This volume reviews our current knowledge of the processes governing the formation of stars, from the collapse and fragmentation of cold molecular gas clouds through to the formation and evolution of disks which can form planets. It provides an especially timely reference for understanding recent discoveries of extrasolar planets and new direct evidence for protoplanetary disks around young stars. Each topic is covered at two levels. A descriptive narrative integrating both observational data and theoretical models is accessible to undergraduates or non-specialists. In addition, each topic is given a rigorous theoretical development with comparison to observations, and is appropriate for first-year graduate students or those desiring a deeper understanding of the physics underlying the story of how stars and accretion disks form and evolve.

This book will be an invaluable reference as the new generation of ground- and space-based telescopes leads to the next level of understanding of the dynamical processes underlying star and protoplanetary disk formation.

ACCRETION PROCESSES IN STAR FORMATION

ACCRETION PROCESSES IN STAR FORMATION

LEE HARTMANN

Smithsonian Astrophysical Observatory

CAMBRIDGE UNIVERSITY PRESS

PUBLISHED BY THE PRESS SYNDICATE OF THE UNIVERSITY OF CAMBRIDGE
The Pitt Building, Trumpington Street, Cambridge, United Kingdom

CAMBRIDGE UNIVERSITY PRESS
The Edinburgh Building, Cambridge CB2 2RU, UK
40 West 20th Street, New York, NY 10011–4211, USA
10 Stamford Road, Oakleigh, VIC 3166, Australia
Ruiz de Alarcón 13, 28014 Madrid, Spain
Dock House, The Waterfront, Cape Town 8001, South Africa

http://www.cambridge.org

First published 1998
Reprinted with corrections 2000

Printed in the United Kingdom at the University Press, Cambridge

Typeface Times 10/12.5pt   *System* LATEX   [UPH]

A catalogue record for this book is available from the British Library

Library of Congress Cataloguing in Publication data

Hartmann, Lee, 1950–
Accretion processes in star formation / Lee Hartmann.
    p. cm.
Includes bibliographical references.
ISBN 0 521 43507 2 (alk. paper)
1. Stars–Formation. 2. Accretion (Astrophysics) 3. Disks
(Astrophysics) 4. Gravitational collapse. I. Title.
QB806.H39 1998
523.8′8–dc21   97-28287   CIP

ISBN 0 521 43507 2 hardback

To my parents

Contents

Preface

The topic of star and protoplanetary disk formation touches almost every area of astrophysics, from galaxy formation to the origin of the solar system. Our understanding of the early evolution of stars has advanced substantially in the last few years as a result of improved observational techniques, particularly in the infrared and radio spectral regions. Although many fundamental problems of star formation remain to be solved, so much has been learned in the last decade about pre-main-sequence accretion processes that an attempt to outline the emerging picture of low-mass star formation seems justified.

In this book I have tried to provide a discussion of accretion in early stellar evolution which can be used at a variety of levels: as an introduction to the subject for advanced graduate students; as a reference for researchers in star formation; and as an overview for scientists in other, related, fields. The text assumes a basic familiarity with astronomical concepts and graduate-level physics, though I have made some effort to include some astronomical definitions and references to fundamental physical equations needed for my development. I have adopted a point of view close to that of my own research, which is generally near the interface between theory and observation, and so have tried to discuss basic physical concepts in relation to observational results. Many plausible and even aesthetically pleasing theories have been constructed which have failed to meet observational tests. Conversely, observations by themselves are not very meaningful unless (or until) they are placed into a physical context. I have also tried to include a substantial number of references, but I warn the reader that my selection is necessarily incomplete and probably biased in such a rapidly-evolving subject; my intent is mainly to provide entry points into the literature for further research.

I especially hope to stir the interest of specialists in other fields where accretion disks are important. There are of course direct applications of pre-main-sequence disk physics and evolution to the study of planet formation, which has taken on added importance with the discovery of extrasolar planets. Beyond this, much of what we currently know about astrophysical disks is based on studies of accreting binary systems, and the accretion disks probably present in active galactic nuclei may have points of similarity to protostellar disks, in that they both exhibit powerful jets and dusty infalling envelopes. One hopes that the similarities among and differences between astrophysical accretion disks will yield further insight into accretion processes.

In writing this book I found myself continually revising material to take current important developments into account. Although this can be problematic, since very recent ideas or results may not have been fully tested, it is difficult to avoid incorporating new material in such a rapidly-developing field. Specialists will recognize that many new results have not been included, and I can only ask for their understanding in view of the rate at which the literature is expanding.

I have tried to express a point of view which is not always that of the 'standard models' if the observations do not support these models. In discussing these matters I have tried to be evenhanded and to provide enough information for the reader to make his or her own judgements, though this can be difficult in addressing current areas of contention. Although some of the issues of today may be of transient importance, the conflict between opposing views seems to me to be part of the excitement of science, and the means by which we sharpen our understanding of astrophysical objects. I have also indulged in a little historical discussion to provide a faint hint of how science is actually done. In an area of rapidly-evolving research such as star formation, texts such as the present one serve not simply to outline the current state of knowledge, but to challenge readers to do better.

Cambridge, Massachusetts   L.H.
October 6, 1997

Acknowledgements

This book could not have been written without the generous support and encouragement of many people and institutions. I gratefully acknowledge the support of the Smithsonian Institution for many years in my astronomical career. I also must thank the Centro de Investigaciones de Astronomía, Mérida, Venezuela, for accommodating many stays during which most of this book was researched and written. In particular, I wish to thank Gustavo Bruzual, Nuria Calvet, and Gladis Magris for providing me with a very hospitable and supportive climate in which to do research.

Many friends and colleagues at the Center for Astrophysics and elsewhere with whom I have collaborated or argued against have contributed in many ways. In particular, I have learned much from my collaborations with three extremely able scientists over the years – Keith MacGregor, John Stauffer, and Scott Kenyon – and I am grateful for the opportunity to have worked with them.

Colleagues at the Harvard-Smithsonian Center for Astrophysics and elsewhere have kindly read drafts of portions of this manuscript, and provided very useful comments and suggestions. In particular, I want to thank Nuria Calvet, Neal Evans, Charles Gammie, Erik Gullbring, Phil Myers, Cesar Briceño, and James Muzerolle, who of course bear no responsibility for any remaining errors and infelicities.

I also wish to thank Doug Lin for suggesting that I write this book, and the people at CUP for their patience and hard work. As a practical matter, I also wish to cite the Astrophysics Data System's abstract service, which was extremely helpful in exploring the literature. Several people, including Phillippe André, Claude Bertout, Chris Burrows, Tom Dame, Suzan Edwards, Neal Evans, Phil Myers, Motohide Tamura, Hans Ungerechts, and Derek Ward-Thompson kindly gave permission to use their figures and/or provided copies.

Figures 1.4, 1.5, 2.1, 2.2, 2.3, 3.3, 4.7, 4.8, 4.9, 4.10, 4.11, 4.13, 6.4, 6.6, 6.11, 6.13, 7.1, 7.9, 7.10, 8.2, 8.4, 8.5, 8.6, 8.12, 9.2, 9.3, 9.4, and 9.6 are reproduced by permission of, and are copyright by, the American Astronomical Society. Figures 7.4, 7.6, 7.7, and 7.8 are reproduced with permission from the *Annual Review of Astronomy and Astrophysics*, Vol. 34, copyright 1996, by Annual Reviews Inc.

Finally, I must acknowledge my debt to Nuria Calvet: first, as a collaborator in many of the topics discussed in this book; second, as a supporter who encouraged me when the task seemed impossible; and third, as a taskmaster to force me to keep to some vague semblance of a schedule. Without her this book would never have been finished. Whatever credit attaches to this work belongs in large measure to her; whatever blame, to me alone.

1

Overview

Much of what we presently know or surmise about the physical processes involved in star formation is derived from the detailed study of a few nearby molecular cloud 'nurseries'. Stars in the solar neighborhood are formed from the gravitationally-induced collapse of cold molecular gas. Typical molecular gas clouds must contract by a factor of a million in linear dimensions to form a star. Because of this dramatic reduction in size, any small initial rotation of the star-forming cloud can be enormously magnified by conservation of angular momentum during rapid collapse. In this way a modestly-rotating gas cloud produces a rapidly-rotating object – a disk – in addition to a small, stellar core at the end of gravitational collapse. Probably most of the material of a typical star is accreted through its disk, with a small amount left behind to form planetary systems.

Advances in observational techniques spanning the electromagnetic spectrum have been essential in developing our present understanding of star formation. Improved infrared detector technology, along with the launch of the Infrared Astronomy Satellite (*IRAS*) in 1983, led to the recognition that dusty disks commonly surround many young stars. Radio-wavelength interferometry with high spatial resolution and sensitivity is now providing a closer look at the distribution of circumstellar material in these disks. Optical studies using large ground-based telescopes are producing more systematic estimates of stellar masses and ages, as well as providing new insights into the accretion flows of young stars. The concurrent development of theory has played an essential role in the maturing of the subject by suggesting a physical framework into which the wide variety of observational phenomena may be placed.

While a coherent picture of star formation appears to be emerging, many mysteries remain. We do not know what determines the masses of stars, nor is it clear whether the paradigm of low-mass star formation applies to massive stars. Although there are several possibilities for initiating protostellar cloud collapse, the relative importance of these alternatives is not known. We are still far from understanding how binary stars, which constitute the majority of stellar systems, are produced. Disk accretion is clearly an important part of the star formation story; although progress has been made lately in understanding some angular momentum transport mechanisms which can drive accretion, theories do not yet have predictive power. The powerful, highly-collimated outflowing jets observed from young stars were totally unexpected, and additional surprises are probably yet in store.

This book attempts to survey our present understanding of the accretion processes involved in low-mass star formation. Given the rate at which instrumental techniques are currently advancing, the observational picture presented here can only serve as an introduction to the subject. In particular, the subject of the formation of dense stellar clusters is left to future treatments. This chapter contains a brief outline of the star formation processes discussed in more detail in the rest of the book.

1.1   Molecular clouds

Young stars are not distributed at random in the Milky Way, but are generally found near the plane of the galactic disk, close to or within clouds of relatively dense molecular gas (Figure 1.1). The sizes and masses of molecular clouds in the Milky Way span a large range, from giant star-forming molecular cloud complexes of masses $\sim 10^6\,M_\odot$ and sizes ~ 100 pc to clouds of $\lesssim 10^1\,M_\odot$ and $\lesssim 1$ pc and smaller. Three of the best-studied (and closest) molecular cloud complexes forming stars are: Taurus, at a distance of ~ 140 pc, with a mass $\sim 10^4\,M_\odot$ and an extent ~ 20 pc (Ungerechts & Thaddeus 1987; Figure 1.1); Ophiuchus, at a distance similar to that of Taurus, and with a similar mass and overall size, but with much denser concentrations of gas (DeGeus, Bronfman, & Thaddeus 1990; Loren, Wooten, & Wilking 1990); and Orion, at a distance of ~ 450 pc, with a mass $\sim 10^5\,M_\odot$ spread over a region ~ 100 pc, and also possessing very dense star-forming regions (Bally *et al.* 1987; Genzel & Stutzski 1989, and references therein). The nearby Chamaeleon and Lupus molecular cloud complexes (cf. Dame *et al.* 1987) are also forming stars in the solar neighborhood (see references in Herbig & Bell (1988)). Results for the Taurus complex are emphasized in this book, not because it is necessarily very representative, but because its stellar population has been particularly well studied.

Most of the large molecular clouds and cloud fragments are roughly consistent with being gravitationally bound. Supersonic ($\sim$ few $\mathrm{km\,s^{-1}}$) velocity dispersions are found on large spatial scales; on small scales, in the densest regions, thermal pressure of the cold gas (with typical temperatures of $T \sim 10$–100 K), along with magnetic fields, may provide the primary support against gravity.

The formation of a star from a molecular cloud of gas requires that gravity overcome the resisting forces of thermal gas pressure, turbulent motions, and magnetic fields. It is therefore not surprising that stars are formed in dense, cold molecular cloud cores, which at least in some cases exhibit small turbulent motions. The substantial dust extinction in many molecular cloud complexes (Cernicharo 1991) may play an important role in enhancing star formation (e.g., McKee 1989) by shielding dense gas from the heating effects of luminous stars and therefore lowering the internal temperature. Dust absorption of external photoionizing radiation fields may be especially important in reducing levels of ionization in molecular clouds; this may enable gas to diffuse across magnetic field lines, reducing the magnetic forces which can counteract gravity (Shu, Adams, & Lizano 1987).

The gas from which many stars are forming is structurally complex and highly filamentary in many places. The origin of this structure is not well understood at present (Scalo 1990); it may be produced in part by magnetic fields (Parker 1966). Filament and sheet formation may be necessary to the fragmentation of molecular clouds into clumps of order the final stellar mass, although the relation between initial cloud mass and the mass of the resulting star is unclear. Observational estimates suggest that the number of molecular clouds with mass M_c is $N_c(M_c) \propto M_c^{-1.5}$ (Blitz 1991). This cloud mass spectrum differs strongly from the initial mass function (IMF)

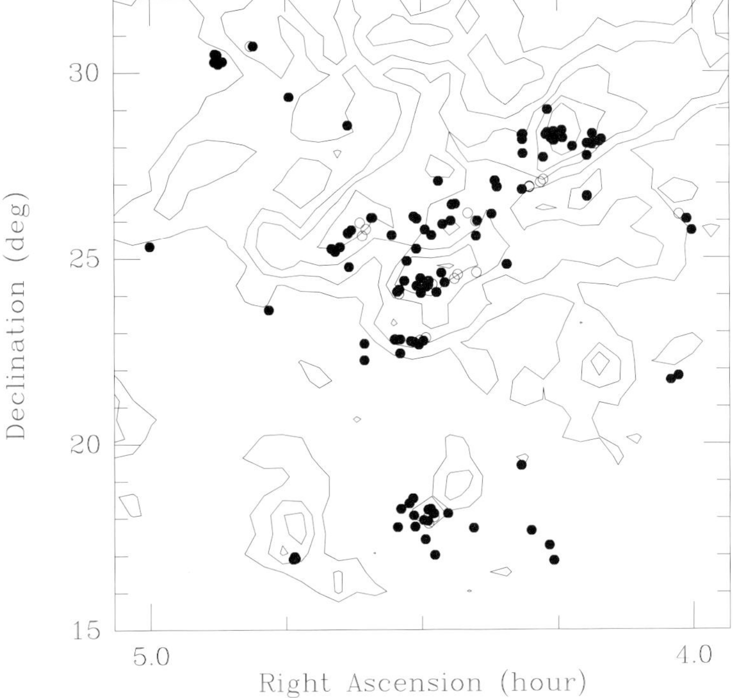

Fig. 1.1. The Taurus-Auriga molecular cloud complex. Contours of integrated ^{12}CO emission are taken from Ungerechts & Thaddeus (1987), with the positions of known young stars (ages $\sim$ 1 Myr) superimposed. The pre-main-sequence stars are generally clustered near regions of high CO column density.

of stars; the latter exhibits an approximate power-law form $N(M) \propto M^{-n}, 2 \lesssim n \lesssim 2.4$ for stars with masses $M \gtrsim 0.5\,M_\odot$, and is much flatter (smaller n) for lower-mass stars (Miller & Scalo 1979; Scalo 1986). Comparisons between cloud and stellar masses are made difficult by uncertainties in the defining boundaries of actual star-forming clouds (Williams, DeGeus, & Blitz 1994).

Although it seems likely that the differing environments of molecular clouds should affect the star formation process, it has proved surprisingly difficult to demonstrate such relationships convincingly. The highest-mass stars form only in regions of high gas/star density; however, because massive stars are generally quite rare, and low-density star-forming regions have relatively small numbers of stars, it is hard to show that the probability of forming high-mass stars varies substantially from region to region. Low-mass stars seem to be formed in large numbers in all environments (e.g., Herbig & Terndrup 1986).

Completely isolated star formation seems to be the exception rather than the rule. For example, the Orion Nebula cluster (centered on the Trapezium group of O stars) contains $\gtrsim 2 \times 10^3$ young stars, with a central density of $\gtrsim 2 \times 10^4$ stars pc^{-3} inside a "core" radius of about 0.15 pc (Jones & Walker 1988; McCaughrean & Stauffer 1994; Ali & DePoy 1995; Hillenbrand 1997; Hillenbrand & Hartmann 1998). Even in a low-density region of star formation like Taurus, young stars are still noticeably clustered (Figure 1.1), with groups of ~ 10–20 stars concentrated in regions ~ 1 pc in size (Gomez *et al.* 1993). In addition, most young stars are members of binary or multiple stellar systems (Ghez, Neugebauer, & Matthews 1993; Leinert *et al.* 1993; Reipurth & Zinnecker 1993; Simon *et al.* 1995; Mathieu 1994).

Many solar-type stars (at least those in low-density star-forming regions like Taurus) are thought to be formed in 'molecular cloud cores', condensations in molecular gas with densities $\gtrsim 10^3$ cm^{-2}. These cores are typically found from radio-frequency surveys of spectral lines of NH_3 and other molecules which are excited only at relatively high gas densities. In Taurus, cloud cores found in NH_3 have densities $\sim 10^4$ cm^{-3} and are typically a few times 0.1 pc in size (Myers & Benson 1983; Myers & Goodman 1988a,b). At such high densities, the expected gravitational collapse timescales are only a few hundred thousand years, so that star formation could proceed quite rapidly from these cores. This expectation is borne out by observations which indicate that almost half of these cloud cores in Taurus have heavily-extincted (presumably recently-formed) stars within their boundaries (Beichman *et al.* 1986).

Bok globules, visually-opaque clouds found by their extinction of background stars, were long ago proposed as sites of low-mass star formation (Bok & Reilly 1947). The globules vary in properties, but typically have masses of a few times that of the Sun, with radii of order a few tenths of a pc in size (Clemens & Barvainis 1988). Many of these globules have young stars within them (Yun & Clemens 1990) and some may be undergoing collapse (Wang *et al.* 1995). Molecular cloud cores may be essentially equivalent to dense Bok globules, but have not been catalogued as such because they are superimposed upon regions of generally large visual extinction, and so are difficult to detect optically.

1.2   Young stars

Young low-mass stars were originally recognized as a subset of optical emission-line objects, typically exhibiting strong hydrogen (Balmer α or Hα) line emission at 6563 Å $= 0.6563\,\mu$m. The youth of these stars was suggested by their spatial correlation with reflection nebulae and dark clouds (Joy 1945), which could be remnant natal material, and by their concentration near associations of high-mass stars which must necessarily be young (Ambartsumian 1947). This grouping is now understood as the result of star formation in molecular clouds; stars of ages $\lesssim 10^6$ yr cannot travel far from their formation sites with typical small velocities ($\lesssim 1$–2 km s^{-1}) relative to the molecular gas (Herbig 1977a; Jones & Herbig 1979; Hartmann *et al.* 1986).

Objective prism surveys for stars with strong emission lines (mostly Hα) led to the first extensive catalogs of young stars. With the advent of infrared techniques, many additional objects (heavily obscured by dust at optical wavelengths) were found. Some of the densest known clusters of newly-formed stars were only detected once infrared arrays were employed to image molecular cloud complexes (Lada *et al.* 1991). Despite the substantial extinction of remnant circumstellar material, X-ray emission has also been used to identify additional young objects, especially stars which do not exhibit strong excess infrared or optical emission (e.g., Montmerle *et al.* 1983; Feigelson *et al.* 1987; Walter *et al.* 1988).

Stellar luminosities and effective temperatures, or equivalently positions in the Hertzsprung–Russell (HR) diagram, provide important clues to the evolutionary states of young stars. Effective temperatures are determined from spectroscopic measurements of optical (and now infrared) absorption line strengths; usually, spectral line ratios are used to determine the stellar spectral type and luminosity class, which can be associated with the stellar surface effective temperature and surface gravity. Stellar luminosities are determined from observed optical and infrared fluxes corrected for the dimming and reddening effects of intervening dust. Using these methods, many surveys of the properties of young stellar populations have been made (e.g. the seminal work of Cohen & Kuhi (1979)).

Many low-mass stars associated with molecular clouds lie above the hydrogen-fusion main-sequence in the HR diagram (Figure 1.2); their low masses M_* and moderately large radii R_* indicate central temperatures too low to fuse hydrogen into helium (since the internal temperatures tend to scale as M_*/R_*). Deuterium fusion can occur at lower temperatures than hydrogen fusion; but because the deuterium abundance is relatively low, the luminosities of typical young stars can be powered by deuterium fusion for only $\sim$ a million years. Without fusion energy release, the young star must contract, generating gravitational potential energy to replace the energy lost by the radiation of the stellar photosphere. This contraction corresponds to moving downward in the HR diagram (Figure 1.2), and thus the positions of young stars in the HR diagram can be used to estimate ages. This contraction stops when stars arrive on the main sequence, at which point hydrogen fusion energy release replaces the energy radiated into space by the stellar photosphere. Thus the ages of stars near the 'zero-age main sequence' (ZAMS) are difficult to determine, especially when uncertainties in stellar luminosities due to distance and dust extinction errors are taken into account.

Low-mass ($M \lesssim 2\,M_\odot$), pre-main-sequence stars, having stellar spectral types F–M (corresponding to surface effective temperatures $\sim$ 7000–3000 K), are called 'T Tauri' stars (Joy 1945; Herbig 1962; Bertout 1989); higher-mass $\sim$ 2–10 $M_\odot$ pre-main-sequence stars are labeled 'Herbig Ae/Be' stars (Herbig 1960) to distinguish them from other types of emission-line A-B stars (effective temperatures $\sim$ 8000–30 000 K) which are presumably more evolved. (The mass ranges are not exact because the classification depends upon the optical stellar spectral type or effective temperature, which can vary substantially as a $\sim 2\,M_\odot$ star evolves; see Figure 1.2). The rapidity with which massive stars evolve make study of their pre-main-sequence evolution more difficult. Stars with masses $\gtrsim 2\,M_\odot$ are typically found near the ZAMS (Figure 1.2; Hillenbrand *et al.* 1992); age estimates for these objects are especially uncertain (cf. Palla & Stahler 1992).

The principal systematic catalog of pre-main-sequence stars is still that of Herbig & Bell (1988 = HBC), which includes spectroscopic information for many objects. There has been an explosion in the number of young stars discovered since the HBC, which is now quite incomplete. Many young stars have only been detected as infrared sources, so their effective temperatures and masses, and often luminosities, are poorly known.

The term 'young stellar objects', or YSOs (Strom 1972), was coined partly in the recognition that the appearance of young stars may be strongly affected or altered by their circumstellar material. It also serves as relatively neutral term to identify objects whose underlying nature – T Tauri star, Herbig Ae/Be star, protostar (§1.3), pre-main-sequence accretion disk (§1.4) – is not yet understood. In this book the term

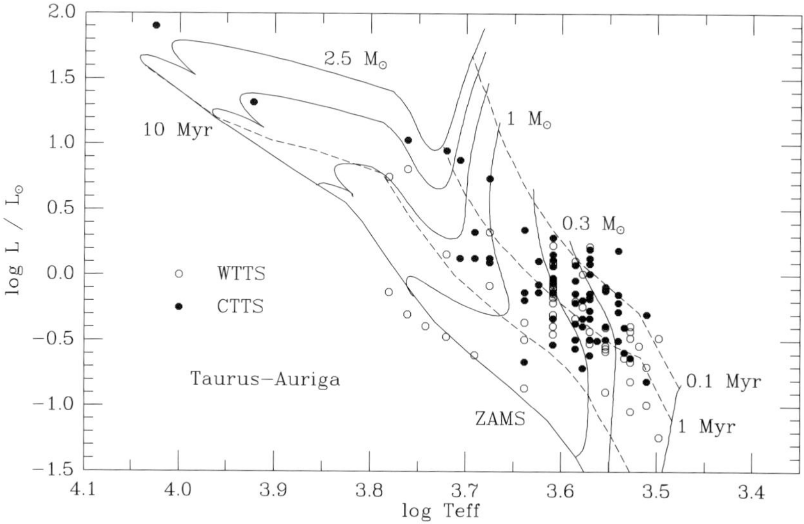

Fig. 1.2. HR diagram positions of young stars lying within the Taurus–Auriga molecular cloud complex (Figure 1.1). For comparison, theoretical evolutionary tracks for pre-main-sequence stars of masses 2.5, 2.0, 1.5, 1.0, 0.5, 0.3, and 0.1 $M_\odot$ are shown. The dashed lines are isochrones for ages of 10^5, 10^6, and 10^7 yr (0.1, 1, and 10 Myr), with the hydrogen-fusion 'zero-age main sequence' or ZAMS shown as the lowest line running from upper left to lower right. The open circles refer to weak-emission T Tauri stars (WTTS; see text), while the filled circles denote the positions of the classical T Tauri stars (CTTS). Stellar properties taken from Kenyon & Hartmann (1995); evolutionary tracks are from D'Antona & Mazzitelli (1994).

YSO is used sparingly, since it does not discriminate clearly between objects which can be intrinsically quite different.

The T Tauri stars originally were identified as late-type stars with strong emission lines and irregular light variations associated with dark or bright nebulosities (see Bertout (1989) for a discussion). Since that time, the term 'T Tauri star' has come to be synonymous with low-mass pre-main-sequence stars, whether or not they are associated with nebulosity or have strong emission lines. Most of the variable stars first identified as T Tauri stars are currently called 'strong-emission' or 'classical' T Tauri stars (CTTS), to distinguish them from 'weak-emission' pre-main-sequence stars (WTTS). The distinction between strong and weak emission is usually made on the basis of Hα emission; if the Hα equivalent width is smaller than 10 Å the object is usually termed a WTTS (HBC). Although this classification is generally useful for K–M stars, it can be misleading when applied to the hotter F- and G-type stars because their photospheric radiation fields tend to suppress Hα emission. It appears that the excess emission of many WTTS can be explained in terms of enhanced solar-type magnetic activity (Walter *et al.* 1988), while the extreme levels of excess emission at optical and infrared wavelengths of many CTTS require an external energy source. *Accretion* from a circumstellar disk appears to supply the energy needed for the non-photospheric emission of CTTS (§1.4).

1.3 Protostars

Dramatic improvements in infrared detector technology have enabled astronomers to find many YSOs that are hidden by dust absorption at optical wavelengths. The measurement of far-infrared emission, which can penetrate through the most opaque dust clouds, was enormously enhanced by the launch of the *IRAS* satellite. In orbit, *IRAS* could be more effectively shielded from the intense thermal emission of the Earth than was possible with ground-based telescopes, and it was actively cooled to liquid He temperatures. *IRAS* produced a survey of the sky at wavelengths of 12, 25, 60, and 100 μm with enough sensitivity to detect many pre-main-sequence stars; the launch of the *ISO* satellite promises further advances.

One of the major results of *IRAS* was the discovery of many infrared sources with luminosities typical of T Tauri stars, but with spectral energy distributions peaking at 60–100 μm (Beichman *et al.* 1986). These objects are now generally identified as stars still surrounded by much of the material from their parent molecular cloud core. The dusty envelopes of these objects absorb energy from the central star at short wavelengths, and re-emit this energy at far-infrared wavelengths, where the envelope is sufficiently transparent for the radiation to escape.

The protostellar phase of star formation is thought to involve the free-fall collapse of dusty envelopes to stellar dimensions, resulting in very large extinctions towards the central energy source (e.g., Larson 1969a,b; Appenzeller & Tscharnuter 1974). Thus, it is plausible to identify some of the heavily-extincted *IRAS* sources as protostars, i.e. very young stars that still are accumulating substantial amounts of mass from their surrounding envelopes. Theoretical models of the dust emission from protostellar envelopes in gravitational free-fall can reproduce the observed infrared emission of these heavily-extincted young stellar objects (Adams, Lada & Shu 1987 = ALS; Butner *et al.* 1991; Kenyon, Calvet, & Hartmann 1993). It has proved difficult to establish with great certainty that these dusty envelopes are actually falling in, although continuing efforts to observe small velocity shifts in radio-wavelength emission lines have now provided suggestive indications of collapse (Walker *et al.* 1986; Zhou *et al.* 1993; Zhou & Evans 1994; Hayashi, Ohashi, & Miyama 1993; Mardones *et al.* 1997).

IRAS surveys of Taurus (Beichman *et al.* 1986; Cohen, Emerson, & Beichman 1989; Kenyon *et al.* 1990, 1994; Beichman, Boulanger, & Moshir 1992) indicate that the number of protostellar sources is $\approx 10\%$ of the total pre-main-sequence population. This suggests that the time for protostellar collapse is similarly about 10% of the age of the pre-main-sequence stars in the region. The resulting estimate of $\sim 10^5$ yr for the protostellar infall phase is roughly consistent with the expected timescale of collapse for the formation of a typical low-mass T Tauri star (Larson 1969a,b; Shu 1977).

1.4 Long-wavelength emission: dusty envelopes and disks

YSOs are frequently classified in the literature based on the shape of the emitted spectrum from near- to far-infrared spectral regions, which emphasizes the properties of circumstellar dust (Lada & Wilking 1984; Lada 1987). The emitted spectrum is usually discussed in terms of the 'spectral energy distribution' or SED, which is a

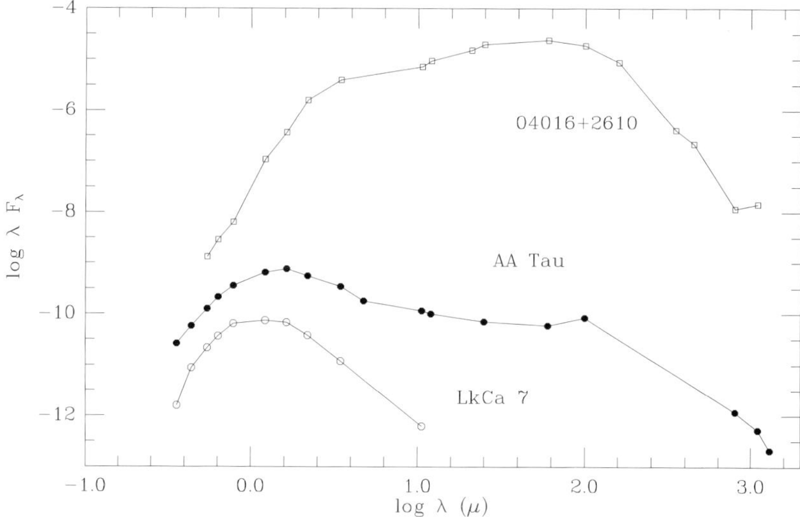

Fig. 1.3. Spectral energy distributions (SEDs) of three YSOs which typify the infrared classification system. The vertical axis is the flux at the Earth in arbitrary units. The Class I object IRAS 04016+2610 is probably a protostar hidden at optical wavelengths by its dusty infalling envelope; the dust absorbs the radiation from the central regions and re-emits it in the far-infrared. The CTTS or Class II object AA Tau is optically visible, but exhibits long-wavelength dust emission generally attributed to a circumstellar disk (see text). The WTTS or Class III object LkCa 7 does not exhibit detectable dust emission; its SED is nearly that of a single-temperature blackbody, and is typical of the photospheric emission of low-mass pre-main-sequence stars. Data from Kenyon & Hartmann (1995).

frequently used shorthand for either the flux distribution λF_λ observed at the Earth, or the luminosity distribution λL_λ, depending upon the context.[*]

The infrared classification scheme depends upon the spectral index s of the emitted flux F of the object, $\nu F_\nu = \lambda F_\lambda \propto \lambda^s$, typically measured between $\lambda \sim 2\,\mu$m and ~ 50–$100\,\mu$m. Class I sources correspond to objects with $s > 0$, i.e. the SED rises toward long wavelengths. An example of such a Class I source is the Taurus YSO IRAS 04016+2610, as shown in Figure 1.3); this heavily-extincted object is one of the protostar candidates discussed in the previous section. The SEDs of such objects can be explained with emission from dusty infalling envelopes.

Class II sources have $-4/3 \lesssim s \lesssim 0$, which is frequently identified as the spectrum produced by a dusty circumstellar disk. A typical example of a Class II source is the T Tauri star AA Tau, shown in Figure 1.3. The CTTS AA Tau exhibits much more infrared radiation than the WTTS Taurus object LkCa 7 (Figure 1.3); the latter is a Class III source, because it has the Rayleigh–Jeans distribution $s \sim -3$ expected at long wavelengths for a star without infrared excess emission from circumstellar dust.

Since this classification scheme was introduced, a further modification has been suggested by the introduction of the 'Class 0' sources (André, Ward-Thompson, & Barsony 1993). Class 0 sources are very red objects, with large amounts of sub-mm

---

[*] In this book we follow the typical astronomical convention of using cgs units. Thus, the flux λF_λ has units of $\mathrm{erg\,cm^{-2}\,s^{-1}}$, while the luminosity λL_λ has units of $\mathrm{erg\,s^{-1}}$.

emission relative to their total luminosities. Evidence is accumulating that the Class 0 sources have especially large amounts of gas and dust in their immediate environs, as might be expected if they represent a generally earlier phase of protostellar evolution than Class I sources (e.g., Bontemps *et al.* 1996).

ALS suggested that the infrared classification scheme can be interpreted in terms of an evolutionary sequence. Protostars are Class I (or 0) sources, surrounded by (roughly spherical) dusty infalling envelopes. The angular momentum of the infalling material causes it to pile up in a rotating, flattened disk. After envelope infall has ceased, the dusty disk can still produce substantial infrared emission, as it can be heated externally by radiation from the central star and internally by viscous dissipation (§1.5); however, because the disk is flat, it does not intercept most lines of sight to the central star, and thus the latter is optically visible. This combination of large amounts of dust in a disk geometry produces a Class II source. Eventually, the disk dust is dissipated or coagulated, and one is left with only detectable emission from the central star – a Class III object.

Surveys of the Taurus molecular cloud suggest that roughly half of the pre-main-sequence stars have dusty circumstellar disks at a typical age of 10^6 yr (Strom, K.M. *et al.* 1989; Strom, S.E. *et al.* 1993; Beckwith *et al.* 1990; Skrutskie *et al.* 1990). There is some indication that binary T Tauri stars exhibit less long-wavelength dust emission than single stars, which may be the result of disk disruption by the tidal forces in the binary system (Jensen, Mathieu, & Fuller 1994, 1996; Osterloh & Beckwith 1995).

In principle, the optically-thin dust emission observed at wavelengths ~ 1 mm can be used to calculate the dust mass surrounding Class II sources. In practice, masses estimated in this way are uncertain for several reasons, principally because the dust opacity is not well understood; this is particularly a problem for circumstellar disks, where appreciable evolution of dust grain sizes is expected to occur (e.g., Weidenschilling & Cuzzi 1993). Current estimates (Beckwith *et al.* 1990; Osterloh & Beckwith 1995) suggest typical disk masses within an order of magnitude of 10^{-2} M$_\odot$. This is comparable to the so-called 'minimum mass solar nebula', i.e. the total mass of solar composition material that would be needed to produce the observed (condensed) material in the planets.

1.5   Imaging of disks

Imaging observations provide the most direct demonstration of circumstellar disks. Unfortunately, imaging the disks of YSOs is often difficult; even in the nearest star-forming regions at a distance ~ 140 pc, a disk of solar system radius (40 AU) has an angular diameter less than 0.6 arcsec, straining the current technological limits of imaging. Disks are also expected to be faint relative to the illuminating central star at optical and near-infrared wavelengths, so that there are difficulties of contrast in discerning a faint extended disk around a much brighter point source. In addition, there may be difficulties in separating disk material from other gas and dust in the region which is either left over from star formation, or even falling in to form the disk. For these reasons, the first clear image of a circumstellar disk was that of the remnant dust disk around the relatively nearby main-sequence star β Pic (e.g., Smith & Terrile 1984). Progress in radio-wavelength interferometry has led to the detection of modest-sized ($\gtrsim 100$ AU) dusty structures around young stellar objects which are probably disks (Koerner, Sargent, & Beckwith 1993; Lay *et al.* 1994; Dutrey *et al.*

1995; Koerner & Sargent 1995; Koerner, Chandler, & Sargent 1995; Mundy *et al.* 1996).

One way of avoiding the difficulty of discerning a faint disk against a bright central point source is to use some much more powerful but distant illuminating source to enhance disk emission. In dense clusters of young stars like the Trapezium in Orion, the hot luminous central stars can illuminate – and even evaporate – the disks expected to form around low-mass stars in the cluster. A variety of optical and radio-wavelength observations of very compact H II regions in the Trapezium were initially interpreted as evaporating disks with sizes $\sim 10^2$ AU around T Tauri stars in the region (cf. Churchwell *et al.* 1987 and references therein); recent results from the Hubble Space Telescope (*HST*) have provided strong support for this picture (O'Dell, Wen, & Hu 1993; O'Dell & Wen 1994). These objects appear to be compact, dense gas clouds which are being photoionized by the hot luminous stars in the Trapezium (mostly θ^1C Ori). The lifetimes required by the source statistics, coupled with studies of the rate of evaporation of material (e.g., McCullough *et al.* 1995), suggest that these objects are gaseous disks with masses of order 10^{-1}–10^{-2} M$_\odot$ (Churchwell *et al.* 1987). Some objects have been detected which are too far away from θ^1C to be heated and ionized, but can be seen in silhouette against the bright nebular background (McCaughrean & O'Dell 1996). The morphologies of these absorption structures are very suggestive of circumstellar disks, with faint, low-mass T Tauri stars at their centers.

Another way of improving the contrast between the disk and the illuminating central object is to use the disk to occult the central star. This should be an extremely rare occurrence, but *HST* observations of the YSO HH 30 apparently show such an edge-on disk (Burrows *et al.* 1996; Figure 1.4). The scattering surfaces of the disk are curved away from the disk midplane, as expected on general principles (Kenyon & Hartmann 1987), which creates an extended region of dust which scatters light from the central star, while providing an absorbing screen to block out direct light from the central star (Whitney & Hartmann 1992). The *HST* images also show a highly-collimated, rapidly-moving bipolar jet, which appears to be a common feature of low-mass star formation (e.g., Edwards, Mundt, & Ray 1993).

1.6   Disk accretion

Relatively small amounts of dust in a T Tauri disk can efficiently absorb a substantial amount of light from the central star, and reradiate this energy effectively at infrared wavelengths. These so-called 'reprocessing' or 'irradiated' disks (sometimes called 'passive' disks) can emit long-wavelength radiation simply as a result of heating by the T Tauri star, without any internal energy source, and models of irradiated disks can account for the infrared emission of many CTTS (Kenyon & Hartmann 1987). However, some disks are much more luminous than can be explained by the amount of irradiation from the central star, and require their own energy source. Many of these T Tauri stars also exhibit strong excess optical and ultraviolet continuum emission along with broad emission lines (Bertout, Basri, & Bouvier 1988; Bertout 1989; Figure 1.5). This optical and ultraviolet continuum emission is dominated by hydrogen bound-free and free-free emission at temperatures $\sim 10^4$ K.

Accretion of mass from a circumstellar disk is almost certainly responsible for the excess emission of the CTTS (Lynden-Bell & Pringle 1974). The dissipation

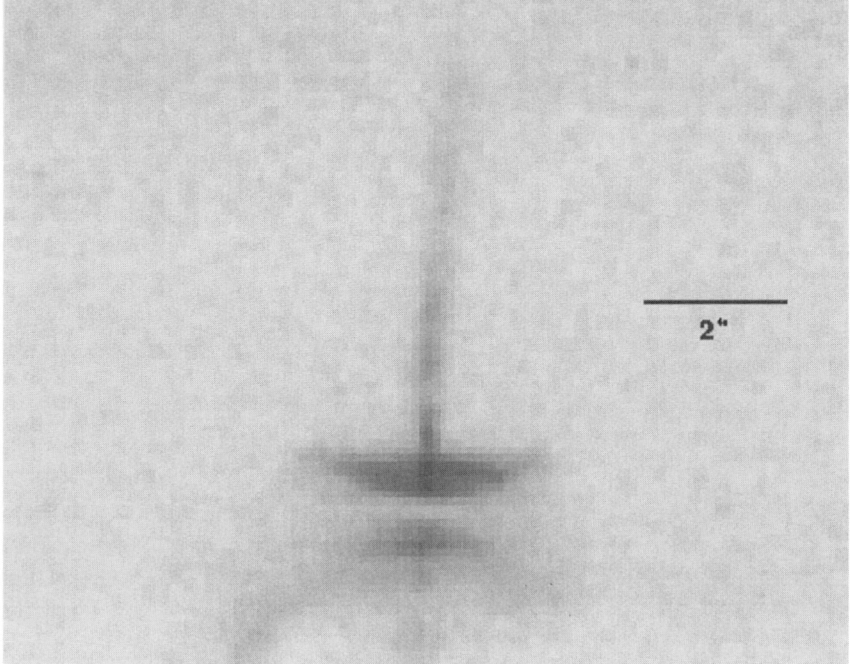

Fig. 1.4. *HST* optical image of the YSO HH 30, shown as a negative image (dark structures are intrinsically bright). The double-concave reflection nebula is produced by light from a hidden central star scattering off the upper surfaces of a dusty disk seen nearly edge-on (see text). The dust absorption in the disk plane (central lane between the upper and lower reflection nebulae) completely obscures the central star. The roughly vertical linear structure is a bipolar high-velocity jet, which is observed in shock-excited emission lines. At the distance of Taurus (140 pc), 2 arcsec corresponds to 280 AU. From Burrows *et al.* (1996).

of accretion energy in the disk enhances the infrared emission, while the infall of material onto the central star produces the high-temperature optical and ultraviolet continuum emission and strong optical emission lines. This conjecture is supported by the strong correlation between infrared and ultraviolet excess emission (Hartigan *et al.* 1990; Edwards 1995); YSOs with substantial near-infrared excess emission (Class II sources) generally show strong optical emission lines and continuum (i.e. are CTTS); stars without near-infrared excess emission (Class III sources) do not show strong optical excess emission (are WTTS).

In some cases the optical and ultraviolet excess continuum emission far exceeds that of the stellar photosphere. This applies to roughly 5–10% of the well-studied T Tauri stars in the Taurus molecular cloud region (Kenyon & Hartmann 1995). In these objects, the accretion energy release is the dominant luminosity source of the system. Our understanding of the evolutionary states of strong-emission CTTS is poor, because the strong hot continuum emission extends to optical wavelengths, making it very difficult to determine reliable positions of these stars in the HR diagram.

Originally, it was thought that the high-temperature emission was produced in

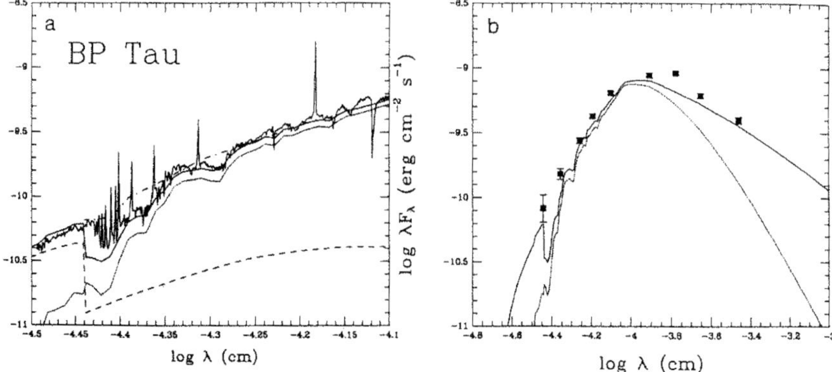

Fig. 1.5. Spectrum of the CTTS BP Tau, showing evidence for excess optical and infrared emission. The left-hand panel shows the observed spectrum (upper heavy line), which is made up of several components; a hot continuum (dashed line), the stellar photospheric emission (second curve from bottom), plus Balmer, Ca II, etc. emission lines. The right-hand panel illustrates broad-band photometry (points) of BP Tau, which indicates both optical and infrared excess emission above that expected for a normal stellar photosphere (bottom curve). The upper curve is a model including hot continuum emission and infrared disk emission. Reproduced from Basri & Bertout (1989).

a shear boundary layer between the rapidly-rotating disk and the more slowly-rotating star (Lynden-Bell & Pringle 1974). It now appears more likely that the inner disk is disrupted by the magnetic field of the central T Tauri star; this results in magnetospheric accretion, in which the disk material is channelled along magnetic field lines to crash into the star (Königl 1991; Camenzind 1990; Ostriker & Shu 1995). The continuum emission is thought to arise from the shock at the base of the magnetospheric accretion column, near the stellar surface, while the emission lines arise in the fast-moving, essentially freely-infalling magnetospheric gas, which is heated to temperatures ~ 8000 K (Calvet & Hartmann 1992; Hartmann, Hewett, & Calvet 1994). The large velocity widths of the strong emission lines in CTTS apparently result from the central star's gravitational acceleration of the infalling gas lifted from the Keplerian disk. Mass accretion rates needed to explain the observed ultraviolet, optical, and infrared emission excesses of CTTS range from $\sim 10^{-9}\,\mathrm{M}_\odot\,\mathrm{yr}^{-1}$ to $\lesssim 10^{-6}\,\mathrm{M}_\odot\,\mathrm{yr}^{-1}$ (Basri & Bertout 1989; Hartigan *et al.* 1991; Valenti, Basri, & Johns 1993; Hartigan, Edwards, & Ghandour 1995; Gullbring *et al.* 1997).

In some pre-main-sequence objects the mass accretion rate increases by orders of magnitude for short periods of time (Herbig 1977b). During these 'FU Orionis' outbursts, the accretion disk becomes 2–3 orders of magnitude more luminous than the central star (Hartmann & Kenyon 1996). Event statistics suggest that the average low-mass star may undergo several FU Ori outbursts during its evolution, during which perhaps 10% or more of the final stellar mass is accreted.

Disk accretion appears to produce the powerful highly-collimated jets and winds which commonly emanate from young stellar objects (Figures 1.4, 1.6). These jets can have velocities of several hundreds of $\mathrm{km\,s}^{-1}$, with collimation angles of only a few degrees, and involve mass ejection at rates as much as 10% of the accretion rate

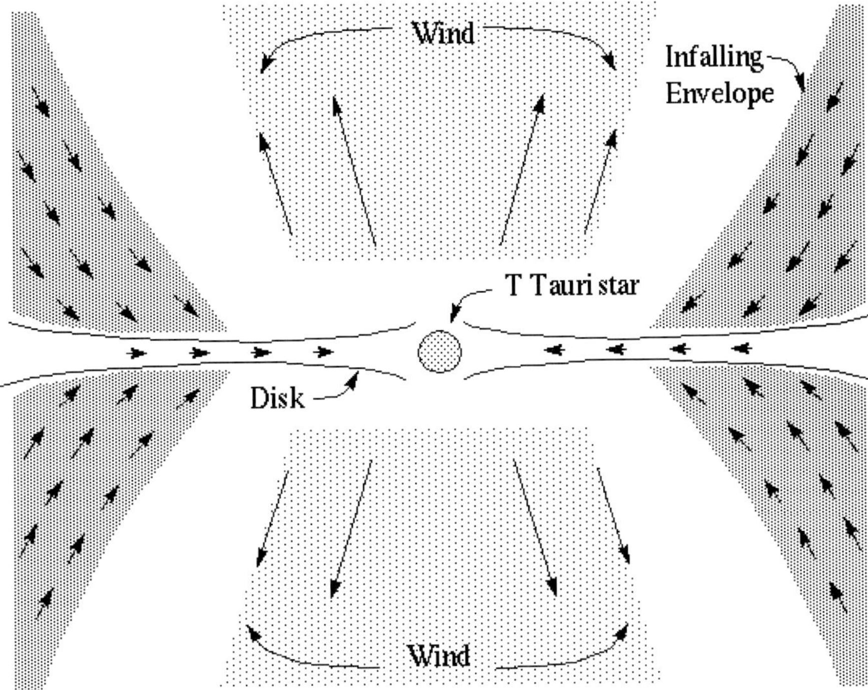

Fig. 1.6. Schematic picture of stellar accretion. Mass is fed into a circumstellar disk by the collapsing protostellar envelope with an infall rate $\sim 10^{-5}\,\mathrm{M_\odot\,yr^{-1}}$. The disk generally accretes at $\sim 10^{-7}\,\mathrm{M_\odot\,yr^{-1}}$ during the T Tauri phase, but during FU Ori outbursts this accretion rate may increase to $\sim 10^{-4}\,\mathrm{M_\odot\,yr^{-1}}$. The disk ejects roughly 1–10% of the accreted material in a high-velocity wind.

(Dopita, Schwartz, & Evans 1982; Mundt & Fried 1983; Lada 1985; Hartigan, Morse, & Raymond 1994; Hartigan *et al.* 1995). Such jets are only observed when there is some evidence for accretion disks; WTTS do not exhibit jets or massive outflows. The collimation of the jets is probably the result of magnetic acceleration from the surface of the Keplerian disk (Pudritz & Norman 1983; Heyvaerts & Norman 1989), although the details of this process are uncertain (Königl 1989; Pelletier & Pudritz 1992; Lovelace, Romanova, & Contopoulos 1993; Shu *et al.* 1994; Najita & Shu 1994).

1.7   A picture of stellar accretion

The (mostly observational) findings outlined above have led to the following scenario for low-mass stellar accretion (Figure 1.7). Cold, dark molecular gas fragments by some unknown process into self-gravitating cloudlets of a few solar masses. These cloudlets or cores evolve into a critical configuration where they cannot support themselves against gravity, and collapse at nearly free-fall. Because of the rapidity of this collapse, any angular momentum transfer must be relatively inefficient, and any initial rotation of the cloud results in collapse to a multiple star system, or to a disk, or both. Since it seems rather unlikely that the initial angular momentum will be so small as to permit collapse directly to stellar dimensions, given the large difference in

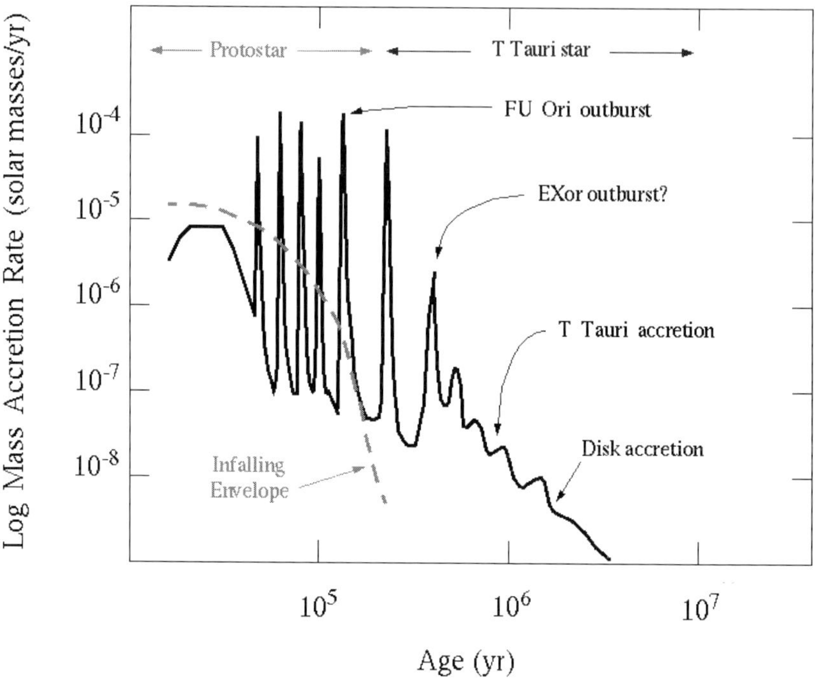

Fig. 1.7. Outline of estimated mass accretion rates during the formation of a typical low-mass (solar-type) star. In relatively cold, isolated regions of star formation, the collapse of the parent molecular cloud of a $\sim 1\,M_\odot$ star is thought to take approximately 0.1–0.2 Myr. The result of this infall is to build up a stellar core and a circumstellar disk. The disk accretes steadily onto the central star at low rates, punctuated by very brief FU Ori outbursts of rapid disk accretion. It is thought that the FU Ori events occur preferentially during the earliest phases, in which mass is still falling onto the disk from the protostellar envelope, thus replenishing the accreted material; during the outburst, as much as $10^{-2}\,M_\odot$ may be dumped onto the central star ($\dot{M} \sim 10^{-4}\,M_\odot\,\mathrm{yr}^{-1}$ for $t \sim 100$ yr). Well after the protostellar envelope has stopped adding mass to the circumstellar disk, the disk continues to accrete and evolve during the T Tauri phase. The figure indicates that disk accretion slowly decays and eventually ceases over periods of a few million years, but this timescale is uncertain, and probably varies substantially from star to star. Many of the current WTTS probably finished accreting from their disks in timescales $\lesssim 1$ Myr.

size between a star and its parent gas cloud, it is plausible to suppose that most of the mass initially lands on the disk(s). Disks are engines for the outward transfer of angular momentum, allowing the accretion of mass to the central star.

Disk accretion rates during early stellar evolution vary widely for typical low-mass stars. During or immediately after the protostellar phase of infall to the disk (Figure 1.7), disk masses are likely to be relatively large, and such disks could be subject to gravitational instabilities which would cause rapid accretion. The rate at which infall adds mass to the disk generally may not be the same as the natural accretion rate of the disk; a mismatch between these rates could explain the FU Ori outbursts, if matter piles up in the disk until it can be discharged in rapid accretion events.

Eventually, infall to the disk stops. At this point the disk slowly evolves and eventually becomes depleted in mass; the processes by which this occurs are not clear. Much of the disk may be accreted into the central star; some material may be driven off by stellar wind ablation, or by evaporation by high-energy stellar photons. Coagulation or accretion of disk material into planets may be the final stage in disk clearing.

These stages of infall and disk accretion are schematically indicated in Figure 1.7. It is difficult to set a precise boundary between the protostar and T Tauri phases, because infall from the envelope may not cease instantaneously. During the main infall phase the central protostar may accrete from its disk at generally similar rates as T Tauri stars – which suggests that, if the protostar could be observed directly, it might appear quite similar to a T Tauri star. Moreover, if one were to try to define a protostar as an object which has not finished accreting its final mass, then T Tauri stars would also be protostars. Definitions based on the relative amount of mass remaining in the disk/envelope vs the amount of mass already in the stellar core are difficult to implement given measurement uncertainties. Therefore, in this book the term 'protostar' is used to refer to phases where a substantial (in the sense of extinction) infalling envelope surrounds the central star (i.e. a Class 0 or Class I source).

There is reason to believe that the FU Ori outbursts are generally concentrated to early phases of evolution, while infall is still occuring, but the frequency and duration of such outbursts are poorly understood; and little is currently known about the so-called 'EXor' outbursts of T Tauri stars (Herbig 1977b). Similarly, there is a wide range of accretion rates among T Tauri stars. The disk accretion rates at later times in Figure 1.7 are meant only to refer to the CTTS; the WTTS may have accreted their disks much faster, assuming that they initially did possess disks.

This picture of stellar accretion is mostly the result of extensive empirical studies over the last two decades, supported by simple theoretical models. In the following chapters we will explore the physical arguments and observational evidence for this scenario of low-mass star formation.

2

Initial conditions for protostellar collapse

Low-mass stars like the Sun are thought to have their origin in dense molecular cloud condensations (cores). In this chapter we present a necessarily brief overview of the complex subject of core formation and equilibrium. The focus is mainly limited to those properties of cores which establish the initial conditions for subsequent stellar accretion processes.

Although supersonic motions are observed in molecular cloud complexes, velocity dispersions approach thermal values on small scales, at least in relatively low-density star-forming regions like the Taurus cloud complex. If the molecular cores are gravitationally bound, as must be the case for star formation to proceed, and if thermal pressure is the principal force which counteracts gravity, then the relationship between the mass M_{cl} and radius R_{cl} of a core is roughly

$$\frac{GM_{cl}}{R_{cl}} \sim c_s^2 = \frac{kT}{\mu m_H},\tag{2.1}$$

where c_s is the sound speed and m_H is the mass of the hydrogen atom. Taking a mean molecular weight $\mu = 2.3$, appropriate for molecular hydrogen plus helium, and a typical cold molecular cloud temperature of $T = 10$ K, equation (2.1) implies that a solar mass molecular cloud core must have a radius $R_{cl} \sim 0.1$ pc. Encouragingly, observations suggest that some molecular cloud cores have roughly these properties. YSOs are detected as infrared sources within many cores, suggesting that star formation proceeds rapidly once cores are formed.

Whether this quasi-equilibrium, thermally-supported core picture also applies to dense, clustered regions of star formation is not clear. Such dense regions exhibit large turbulent velocities, suggesting that the sound speed in equation (2.1) should be augmented by turbulent motions; however, it may not be appropriate to assume hydrostatic equilibrium as a starting point. One should keep in mind that dense cluster regions may be the typical sites of star formation, rather than low-density and relatively quiescent molecular cloud complexes such as Taurus.

Just what triggers a cloud to form a star, or how the gas fragments to make individual stars or multiple systems, are matters of current debate. Theory and observations both suggest that cloud cores cannot heat up substantially if they are compressed; if a core close to hydrostatic equilibrium is perturbed for some reason to a smaller radius than that given by equation (2.1), thermal pressure cannot balance gravity, and gravitational collapse will ensue. The inclusion of magnetic fields modifies this picture, but even with magnetic pressure support there is a well-defined limit to

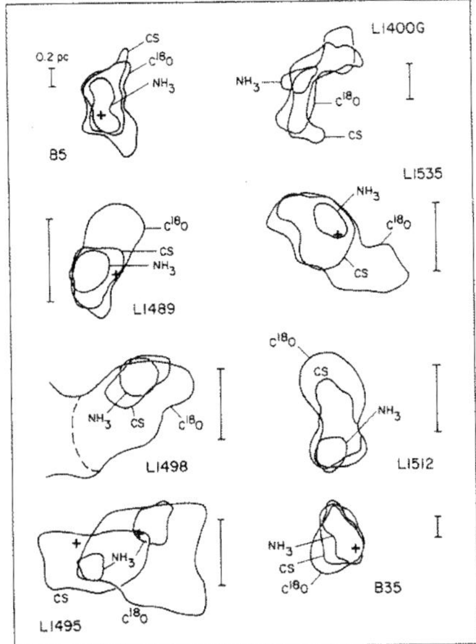
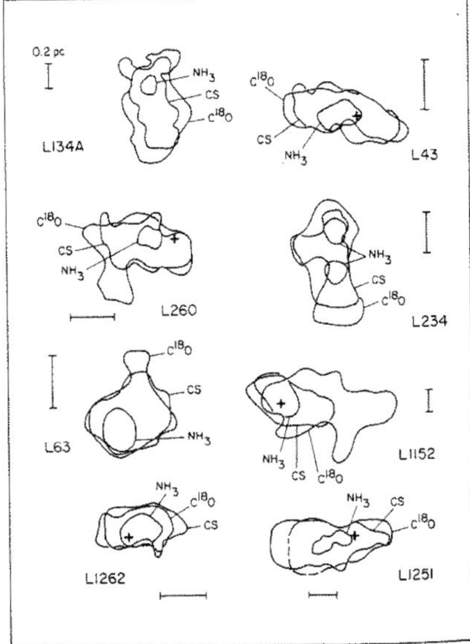

Fig. 2.1. Half-maximum intensity contours for a sample of dense molecular cloud cores, as observed in a variety of molecular species. For each core a linear dimension of 0.2 pc is indicated. Crosses indicate associated star (*IRAS* source). From Myers *et al.* (1991).

the size of the cloud core below which collapse occurs. The predicted slow inward diffusion of gas across magnetic field lines may help initiate gravitational collapse; but whether magnetic field diffusion is necessary for star formation to occur is not at all clear, especially since dense star clusters suggest formation times much shorter than magnetic field diffusion times.

2.1 Observations of molecular cloud cores

The properties of dense molecular gas clouds have mostly been inferred from observations of cm- and mm-wavelength spectral lines. Emission from the ^{12}CO molecule provides the most extensive and sensitive tracer of molecular gas, because the (much more abundant) H_2 molecule is difficult to observe directly except when heated by shocks or ultraviolet radiation fields. The lowest rotational spectral line transitions of ^{12}CO are usually extremely optically thick in many molecular cloud regions, making it difficult to use these lines to study the densest regions of gas. It is necessary to use other tracers whose transitions are more strongly excited at high densities, such as NH_3 or CS, or to use rare isotopes of CO, which are less optically-thick, to study dense molecular cloud cores.

Observations of molecular cloud cores in the NH_3 (J,K) = 1–1 inversion transition in nearby star-forming regions (Benson & Myers 1989) indicate typical molecular hydrogen number densities between 2×10^3 and $2 \times 10^5\,cm^{-3}$, velocity dispersions between 0.2 and 0.9 $km\,s^{-1}$, and median masses $\sim 10\,M_\odot$. On the smallest scales, the velocity dispersions of these cores approach thermal values. The median size of these objects is a few $\times\,0.1$ pc (Figure 2.1).

The properties of cores depend somewhat on which molecular tracer is being used to define the object. Different species and transitions reflect differing density and temperature ranges (Zhou *et al.* 1989; Fuller & Myers 1993; Butner, Lada, & Loren 1995), and yield somewhat different results for velocity widths and sizes (e.g., Figure 2.1).

Magnetic fields are difficult to measure in cloud cores; most observations are made in lower-density environments. For example, in one dark cloud in the Perseus molecular complex, B1, with a size ~ 1 pc and an estimated average density $\sim 10^3 \, \mathrm{cm}^{-3}$, the measured magnetic field is $\sim 30 \, \mu\mathrm{G}$ (Goodman *et al.* 1989). Measurements are often not available for dense cores, but values for the nearby environment suggest that magnetic pressures $B^2/8\pi$ are often comparable to thermal pressures (see Heiles *et al.* (1993); Myers and Khersonsky (1995); Troland *et al.* (1996), and references therein). More generally, it is suggested that the gravitational, magnetic, and thermal plus 'turbulent motion' energy densities are roughly comparable (Myers & Goodman 1988a,b).

Observations suggest that some of the smallest, least turbulent, protostellar molecular cloud cores are nearly in hydrostatic equilibrium (Myers & Goodman 1988b; Fuller & Myers 1992; Figure 2.2). It is difficult to prove this, however, since the difference in energy between static and unbound systems is only a factor of two, which is within the uncertainties in estimating masses, magnetic field pressures, appropriate velocity moments, appropriate geometry, etc.

Current surveys suggest that, at least in the Taurus molecular cloud complex, many known cores are associated with heavily-extincted YSOs (Beichman *et al.* 1986). The high frequency of association of known cores with young stars indicates that many cores may be observed in a stage of collapse (e.g., Figure 2.2), because the star is presumably the result of some core material having already collapsed (Fuller 1994). It is not known whether all cores without stars will eventually form stars (e.g., Bonnell, Bate, & Price 1996).

Many molecular cloud cores seem to be elongated rather than spherical. The observed ratios of major to minor axes in Taurus core maps are ~ 2 (Myers *et al.* 1991). The statistical properties of elongation, coupled with the assumption of random inclinations, suggests that cores may be mostly prolate rather than oblate. These observations should be kept in mind when considering the application of theoretical calculations in spherical geometry to real objects. Binary or multiple stars may be formed more easily if the parent cloud cores are elongated (Bonnell & Bastien (1992); see the review of binary formation by Boss (1995)).

On small scales in dense cores, the mass motions appear to be dominated by thermal motions and velocity dispersions are small (Myers & Goodman 1988b; Figure 2.3). In contrast, on large scales, velocity dispersions can be dominated by non-thermal motions (Larson 1981).

Most molecular cloud cores exhibit modest projected spatial velocity gradients. Typical angular velocities are $\Omega \sim 10^{-14}$–10^{-13} rad s^{-1}, corresponding to ~ 0.3–3 km s^{-1}pc^{-1} (Goodman *et al.* 1993). At the upper end of this range, the rotation at the outer edge of a cloud core of radius 0.1 pc would be ~ 0.3 km s^{-1}, comparable to the thermal support velocity, but most clouds surveyed seem to have slower rotation than this. In general, rotation does not appear to provide much support for molecular cloud cores against their self-gravity.

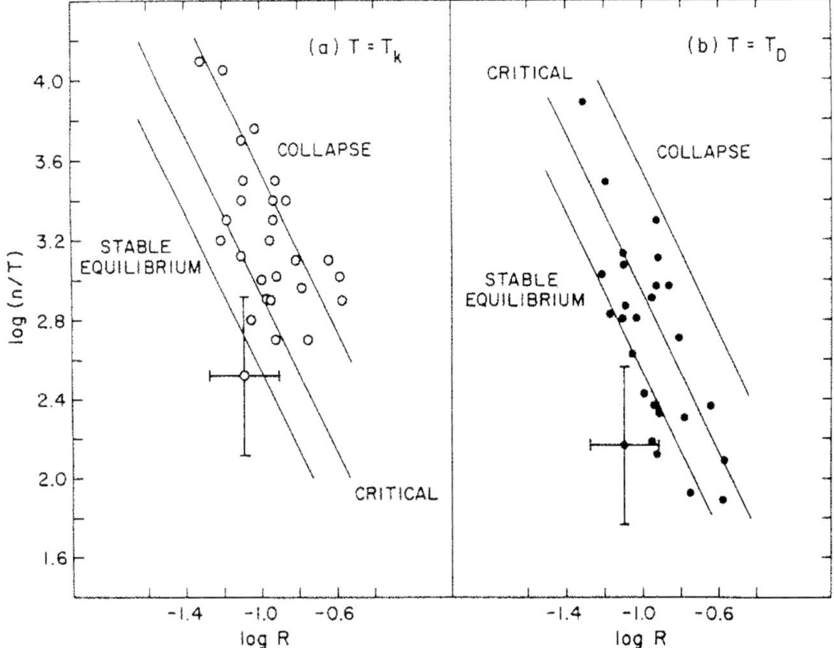

Fig. 2.2. Comparison of cloud core density, temperature, and radius with the requirements for hydrostatic equilibrium. The left-hand panel assumes that only thermal pressure supports the cloud, while the right-hand panel incorporates a pressure due to non-thermal motions indicated from the velocity widths of the spectral line profiles. The solid lines are based on a model of an isothermal pressure-bounded equilibrium sphere. Points above the upper line correspond to conditions for which no hydrostatic equilibrium is possible; the equilibrium is unstable if the data point lies above the middle line (see discussion in §2.3). Typical estimated uncertainties for the data points are shown by the error bars. The results suggest that most cores are near equilibrium, especially if turbulent support is taken into account. From Myers & Benson (1983).

2.2 Fragmentation and cores

Molecular cloud cores have presumably condensed out from lower-density regions of the molecular cloud. How this happens is not clear, but the development of self-gravitating molecular cloud cores may be related to gravitational instabilities. The so-called 'Jeans length' or 'Jeans mass' can provide a useful guide to the scales over which gravitational instabilities can grow and produce high-density regions.

The physical basis of the Jeans instability can be understood from the following simple argument (Binney & Tremaine 1987). Imagine an isothermal, initially uniform gas with sound speed c_s, density ρ_o, and pressure P. In this state, there are no net pressure or gravitational forces. Suppose a spherical region of this gas is slightly compressed so that its density is higher by an amount $\chi\rho_o$ within a radius r. There will now be an outward pressure force, of order of magnitude per unit mass

$$F_P \sim \nabla P/\rho \sim \chi c_s^2/r. \tag{2.2}$$

The higher density leads to an inward gravitational force per unit mass of

$$F_G \sim GM\chi/r^2 \sim G\rho_o\chi r, \tag{2.3}$$

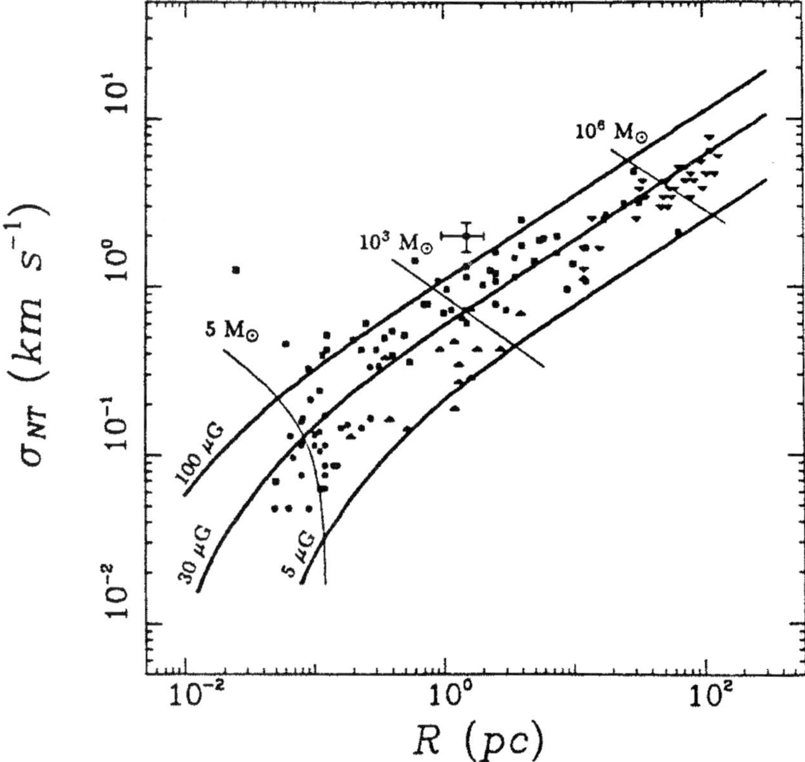

Fig. 2.3. Non-thermal velocity dispersion vs size of cloud cores. The curves show predictions for virial equilibrium models for an assumed gas temperature of 10 K (see following sections). Contours of constant mass and assumed magnetic fields are indicated. From Myers & Goodman (1988b).

where M is the mass of the cloud. Thus, gravity wins on large scales such that

$$r^2 \gtrsim \frac{c_s^2}{G\rho_\circ}.$$ (2.4)

A detailed analysis (Appendix 2) leads to a critical Jeans length

$$\lambda_J = \left(\frac{\pi c_s^2}{G\rho_\circ}\right)^{1/2},$$ (2.5)

and a critical Jeans mass

$$M_J = \lambda_J^3 \rho_\circ = \left(\frac{\pi c_s^2}{G}\right)^{3/2} \rho_\circ^{-1/2}.$$ (2.6)

According to this analysis, length scales larger than the Jeans length, or masses greater than the Jeans mass, will be unstable to gravitational collapse. Assuming a pure molecular hydrogen gas,

$$M_J \sim 750\,\mathrm{M_\odot}\,T_{10}^{3/2}\,N_{H_2}^{-1/2},$$ (2.7)

where T_{10} is the temperature in units of 10 K and N_{H_2} is the molecular hydrogen density in units of cm^{-3}.

Whether these simple considerations apply directly to real molecular clouds is uncertain. Molecular clouds exhibit supersonic motions on large scales (cf. Larson 1981; Figure 2.3), whereas the background medium is assumed to be at rest in the Jeans analysis. If the supersonic motions are appropriately 'turbulent' so that they exert an isotropic pressure equivalent to the thermal gas pressure at a higher temperature, an effective sound speed might be defined to carry through the analysis in a very approximate way. For example, the Taurus molecular cloud complex contains several large dark clouds of average density $\sim 10^3$ cm^{-3}, sizes of order $\sim$ 1–2 pc, and velocity 'dispersions' ~ 0.6 km s^{-1}. Taking the velocity dispersion to correspond to a 'sound speed', the resulting 'Jeans length' is comparable to the observed sizes, suggesting that gravity is important in confining these clouds. Since the nature of the 'turbulence' is not well understood, and may include macroscopic flows, the application of this analysis to the observed structures is uncertain.

Supersonic motions probably must be reduced by shock dissipation or some other process before low-mass star formation can proceed. Assuming that this is accomplished by some means (see McKee *et al.* (1993) for a discussion of some possibilities), equation (2.6) implies that as gas condenses to higher densities, the Jeans mass decreases. This dependence of the Jeans mass on density suggests the possibility of *fragmenting* ever smaller masses out of the original larger self-gravitating cloud, a scenario for fragmentation into star-sized clouds originally proposed by Hoyle (1953). However, there are difficulties with this idea, as discussed by Larson (1985). The basic problem is illustrated by the dispersion relation for the Jeans analysis (Appendix 2). Assuming density perturbations of the form $\delta\rho \propto \exp[i(\omega t - kx)]$, the dispersion relation is

$$\omega^2 = c_s^2 (k^2 - k_J^2), \quad k_J^2 = 4\pi G \rho_{\circ}/c_s^2. \tag{2.8}$$

When the wavenumber $k = 2\pi/\lambda$ is smaller than the critical wavenumber k_J, i.e. the wavelength of the perturbation is larger than the Jeans length $\lambda_J = 2\pi/k_J$, ω is imaginary, and there is an exponentially growing (unstable) mode. The growth rate $-i\omega$ increases monotonically with decreasing k, and is *largest* as $\lambda \to \infty$. In other words, the largest scales have the fastest growth rates and thus the fastest collapse times. This faster collapse of the larger scales makes it very hard to fragment smaller pieces out (Larson 1985).

Things change qualitatively if the background medium is not uniform, as suggested by observations of molecular clouds. Consider the next most complicated case, that of an isothermal thin sheet with a surface mass density $\Sigma_{\circ}$. As shown in Appendix 2, the dispersion relation indicates that surface density perturbations can grow exponentially in time at a rate Γ, where

$$\Gamma^2 = 2\pi G \Sigma_{\circ} k - c_s^2 k^2. \tag{2.9}$$

There is a critical wavenumber,

$$k_c = 2\pi G \Sigma_{\circ}/c_s^2, \tag{2.10}$$

above which no exponential growth is possible; i.e. there is a minimum wavelength (a 'Jeans' length) for gravitational instability. Now there is a wavenumber with a maximum growth rate, because the growth rate is zero at both k_c and $k \rightarrow 0$. Differentiating equation (2.9) with respect to k, we find the wavenumber at which the exponential growth is fastest,

$$k_f = \pi G \Sigma_\circ / c_s^2 = k_c/2. \tag{2.11}$$

This result suggests that the sheet will break up into fragments of preferred mass

$$M_f \sim \lambda_f^2 \Sigma = 4 c_s^4 / G^2 \Sigma_\circ, \tag{2.12}$$

where $\lambda_f = 2\pi/k_f$.

Consider dark clouds (cores) in the Taurus complex with $A_V \sim 5$ mag, and suppose that the extinction arises in a single (face-on) sheet. With the usual calibrations of dust extinction per mass column density (Mathis 1990; Figure 4.1), the surface density is $\Sigma_\circ \approx 3.2 \times 10^{-2} \, \mathrm{g\,cm^{-2}}$. For a pure molecular hydrogen gas at $T = 10$ K, $c_s = 0.19 \, \mathrm{km\,s^{-1}}$; then the critical wavenumber from (2.10) corresponds to a critical wavelength $\lambda_c \sim 0.05$ pc. The fastest growth rate occurs for perturbations of wavelengths $\lambda_f \sim 0.1$ pc, corresponding to $M_f \sim 2\,\mathrm{M_\odot}$. These results suggest that gravitational fragmentation may produce some molecular cloud cores in Taurus of appropriate sizes to make low-mass stars.

Although the assumption of an infinitely thin sheet is unrealistic, analysis of an isothermal, self-gravitating sheet in hydrostatic equilibrium also shows similar instability. The gas layer has a finite thickness in the direction z perpendicular to the midplane,

$$\rho(z) = \rho(0) \operatorname{sech}^2(z/H), \tag{2.13}$$

where the scale height H is defined by (Spitzer 1978)

$$H = \frac{c_s^2}{\pi G \Sigma_\circ}. \tag{2.14}$$

The critical and preferred wavenumbers of this layer are half of the corresponding values for the infinitely thin sheet (Larson 1985). This result is particularly interesting because it corresponds to a true static initial state of the background medium, unlike the Jeans analysis for the homogeneous medium. The critical wavenumber is the inverse of the scale height of the medium,

$$k_c = \pi G \Sigma_\circ / c_s^2 = H^{-1}. \tag{2.15}$$

Larson (1985) showed that similar results obtain even with added complicating factors such as rotation, magnetic fields, or filamentary geometry. In the case of equilibrium filaments, the critical wavelengths are of the order of characteristic thicknesses of the filaments. Rotation tends to suppress collapse, especially on the largest scales (Chapter 5). Magnetic fields can also prevent collapse if they are strong enough (Nakano & Nakamura 1978); treatments of stability with magnetic forces are considered in §2.6.

To summarize, in circumstances where thermal pressure provides the dominant support against gravity, and the background medium is in hydrostatic equilibrium, there is a minimum scale (the 'Jeans length') over which gravitational instabilities can grow. If the background medium is non-uniform, for example the gas is distributed in a sheet or in a filament, there is a preferred scale for instability, which is usually a few times larger than a characteristic length of the background, such as the scale height for a gaseous equilibrium sheet. Whether these considerations apply to real molecular clouds depends upon the level of turbulent supersonic motions (whose nature is not well understood) and the manner in which these motions decay or are suppressed in dense regions (e.g., McKee *et al.* 1993).

2.3 Virial theorem and cloud stability

A somewhat different, and generally older, view of gravitational instability starts from clouds with reasonably well-defined outer boundaries. Then the problem of star formation becomes the analysis of the stability of these clouds, which one might identify with molecular cloud cores. (This analysis could also be applied to isolated Bok globules, but modifications would have to be made to account for heating and photodissociation processes in their outer layers, which are relatively more exposed to the radiation fields of hot stars.)

In its simplest form, the analysis proceeds using the virial theorem. The equations of momentum and mass conservation (e.g. Shu 1992; Appendix 1) can be combined to yield

$$\rho \frac{D\mathbf{v}}{Dt} = -\nabla P - \frac{1}{8\pi} \nabla B^2 + \frac{1}{4\pi} (\mathbf{B} \cdot \nabla) \mathbf{B} - \rho \nabla \phi. \tag{2.16}$$

Since

$$\frac{D^2 r^2}{Dt^2} = \frac{D^2 \mathbf{r} \cdot \mathbf{r}}{Dt^2} = 2 (\dot{\mathbf{r}} \cdot \dot{\mathbf{r}} + \mathbf{r} \cdot \ddot{\mathbf{r}}), \tag{2.17}$$

taking the dot product of the momentum equation with the position vector $\mathbf{r}$, and integrating over a volume V which has a closed surface S, the left-hand side becomes

$$\frac{1}{2} \frac{D^2}{Dt^2} \int r^2 dm - \int dm \, v^2 = \frac{1}{2} \frac{D^2 I}{Dt^2} - 2E_k, \tag{2.18}$$

where E_k is the bulk kinetic energy of the material and the D/Dt operator is the time derivative following the motion. The generalized moment of intertia, I, is the integral of r^2 over the mass element $dm = \rho dV$.

Because translational motion does not affect the basic analysis, we set $E_k = 0$ and consider the volume to be at rest. After further manipulation and use of vector identities, we have the virial theorem,

$$\frac{1}{2} \frac{D^2 I}{Dt^2} = \int_V dV \, 3P + \int_V dV \frac{B^2}{8\pi} - \int_S \left(P + \frac{B^2}{8\pi} \right) \mathbf{r} \cdot d\mathbf{S}$$

$$+ \frac{1}{4\pi} \int_S (\mathbf{r} \cdot \mathbf{B})(\mathbf{B} \cdot d\mathbf{S}) - \int_V dV \rho \mathbf{r} \cdot \nabla \phi. \tag{2.19}$$

The integrals are taken over the closed surface S or the volume V which is enclosed by S; the unit vector normal to the surface, $d\mathbf{S}$, is taken to be pointing outward. The first term on the right hand side is a volume integral of the internal thermal energy; the second term is the magnetic energy; and the last term is the gravitational potential energy. The third and fourth terms constitute surface pressure effects; the magnetic field enters in two different ways because the force involved can be separated into pressure and tension terms (e.g., Priest 1984).

To see how this result can be used to study cloud stability, consider the case of a spherical, isothermal, uniform, unmagnetized cloud in virial equilibrium, i.e.

$D^2 I/Dt^2 = 0$. The cloud has a radius R_{cl} and mass M_{cl}, and the external medium exerts a pressure P_o at its surface. With these assumptions equation (2.19) reduces to (Spitzer 1978)

$$4\pi R_{cl}^3 P_o = 3c_s^2 M_{cl} - \frac{3}{5}\frac{GM_{cl}^2}{R_{cl}}. \tag{2.20}$$

If the mass of the cloud and its internal temperature are fixed, equation (2.20) determines the equilibrium relationship between parameters. In the limit $R_{cl} \to \infty$, gravitational forces become unimportant and equilibrium is established by a balance of the internal and external pressures. Decreasing R_{cl} increases the importance of gravity. For a given M_{cl} and c_s, there is a minimum radius

$$R_{min} = \frac{1}{5}\frac{GM_{cl}}{c_s^2}, \tag{2.21}$$

below which there is no possible equilibrium state. This equation simply expresses the inability of thermal pressure forces to support a cloud against gravity if it is too small for its mass (e.g. equation (2.1)).

In practice this minimum radius is not the critical condition, because not all equilibria are stable. This can be seen by considering the relationship between the external pressure and the cloud radius in more detail. In equilibrium, the derivative of P_o as a function of radius is

$$\frac{dP_o}{dR_{cl}} = \frac{1}{4\pi}\left(-\frac{9c_s^2 M_{cl}}{R_{cl}^4} + \frac{12}{5}\frac{GM_{cl}^2}{R_{cl}^5}\right). \tag{2.22}$$

At large R_{cl}, $dP_o/dR_{cl} < 0$, so an increase in external pressure causes the radius of the cloud to shrink. However, there is a maximum external pressure in equilibrium, because equation (2.22) changes sign at a critical radius. The maximum pressure, from equation (2.22), occurs when the cloud has an equilibrium radius slightly larger than R_{min},

$$R_{crit} = \frac{4}{15}\frac{GM_{cl}}{c_s^2}; \tag{2.23}$$

the corresponding maximum pressure from equation (2.20) is

$$P_{crit} = 3.15\frac{c_s^8}{G^3 M_{cl}^2}. \tag{2.24}$$

There is no possible equilibrium for pressures larger than this critical value. At lower pressures collapse may still ensue if $R_{cl} < R_{crit}$. In this case, equation (2.22) shows that the equilibrium pressure must decrease with decreasing radius. This is an unstable equilibrium; for a fixed pressure, a perturbation decreasing the cloud radius will result in the external pressure being larger than the maximum equilibrium value, which will cause the cloud to contract further, etc. The process runs away; the gas pressure becomes increasingly less important than gravity, and free-fall collapse eventually ensues, i.e. gravitational forces are essentially unopposed by pressure forces (Chapter 3).

If we recast this critical condition in terms of the density ρ_o of the uniform sphere and a 'critical mass' $M_{crit} = 4\pi R_{cl}^3 \rho_o / 3$,

$$M_{crit} = \left(\frac{3}{4\pi}\right)^{1/2} \left(\frac{15}{4}\right)^{3/2} \left(\frac{c_s^2}{G}\right)^{3/2} \rho_o^{-1/2}. \tag{2.25}$$

Thus the stability analysis can be transformed into a relationship between the mass of cloud and its internal density and temperature, in a form which differs only by constants of order unity from the Jeans mass relation (2.6).

The existence of a maximum pressure for a cloud of given mass, or a minimum radius, arises from the assumption that the cloud internal temperature remains constant, and does not increase under contraction. Observations indicate that most molecular cloud cores in Taurus have roughly the same temperature (e.g., Cernicharo 1991). The cloud temperature depends upon the balance of heating and cooling processes; for typical physical conditions in Taurus, where heating of the cloud is probably dominated by cosmic rays (except for outer, exposed regions where heating by ultraviolet radiation may be important (Hollenbach & Natta 1995)), and the cooling by rotational bands of molecules (primarily CO), the temperature is relatively independent of density over a wide range of parameters (Goldsmith 1988; Boland & DeJong 1984). These results suggest that the above simple stability analysis is relevant to real molecular clouds.

Equation (2.23) may be converted into a more suitable form for comparison with observations,

$$\frac{\rho_o}{c_s^2} = \frac{45}{16\pi G R_{crit}^2}. \tag{2.26}$$

The lines labeled 'critical' in Figure 2.2 approximate this relation. The observations of cloud cores suggest that many are near the critical equilibrium point (Myers & Benson 1983; Myers & Goodman 1988a,b; Benson & Myers 1989).

As is the case with the Jeans analysis, the observation of 'turbulent' motions which can be important or even dominate thermal motions on large scales poses a difficulty for the application of the virial equilibrium. Hydrostatic equilibrium models have been constructed for cloud cores which include supersonic turbulence as a spatially-varying, isotropic pressure (Lizano & Shu 1989; Myers & Fuller 1992; Caselli & Myers 1995). It may be necessary to include the effects of supersonic motions in some cases, especially in more massive cloud cores, but it is far from clear that the observed supersonic motions really can be treated as exerting a microscopic, isotropic pressure; the discarded kinetic energy terms in equation (2.18) may not be negligible.

2.4   Fragmentation vs 'cloud' stability

Before further discussion of equilibria, it is worth considering the possible distinctions between the 'fragmentation' and 'cloud' virial equilibrium pictures. Beyond some modest differences in geometry, both approaches yield similar results in terms of defining a critical or 'Jeans' mass. The basic distinction between these viewpoints lies in initial conditions. In the cloud scenario, one posits a self-gravitating molecular cloud (core) of a given mass and initial radius in hydrostatic equilibrium. This cloud may or may not be close to the critical condition for collapse. It may require just a small perturbation, for example, in the external pressure for collapse to ensue; or it may be very strongly stable and resistant to its self-gravity. In the gravitational fragmentation scenario, a 'cloud' cannot be distinguished from the background medium *until it has started to collapse*; any self-gravitating cloud fragmenting out of the ambient medium is by definition unstable.

Probably both viewpoints are applicable in different parts of molecular clouds. There is no question that small molecular clumps or clouds exist which are not strongly self-gravitating, and for which external pressures play an essential role in their confinement (cf. Elmegreen 1991); gravitational fragmentation cannot be relevant to the production of such clouds. On the other hand, the molecular cloud cores which are thought to produce stars must be self-gravitating. If gravitational fragmentation produces cores, it is 'too late' to apply virial equilibrium to these objects, because they are already contracting; the instability analysis must be applied at an earlier stage.

It is not clear at present whether molecular cloud cores form 'slowly' or 'quickly' (e.g, Bonnell *et al.* 1996a); i.e. it is not obvious if cores are built up over many dynamical timescales, so that hydrostatic equilibrium is a good approximation, or if cores are formed dynamically, as in the gravitational fragmentation scenario. Detailed observations of cloud cores generally suggest that they are close to virial equilibrium (Benson & Myers 1989; Myers & Goodman 1988a,b); however, the uncertainties in the observations do not rule out the possibility that cores are collapsing (or even expanding).

In principle one might be able to estimate the lifetimes of cores from their numbers. If cores are very stable, and therefore long-lived, there should be many more cores without stars than cores with (young) stars. Initial studies suggested that about half of the known molecular cloud cores in Taurus contain YSOs (Beichman *et al.* 1986; §2.1); a more recent study in Taurus, using IRAS emission to define dusty cores to survey cores more systematically (Wood, Myers, & Daugherty 1994), found a total of ~ 50 cores, of which about 1/4 contained IRAS sources (i.e. YSOs with strong far-infrared emission). These initial results suggest that cores may not last very much longer than the ages of the stars they produce; this would mean that cores have lifetimes $\lesssim 10^6$ yr (cf. Chapter 1), suggesting that cores evolve fairly rapidly (see also Onishi *et al.* (1996)). However, there are many difficulties in attempting this comparison; systematic biases may still remain in core surveys, and the assumption that all cores form stars may be incorrect.

Whether cores can be treated as equilibrium entities may also depend upon the density of forming stars. In regions forming clusters, the collapse process may involve collective effects such as the gravitational potential well of the cluster as a whole (e.g., Zinnecker 1982; Bonnell *et al.* 1996b). Interactions between individual stars probably introduce other effects which are not included in the standard picture of isolated star formation.

2.5   Centrally-concentrated clouds

The virial equilibrium analysis of §2.3 assumed a uniform density cloud for simplicity; however, such clouds are not in hydrostatic equilibrium, because there are no internal pressure gradients to balance gravity. A more appropriate equilibrium initial state can be found from solving the force balance equation. Assuming spherical geometry and no magnetic pressure, the equation of hydrostatic equilibrium is

$$\frac{dP}{dr} = -\rho \frac{GM_r}{r^2}, \tag{2.27}$$

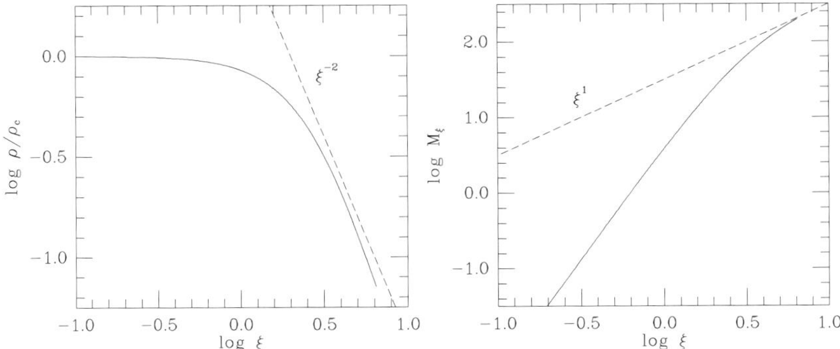

Fig. 2.4. Density and mass distributions for the critical Bonnor–Ebert sphere, i.e. with the maximum central concentration allowed by stability. The left-hand panel illustrates the density distribution with normalized radius ξ (see text), while the right-hand panel illustrates the mass enclosed within ξ. The dashed lines show the power-law dependence of the singular isothermal sphere for comparison.

where

$$\frac{dM_r}{dr} = 4\pi r^2 \rho. \tag{2.28}$$

For an isothermal cloud, these equations can be combined to yield

$$\frac{1}{r^2}\frac{d}{dr}r^2 c_s^2 \frac{d\ln\rho}{dr} = -4\pi G\rho. \tag{2.29}$$

Making the substitution $\ln(\rho/\rho_c) \equiv -u$, one arrives at the Emden equation

$$\frac{1}{\xi^2}\frac{d}{d\xi}\xi^2\frac{du}{d\xi} = e^{-u}, \tag{2.30}$$

where $\xi = r/(4\pi G\rho_c/c_s^2)^{1/2}$ is the non-dimensional radial coordinate.

If ρ_c is taken to be the central density, then the boundary conditions of (2.30) are $u(0) = 0$ and $du/d\xi|_0 = 0$ (by symmetry). There is a family of solutions to this equation, known as Bonnor–Ebert spheres (Ebert 1955; Bonnor 1956), each distinguished by an outer radius ξ_1. The results of integrating the Emden equation are shown in Figure 2.4 for the case of critical stability (see below). The limiting case, where $\xi_1 \to \infty$, is that of the singular isothermal sphere,

$$\rho = \frac{c_s^2}{2\pi G}r^{-2}, \tag{2.31}$$

which corresponds to an infinitely-concentrated configuration. One may verify that this is a solution by direct substitution into equation (2.27). Equation (2.31) provides a convenient initial power-law density distribution which is often used in analyzing the collapse of protostellar clouds (§3.2).

Like the uniform sphere, the Bonnor–Ebert sphere also has a maximum external pressure or minimum stable radius for a given mass and temperature. The variation of P_o with R_{cl} for solutions of equation (2.30) has basically the same form as for the uniform cloud (2.22). The critical radius and pressure for the Bonnor–Ebert sphere exhibit the same parameter dependence as in the case of the uniform sphere, but with slightly different numerical coefficients (Spitzer 1968),

$$R_{crit} = 0.41 \frac{GM_{cl}}{c_s^2},$$ (2.32)

and

$$P_{crit} = 1.40 \frac{c_s^8}{G^3 M_{cl}^2}.$$ (2.33)

Figure 2.4 illustrates the variation of density and mass with radius for this limiting solution. The density variation is similar to that of the the singular isothermal sphere over roughly the outer half of the cloud. In the inner regions, the density approaches a constant. The central density is ~ 5.8 times the mean density; this modest concentration explains why the limiting pressure for the critical Bonnor–Ebert sphere is not much different from that calculated with the simple uniform density assumption. (Note that the singular isothermal sphere is extremely unstable to infinitesimal perturbations, and thus cannot represent a realizable hydrostatic equilibrium.)

For a $1 \, M_\odot$ cloud at $T = 10$ K, we have $R_{crit} = 1.5 \times 10^{17}$ cm ~ 0.05 pc $\sim 10^4$ AU and $P_{crit} \sim 2 \times 10^{-11}$ dyne cm^{-2}. This external pressure is about an order of magnitude higher than typical interstellar medium pressures $P \sim 10^4 k \sim 1.4 \times 10^{-12}$ dyne cm^{-2} (Elmegreen 1991); turbulent motions, or a high density in the external molecular cloud region, may provide additional confining pressure.

2.6   Stability of magnetized clouds

Magnetic fields can help support clouds against their self-gravity. The subject is highly complex, and the effects of magnetic fields are treated here only in a very schematic way.

Returning to the case of the uniform-density spherical cloud for simplicity, we follow Spitzer (1978) and assume that the magnetic field is also uniform within the cloud. It is necessary to allow for decreasing magnetic field strength outside the cloud, which is assumed to be surrounded by a low-density medium. A convenient form to adopt is $B = B_{cl}(R_{cl}/r)^3$, which roughly reproduces the variation of energy density with radius in a dipole magnetic field.

If one takes the bounding surface S to be far outside the cloud at a radius $R_b \gg R_{cl}$, then the magnetic surface integral terms become negligible. Then the terms in equation (2.19) are

$$0 = 3c_s^2 M_{cl} + \int_{R_{cl}}^{R_b} dV \, 3P_\circ - 4\pi R_b^3 P_\circ + \frac{1}{6} B_{cl}^2 R_{cl}^3 + \int_{R_{cl}}^{R_b} dV \, \frac{B^2}{8\pi}$$
$$- \frac{3}{5} \frac{GM_{cl}^2}{R_{cl}} - \int_{R_{cl}}^{R_b} dV \, \rho r \nabla \phi.$$ (2.34)

The very last term in (2.34) can be written as $\int dM(r\nabla\phi)$, and since by assumption the mass exterior to the cloud is negligible, we may set this term to zero. It is straightfoward to show that the integral of the magnetic energy density outside of the cloud equals the volume integral inside the cloud as $R_b/R_{cl} \to \infty$. Finally, the volume integral of the external pressure term outside the cloud is $4\pi P_\circ(R_b^3 - R_{cl}^3)$. With this result, equation (2.34) becomes

$$4\pi R_{cl}^3 P_{\circ} = 3c_s^2 M_{cl} - \frac{3}{5}\frac{GM_{cl}^2}{R_{cl}} + \frac{1}{3}R_{cl}^3 B_{cl}^2. \tag{2.35}$$

Under conditions where the gas is sufficiently conducting, the magnetic flux passing through a given parcel of gas remains constant (e.g., Priest 1984). This 'flux-freezing' condition requires that the magnetic flux $\Phi_B = \pi R_{cl}^2 B_{cl}$ remain constant. Under this condition, the gravitational and magnetic terms in equation (2.35) both vary as R_{cl}^{-1}, i.e. they remain in the same ratio. Thus, with magnetic flux freezing, if magnetic forces do not prevent collapse at one time, they will not be able to prevent collapse at any later time. Conversely, flux-freezing implies that clouds in which magnetic forces dominate (and therefore resist) gravity will always maintain this relationship and therefore gravitationally-induced collapse cannot occur.

The gravitational energy term in equation (2.35) exceeds the magnetic energy term if the mass of the cloud is larger than a critical value, which depends upon the magnetic flux through the cloud,

$$M_B = \frac{1}{\pi}\left(\frac{5}{9G}\right)^{1/2}\Phi_B. \tag{2.36}$$

This result can also be written in terms of the mean density of the cloud ρ_c,

$$M_B = \frac{5^{3/2}}{48\pi^2}\frac{B_{cl}^3}{G^{3/2}\rho_c^2}. \tag{2.37}$$

Gravitational collapse can occur only if $M_{cl} > M_B$. As in the case of the non-magnetized sphere, the external pressure at the critical equilibrium point is

$$P_{crit} = \frac{3.15c_s^8}{G^3 M_{cl}^2 [1 - (M_B/M_{cl})^{2/3}]^3}. \tag{2.38}$$

Comparison with equation (2.24) illustrates the effect of the magnetic field in resisting collapse.

More sophisticated calculations allowing departures from uniformity and sphericity (which must be present since the magnetic field is never isotropic) have been made which exhibit the same basic behavior as this simple model, but differ modestly in numerical coefficients. Calculations of flattened, centrally-condensed, equilibrium clouds suggest that equation (2.36) is more accurately written as (Mouschovias & Spitzer 1976)

$$M_B \sim \frac{0.13}{G^{1/2}}\Phi_B. \tag{2.39}$$

From equation (2.39) we find immediately that the maximum magnetic flux that a $1\,\mathrm{M_\odot}$ cloud core can have and still contract gravitationally is $4 \times 10^{30}\,\mathrm{G\,cm^2}$.

Alternatively, we may write

$$M_B \sim 3.5 \times 10^{-3}\frac{B_o^3}{G^{3/2}\rho_c^2} \approx 10^2\,\mathrm{M_\odot}\left(\frac{B_{cl}}{100\,\mu\mathrm{G}}\right)^3\left(\frac{N_{H_2}}{10^4\,\mathrm{cm^{-3}}}\right)^{-2}, \tag{2.40}$$

where we have assumed that the gas is entirely composed of molecular hydrogen and has a mean molecular weight of 2. At typical cloud core densities $10^4\mathrm{cm^{-3}}$, a $M_{cl} \sim 1\,\mathrm{M_\odot}$ core will be able to collapse only if $B_{cl} \leq 20\,\mu\mathrm{G}$. With this magnetic field, and assuming an internal core temperature of 10 K, the ratio of gas pressure to magnetic pressure is $\beta = \rho_c c_s^2/(B_{cl}^2/8\pi) \sim 1$, i.e. the magnetic energy (pressure) is roughly in equipartion with the thermal energy (pressure).

2.7 The magnetic flux problem(s)

If the magnetic field is frozen to the fluid, and if reconnection or diffusion does not eliminate magnetic flux, then any mass once threaded by specific magnetic field lines is always threaded by those field lines, and the magnetic flux Φ_B remains constant. This assumption suggests difficulties in making solar-type stars. For example, a reasonable initial condition might be derived from properties of diffuse neutral hydrogen clouds of modest visual extinction, from which molecular clouds may condense. Measurements of the Zeeman effect in such H I clouds (Myers *et al.* 1995) suggest magnetic field strengths of order $\sim 10\,\mu$G for gas of approximate density $n_H \sim 10^2\,\mathrm{cm}^{-3}$. A spherical volume encompassing one solar mass would then have a radius of ~ 0.4 pc and a magnetic flux of $5 \times 10^{31}\,\mathrm{G\,cm}^2$. This is an order of magnitude larger than the critical flux for a solar-mass star to collapse (equation (2.39)), suggesting that something must happen to reduce the magnetic flux per unit mass.

Magnetic fields do not exert isotropic forces in general; gas will tend to move *along* magnetic field lines (unless turbulent, probably magnetic, pressure opposes this motion; Gammie & Ostriker (1996)), allowing the gas to concentrate relative to the magnetic field. Mouschovias and collaborators have shown that this effect can be quite important, resulting in a relationship between mean magnetic field strength and gas density $B \propto \rho^n$, where $1/3 \gtrsim n \gtrsim 1/2$ (see Mouschovias (1991) for an exhaustive review on this and related topics). The limit $n \sim 1/2$ corresponds to a balance between thermal and magnetic pressures, $B^2/8\pi \propto \rho c_s^2$, as discussed in §2.6. Observations suggest that, when magnetic field strengths can be measured, they frequently are roughly consistent with *equipartition* strength with the other pressures (Myers *et al.* 1995). This is intuitively reasonable, since one might suppose that the magnetic field can only be compressed by thermal gas (or other) pressures to a comparable level, and vice versa. Using these ideas to scale from the H I clouds, one might infer a magnetic field strength of order 30 μG at a density of order $10^3\mathrm{cm}^{-3}$, consistent with some measurements of dark clouds (Crutcher *et al.* 1994), such as B1 (§2.1; Goodman *et al.* 1989). However, even with this field strength and density, the critical mass (2.40) is still hundreds of solar masses.

The large extinctions through a molecular cloud core may prevent the ultraviolet radiation field of the galaxy (or of the nearby hot stars) from penetrating into the core interior. These radiation fields photoionize species, and the resulting ions help couple the gas to the magnetic field. In a dark cloud interior, where the external ionizing radiation fields do not penetrate, the ion fraction drops to very low values. Although the ions are 'frozen' to the magnetic fields by the Lorentz force (e.g., Priest 1984), the neutral species are not. The neutrals are affected by the magnetic field only indirectly through collisions with ions, which are following the field. If the frequency of collisions between ions and neutrals is sufficiently low, as may occur in dark cloud cores, the neutral gas can 'slip' through the field lines on interesting timescales. In principle, this process of 'ambipolar diffusion' represents a means by which the molecular gas can be condensed further relative to the magnetic field (Mestel & Spitzer 1956). In a cloud core, the gravitational force will pull mass in through the resisting magnetic field, concentrating the cloud as the magnetic field is slowly 'left behind'.

To illustrate this process in an approximate way, we follow Spitzer (1978) in considering an infinite cylinder of uniform-density molecular gas. We distinguish between the neutral particles, which have a mass density $\rho_n = n_n m_H \mu$, where μ is the mean molecular weight, and the ions, which have a number density n_i. Assuming the neutral particles comprise most of the mass, Gauss' law (e.g. Priest 1984) applied to the cylindrical configuration yields the gravitational force per unit mass

$$-\nabla\phi \; = \; 2\pi R \, G\rho_n, \tag{2.41}$$

where ϕ is the gravitational potential and R is the distance measured from the cylinder's axis. For simplicity we assume that the magnetic field is responsible for providing the balancing force against gravity. In this situation the neutrals do not directly feel the magnetic restoring force, but are affected indirectly by collisions with the ions, which are tied to the field. The momentum transfer to the neutrals from the ions is

$$\rho_n \, (n_i \, <u\sigma>) \, w_D, \tag{2.42}$$

where u is the relative velocity (assumed to be a thermal or random velocity) between ions and neutrals, σ is the collisional cross-section, and w_D is the drift velocity of the neutrals relative to the ions. The term in brackets is the net collision rate. Balancing this force against gravity, and solving for a characteristic diffusion time,

$$t_D \; \equiv \; \frac{R}{w_D} \; = \; \frac{<u\sigma>}{2\pi G m_H \mu} \left(\frac{n_i}{n_H} \right), \tag{2.43}$$

where we have scaled the neutral density in terms of the density of hydrogen atoms n_H. Often $<u\sigma>$ varies slowly with temperature. Taking $<u\sigma> \sim 2 \times 10^{-9} \; \mathrm{cm^3 \, s^{-1}}$,

$$t_D \; \sim \; 5 \times 10^{13} \left(\frac{n_i}{n_{H_2}} \right) \; \mathrm{yr}, \tag{2.44}$$

now incorporating a scaling to the number density of neutral hydrogen molecules n_{H_2} since these are likely to be the dominant constituents of regions where ambipolar diffusion may be applicable. It is clear that the ambipolar diffusion process will be important for star formation only if $n_i/n_{H2} \lesssim 10^{-7}$. Typical estimates based on cosmic-ray ionization of the molecular gas suggest $n_i/n_{H2} \sim 10^{-7}(n_{H2}/10^4 \; \mathrm{cm^{-3}})^{-1/2}$ (McKee 1989). Thus it appears that in conditions representative of molecular cloud cores, ambipolar diffusion can remove sufficient magnetic flux over timescales of $\sim 10^7$ yr and this would eventually lead to collapse (see also Umebayashi & Nakano (1990)).

The idea of ambipolar diffusion plays an important role in what might be termed the 'standard' picture of low-mass star formation (Nakano 1984; Mestel 1985). As summarized by Shu *et al.* (1987), the idea that ambipolar diffusion *must* occur before star formation takes place leads naturally to the 'slow' picture of star formation (§2.4). Because the diffusion time is much longer than the free-fall time, magnetic support helps establish hydrostatic equilibrium before the diffusion of magnetic field in the central regions precipitates collapse. The basic picture is attractive because it essentially states that, given enough time, a molecular cloud core will collapse to form a low mass star even if it is initially strongly 'magnetically subcritical', i.e. $M_{cl} < M_B$.

In recent years serious challenges to the ambipolar diffusion picture of star formation have arisen. One problem is that ionization fractions in many star-forming regions may be higher than previously thought, because the photoionization by hot stars has been underestimated, especially in regions of massive star formation (Myers & Khersonsky 1995). Perhaps an even more fundamental problem is the increasing evidence for rapid star formation, not only in low-density regions like Taurus (Hartmann *et al.* 1991), but especially in dense regions where star clusters form (e.g., Herbig & Terndrup 1986). Observations suggest that the bulk of star formation in the Orion Nebula Cluster occurs over periods of 1 Myr (e.g., Hillenbrand 1997); this timescale is uncomfortably short compared with typical estimates of the time required for ambipolar diffusion. The formation of dense clusters of stars probably provides the stiffest challenge to the ambipolar diffusion picture.

The application of the virial theorem to cloud stability which led to equation (2.40) assumed that the magnetic field exterior to the 'cloud' did not encompass any significant amount of mass. The geometry of the external magnetic field and the structure of the external medium may invalidate this assumption. The large critical masses indicated by equation (2.40) may simply refer to a larger cloud which produces a cluster or group of stars, later fragmenting into lower-mass condensations. To emphasize this point, consider the cloud with a magnetic field strength of 30 μG and a density of 10^3 cm^{-3} as discussed above. The *minimum* Jeans mass for this cloud is *also* much larger than a solar mass using equations (2.6) or (2.7). Indeed, if one supplements the thermal velocities at $T = 10$ K by turbulent motions which dominate on larger scales, the 'effective' Jeans mass may be comparable to the magnetic critical mass. The point is that low-density clouds cannot collapse directly under gravity to solar-mass stars. Larger clouds must be involved (presumably the clouds out of which groups and clusters form), so it is misleading to pose the magnetic flux problem in terms of the direct formation of low-mass stars from low-density conditions.

Another way of looking at the magnetic flux problem is to note that the condition $M_{cl} > M_B$ is equivalent to a critical surface density through the cloud, which in turn can be translated into a critical mean visual extinction (Shu *et al.* 1987)

$$A_V > 4 \,\mathrm{mag} \left(\frac{B}{30 \,\mu\mathrm{G}} \right) . \tag{2.45}$$

Using the total column density or extinction integrated through the entire cloud in principle addresses the issue of criticality on large scales without reference to the final desired stellar mass scale. Stars form in a variety of environments, from regions of very high column densities and extinctions in star-forming regions such as the Orion Nebula (Herbig & Terndrup 1986; Hillenbrand 1997), to low-density, low-extinction regions like Taurus. This wide range in initial conditions motivated Shu, Adams, & Lizano (1987) to suggest that there might be two distinct modes of star formation, depending upon whether the initial cloud is magnetically subcritical or supercritical, i.e. whether M_{cl} is less than or greater than M_B (equation (2.39)). Regions like Taurus, which has a visual extinction through its 'envelope' of $A_V \sim 1$–2 mag, and estimated magnetic field strengths of $30 \,\mu$G, could be mostly subcritical; stars then must be formed through the process of ambipolar diffusion (Shu *et al.* 1987). In contrast, regions like the Orion Nebula Cluster are magnetically supercritical, which produces stars with high efficiency and makes it possible to form a bound cluster. Because we now recognize that the Orion Nebula Cluster and other dense regions form low-mass stars in great abundance, in addition to making high-mass stars, the above argument implies that many low-mass stars are formed in magnetically supercritical conditions – immediately calling into question the importance of ambipolar diffusion in typical star formation.

Given these complications, it is worth returning to the original virial equilibrium argument which led to the concept of the magnetic flux problem for cloud collapse in the first place. From equation (2.35), one sees that if the external pressures can be neglected, *a magnetically subcritical cloud cannot be in equilibrium.* By definition the magnetic term for subcritical clouds is larger than the gravitational term; thus, virial equilibrium cannot be maintained. A finite contribution from internal gas pressure simply adds to the expansion. Magnetically subcritical clouds must be confined at an important level by external pressure; otherwise, they will expand (Fiedler & Mouschovias 1993). Conversely, if star-forming clouds near equilibrium are mostly confined by internal gravity, not external or surface pressures on the cloud, then they will be automatically supercritical (because gravity must counteract not only magnetic but thermal and turbulent pressures).

The low efficiency of turning gas into stars in low-density regions of star formation like Taurus is usually attributed to the need for ambipolar diffusion to take place before gravitational collapse occurs, resulting in slowed contraction for all clouds. However, one might also interpret the observations as suggesting that the Taurus complex is not that far from criticality – as should be the case if gravity is the force which concentrates molecular material in opposition to magnetic and turbulent pressure forces. There are regions in Taurus where $A_V \sim 5$, comparable to the limit indicated in equation (2.45); many stars may have emerged from their natal envelopes, so that the *original* column densities of their clouds were higher than $A_V \sim 1$. In addition, the magnetic field structure and strength in the region is not very well known. If some portions of the Taurus cloud are supercritical, while others are subcritical, the supercritical portions can form stars rapidly enough to explain observations (Hartmann *et al.* 1991), while the subcritical portions might be dispersed by stellar heating and winds long before ambipolar diffusion can take place. In such a picture, the efficiency of star formation reflects the relative amounts of supercritical and subcritical gas in the cloud. Since Taurus may only form stars from the molecular gas with an efficiency of a few per cent (Cohen & Kuhi 1979), this interpretation does not require most of the molecular gas in the Taurus molecular complex to be supercritical.

The magnetic flux problem is not the only unresolved issue about the role of magnetic fields in star-forming regions. The efficiency with which gas is turned into stars must also depend upon the rate of dissipation of supersonic (magnetic) turbulence, the nature of which is currently unclear. Manifestly much remains to be understood about the constraints of magnetic fields on star formation.

Sometimes another magnetic flux problem is emphasized. The critical magnetic flux for a $0.5\,\mathrm{M}_\odot$ star, $\Phi_B \sim 2 \times 10^{30}\,\mathrm{G\,cm^2}$, corresponds to a magnetic field of $\sim 3 \times 10^7\,\mathrm{G}$ at the radius of a typical pre-main-sequence star. Current estimates for T Tauri stars suggest much lower stellar surface magnetic fields, $B \lesssim 10^3\,\mathrm{G}$ (Basri, Marcy, & Valenti 1992). In the absence of diffusive or reconnective processes, this stellar field strength implies a very low magnetic field strength at the point of cloud core collapse; once the collapse gets underway, the free-fall time is so much shorter than the diffusion time that the magnetic field is effectively dragged in with the gas (e.g., Galli & Shu 1993a,b). It is implausible that ambipolar diffusion continues until the magnetic flux is three or four orders of magnitude below values that would permit collapse (Mouschovias 1991). Other processes which occur after collapse must be responsible for the low magnetic fields of T Tauri stars. One possibility is that the magnetic field decouples from the infalling material because of the decrease in ionization fraction at high densities; calculations suggest this may occur in the very innermost regions of the collapsing cloud, where $\sim 10^{10}$–10^{12} cm^3 (Umebayashi & Nakano 1990). In addition, magnetic reconnection may play an important or even dominant role in reducing the magnetic flux which remains in pre-main-sequence stars. If star formation proceeds by accretion from a circumstellar disk, the bulk of magnetic field loss could occur through reconnection and diffusion in the disk.

Early theoretical investigations of star formation emphasized the potential role of the magnetic field of the molecular cloud in solving the 'angular momentum problem' (Ebert *et al.* 1960; Mestel 1965). A magnetic field threading a molecular cloud core can transfer angular momentum to the external medium on a timescale comparable to the time it takes for the Alfvén waves generated by the rotating cloud to flow into, and thus transfer angular momentum to, an external region with a moment of inertia comparable to that of the cloud core. This timescale is generally compared with the ambipolar diffusion time, and so could be a very important process, especially for subcritical star formation (see Mouschovias (1991) for an exhaustive discussion). However, many observed molecular cloud cores do appear to retain significant angular momentum, in amounts that are not important in supporting the core against gravity (Goodman *et al.* 1993), but are much larger than would permit collapse to stellar dimensions.

Once dynamical collapse begins, the angular momentum should be retained (because the free-fall time is necessarily comparable to or shorter than the angular momentum transport time; Shu (1995)). This retention of angular momentum during collapse now does not seem to be a problem for two reasons: first, the outcome of collapse must in general be a binary system; and second, we now believe that *accretion disks* are present around many young stars. As discussed in Chapters 5–7, the observed disks appear to provide sufficient angular momentum transport to solve much of the angular momentum 'problem' for star formation.

3

Protostellar cloud collapse

The developments of the previous chapter suggested that molecular cloud cores may evolve to gravitationally-unstable configurations, based on the inability of the cloud core to heat up as it contracts. The onset of gravitational collapse may be aided by the inward diffusion of the mostly neutral gas relative to the magnetic field, reducing magnetic support of the cloud. Instability may also occur because of an increase in the external pressure, or perhaps simply because the core was never in hydrostatic equilibrium to begin with, but gravitationally fragmented from a larger cloud. In any case, all studies indicate that soon after collapse starts, pressure forces become relatively unimportant in impeding the flow, and the cloud falls in nearly freely under the action of gravity.

One may anticipate the general nature of the collapse for a thermally-supported spherical molecular cloud core by estimating that the initial infall velocity should be of the order of the free-fall velocity from the outer cloud radius,

$$v_{in} \sim (GM/R)^{1/2} \sim c_s \, ; \tag{3.1}$$

therefore, the collapse time is of order $t_{in} \sim R/c_s$ and the mass infall rate is

$$\dot{M} \sim M/t_{in} \sim c_s^3/G. \tag{3.2}$$

The collapse can be halted basically for one of two reasons. First, if the gas temperature increases to the point that the sound speed is comparable to the escape velocity, the thermal pressure will become large enough to combat gravity. Numerical simulations indicate that this can occur only on distance scales much smaller than the initial cloud, where the optical depths become large enough to trap the radiant energy released by infall. It seems likely that in many cases a second reason is more important in stopping collapse, namely the angular momentum of the infalling material, which causes it to fall out onto a rotating disk. Most clouds apparently collapse to binary (or multiple) star systems; binary systems may form with disks initially around both objects.

In this chapter we review some simple physical models for collapsing protostellar clouds. In Chapter 4 the predictions of these models are compared with observations of candidate protostars.

3.1 Free-fall collapse of a uniform cloud

The simplest model one might imagine of a molecular cloud core is that of a spherical cloud of uniform density. As discussed in Chapter 2, cloud cores can become gravitationally unstable because the internal pressure forces can no longer balance gravity. As a limiting case, consider the collapse of a uniform, spherical cloud with no gas pressure to counteract gravity. The equation of motion of a shell of material which starts at radius $r_\circ$ is

$$\frac{d^2 r}{dt^2} = -\frac{GM_r}{r^2} = -\frac{4\pi G\rho_\circ r_\circ^3}{3\,r^2}, \tag{3.3}$$

where $\rho_\circ$ is the initial density and M_r is the mass interior to radius r. Here we have taken advantage of Newton's result that, for a spherical mass configuration, only the mass interior to the point in question has any gravitational effect. Multiplying through by dr/dt and integrating once, and then making the substitution $r/r_\circ = \cos^2 \beta$, the resulting solution is

$$\beta + \frac{1}{2}\sin 2\beta = \left(\frac{8\pi G\rho_\circ}{3}\right)^{1/2} t, \tag{3.4}$$

where t is the time from the beginning of the collapse (when $r = r_\circ$).

At a given time t, β is fixed no matter what the original starting radius $r_\circ$ was. Therefore, the shells do not cross, and they all reach the center at the (same) 'free-fall' time,

$$t_{ff} = \left(\frac{3\pi}{32G\rho_\circ}\right)^{1/2} \sim \frac{3.4 \times 10^7}{N_{H_2}^{1/2}}\,\mathrm{yr}, \tag{3.5}$$

where N_{H_2} is the number density of a pure molecular hydrogen gas.

A $\sim 1\,\mathrm{M_\odot}$ cloud core in hydrostatic equilibrium maintained by thermal pressure support at ~ 10 K has a radius of approximately 0.1 pc (Chapter 2). If for some reason the supporting gas pressure were 'turned off', equation (3.5) implies a free-fall time of $\sim 5 \times 10^5$ yr. Even with pressure forces retarding the infall, it is clear that the protostellar collapse phase is likely to be fairly rapid.

The free-fall result also helps provide some insight into the Jeans stability results of Chapter 2 and Appendix 2. The pressureless collapse calculation becomes increasingly relevant on the largest scales, where pressure forces are relatively unimportant. Because the free-fall time scales as $t_{ff} \propto \rho_\circ^{-1/2}$, a (pressureless) uniform density sphere collapses with material from all radii arriving at the center simultaneously. This means that the outermost radii must fall in fastest, i.e. must have the fastest growth rates, as illustrated by the dispersion relation for the standard Jeans analysis.

3.2   Similarity solution for collapse

The simple free-fall collapse discussed above is not applicable to the general proto-stellar collapse situation, because pressure forces are likely to be important – indeed, must be important if the protostellar cloud is nearly in hydrostatic equilbrium to start with. Numerical solutions of protostellar cloud collapse have been presented by Larson (1969a,b), Appenzeller & Tscharnuter (1974), Bodenheimer (1978), Winkler & Newman (1980a,b), Boss & Black (1982), and others. If considered in its full general-ity, the problem is made difficult by the need to treat both the very large scales of the initial cloud along with the very small scales of the resulting stellar core, placing great demands on the accuracy of numerical solutions (see Tscharnuter (1991); Chapter 9). Moreover, the collapse of real clouds must have many complicating factors which are difficult to incorporate. It is therefore worth developing simple models which illustrate the basic features of collapse.

As discussed in §2.5, the density distribution in the outer layers of an isothermal sphere in hydrostatic equilibrium approaches $\rho \propto r^{-2}$. This approach to a power-law density distribution suggests that (ignoring the inner regions) one might find a similarity solution for the collapse. In a similarity solution, non-dimensional functions for properties like the density and velocity field can be used to describe the motion at any time by some appropriate scaling. The tendency of the numerical calculations to adjust to a (roughly) self-similar structure led Larson (1969a), Penston (1969), Hunter (1977), and Shu (1977) to develop similarity solutions for the infall problem. Of course, any similarity solution for protostellar collapse must fail at early and late times, when the influence of inner and outer boundary conditions must be felt; however, the mathematical simplicity of such solutions leads to a basic understanding of many essential physical features. We follow Shu's (1977) elegant development for the singular isothermal sphere to illustrate the basic physics of the problem in a particularly simple and straightforward way.

Before considering the mathematical details it is useful to outline the qualitative properties of the collapse solution. If the cloud density initially has the singular form $\rho \propto r^{-2}$, then the mass interior to r is $M_r \propto r$. The free-fall time (equation (3.5)) scales as the inverse square root of the mean density, so one might guess that the mass M_r will fall in to the center after an elapsed time $t(r) \approx t_{ff} \propto <\rho>^{-1/2} \propto r$. Therefore, the total amount of mass that has fallen to the center should increase linearly with time, i.e. the mass infall rate should be constant. The inner regions should collapse first because they have the shortest free-fall times (highest densities), while the outer parts of the cloud take longer to fall in. Since the mass that has already fallen in increases linearly with time, and the cloud mass increases linearly with the radius r, this 'inside-out' collapse results in a free-fall zone within the original cloud whose radial extent expands linearly with time.

Now consider the detailed similarity solution. The usual equation of mass conservation in spherical symmetry,

$$\frac{\partial \rho}{\partial t} + \frac{1}{r^2}\frac{\partial}{\partial r}r^2\rho u = 0, \tag{3.6}$$

where u is the radial velocity, can be transformed into an alternative form in terms of M_r,

$$\frac{\partial M_r}{\partial t} + u\frac{\partial M_r}{\partial r} = 0, \qquad \frac{\partial M_r}{\partial r} = 4\pi r^2\rho. \tag{3.7}$$

The momentum equation for isothermal flow is

$$\frac{\partial u}{\partial t} + u\frac{\partial u}{\partial r} = -\frac{c_s^2}{\rho}\frac{\partial \rho}{\partial r} - \frac{GM_r}{r^2}. \tag{3.8}$$

The similarity solution assumes that the outer and inner boundary conditions are unimportant, and so the only dimensional quantities in the problem are G, c_s, r, and t. Dimensional analysis gives the similarity variable $x = r/c_s t$. Then one looks for solutions of the form

$$\rho(r,t) = \frac{\alpha(x)}{4\pi G t^2}, \qquad M_r(r,t) = \frac{c_s^3 t}{G}m(x), \qquad u(r,t) = c_s v(x), \tag{3.9}$$

where α, m, and v are non-dimensional functions of the coordinate x, and the solution is subject to the boundary condition that at $t = 0$ the mass of the core $M_r(0, t) = 0$.

The equations of motion can be written in terms of the similarity variable x through suitable transformations. For the radial coordinate, the transformation is

$$\frac{\partial}{\partial r} = \frac{1}{c_s t} \frac{\partial}{\partial x}. \tag{3.10}$$

The partial time derivative for the similarity solution is to be taken at a constant x, which is moving with respect to the Eulerian coordinate system r. Thus, the required transformation is

$$\frac{\partial}{\partial t_r} = \frac{\partial}{\partial t_x} - \frac{x}{t} \frac{\partial}{\partial x}. \tag{3.11}$$

Using these relations, equations (3.7) and can be written as the ordinary differential equations

$$m - (x - v)\frac{dm}{dx} = 0, \quad \frac{dm}{dx} = x^2\alpha, \tag{3.12}$$

or

$$m = x^2\alpha(x - v). \tag{3.13}$$

Equations (3.7) and (3.8) can be transformed to two coupled first-order differential equations,

$$[(x - v)^2 - 1]\frac{1}{\alpha}\frac{d\alpha}{dx} = (x - v)\left[\alpha - 2(x - v)/x\right], \tag{3.14}$$

$$[(x - v)^2 - 1]\frac{dv}{dx} = (x - v)\left[\alpha(x - v) - 2/x\right]. \tag{3.15}$$

An extended description of the properties of these equations is given in Shu (1977). Here we present a limited discussion, focussed on the collapse problem. One exact analytic solution of these equations is given by

$$v = 0 \ , \quad \alpha = 2/x^2 \ , \quad m = 2x. \tag{3.16}$$

This corresponds to the singular isothermal sphere, which is static but unstable. If the velocities are initially small, i.e. if $v \to 0$ as $x \to \infty$, the asymptotic behavior of equations (3.13)–(3.15) is

$$\alpha \sim A/x^2, \ v \sim -(A - 2)/x, \ m \sim Ax \text{ as } x \to \infty. \tag{3.17}$$

Here A is a constant which must be greater than 2 for v to be negative, as required for infall. This asymptotic solution has the same power-law form for the density as the singular isothermal sphere, but if $A > 2$ there is no balance between gravity and pressure forces and so collapse begins everywhere.

One solution of particular interest is $A = 2 + \epsilon$, $\epsilon \ll 1$, so that the cloud is nearly in hydrostatic equilibrium for $x \geq 1$. A numerical solution for $A = 2.0005$ is shown in Figure 3.1, which was obtained by integrating the two equations (3.14) and (3.15) simultaneously inwards. By taking $A > 2$, the outer envelope is not precisely in hydrostatic equilibrium (see Figure 3.1), but the velocities for $x > 1$ are so small as to be negligible. As $\epsilon \to 0$ the external velocities become arbitrarily small, but then a singularity near $x = 1$ must be considered (see below).

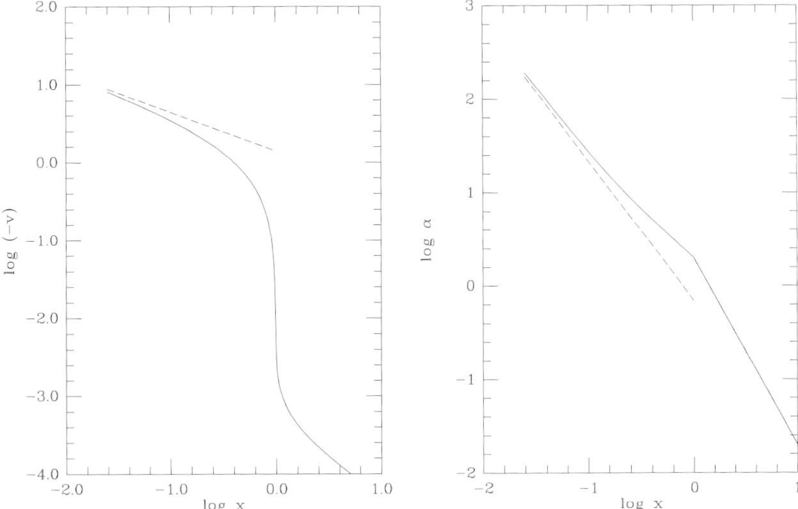

Fig. 3.1. Similarity solution for the collapse of the singular isothermal sphere, adopting the constant $A = 2.0005$ (see text). The left-hand panel exhibits the non-dimensional velocity, while the right-hand panel exhibits the non-dimensional density α. The dashed lines correspond to the case of free-fall (pressureless) collapse at a constant rate onto a core containing all the mass.

The principal properties of this solution can be outlined as follows. At small distances, expansion of the equations shows that

$$m \rightarrow m_\circ, \quad \alpha \rightarrow (m_\circ/2x^3)^{1/2}, \quad v \rightarrow -(2m_\circ/x)^{1/2} \text{ as } x \rightarrow 0. \tag{3.18}$$

For $\epsilon \rightarrow 0$ the core mass is $m_\circ = 0.975$. Since the total mass contained in the region $x \leq 1$ is $m(1) = 2$ (equation (3.16)), about 49% of the mass is in the core, while the rest is still falling in. From the scaling of equation (3.9), the central mass is

$$M_r(0,t) = 0.975 c_s^3 t/G, \tag{3.19}$$

and so the limiting mass infall rate is constant,

$$\dot{M} = 0.975 c_s^3/G. \tag{3.20}$$

Equations (3.18) show that in the innermost regions, the self-gravity of the envelope is negligible in comparison with the gravitational field of the central mass. Because the gas pressure is also relatively unimportant in impeding the (supersonic) collapse at small radii, the infall velocity approaches the free-fall velocity $v_{ff} \approx (2GM/r)^{1/2}$ (indicated as a dashed line in the left-hand panel of Figure 3.1). The density of the infalling material approaches

$$\rho_i \approx \frac{\dot{M}}{4\pi r^2 v_{ff}} = \frac{c_s^3}{4\pi r^2 G v_{ff}}, \tag{3.21}$$

and so $\rho_i \propto r^{-3/2}$ (dashed line in the right-hand panel of Figure 3.1) because the mass infall rate is constant.

These equations have a critical point at $x - v = 1$. In the present case, the limiting

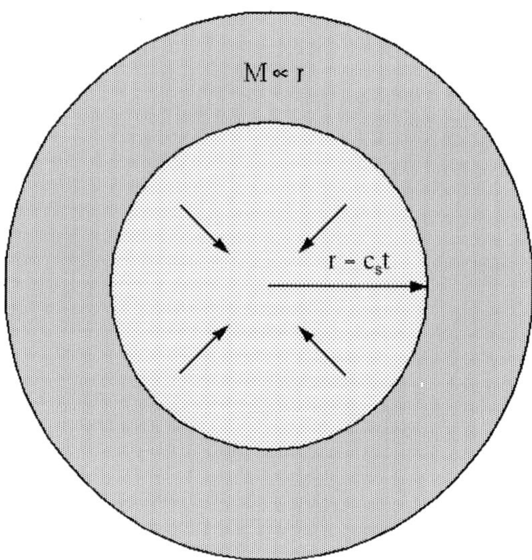

Fig. 3.2. Schematic diagram of purely spherical, 'inside-out' similarity collapse. If the initial cloud is centrally condensed, the innermost regions collapse first because they are densest and have the shortest free-fall times. The outermost regions remain nearly unchanged because they cannot respond on short timescales, and because the spherical collapse of mass to a central point does not affect the gravitational force on large scales. As collapse proceeds, a region of radius $r = c_s t$ becomes evacuated as the material originally in this region falls onto the central mass.

solution approaches a critical point $x \sim 1$. Discussions of the significance of critical points can be found in Shu (1977) and Hunter (1977). In general, the presence of critical points depends upon the specific initial conditions adopted.

This simple similarity solution can be exploited to understand the time dependence of the infall. The variable x is transformed to physical radial distance r by $r = c_s t x$. Therefore, at time t, the position $x = 1$, which corresponds to the boundary between the hydrostatic and infall regions (Figure 3.1), lies at $r = c_s t$ (Figure 3.2). The mass initially contained within this radius r is $M_r = 2c_s^3 t/G$; of this amount, $m_\circ c_s^3 t/G$ has already fallen into the center, and the rest is in the infall region. The physical interpretation of this behavior is as follows: the collapse begins first in the inner regions, because they are densest and therefore have the shortest free-fall times. As the material in the inner region falls in, the pressure support of the overlying layers is removed, allowing these to fall in as well. Because of exact spherical symmetry, the gravitational field seen by the external material is unaffected by the collapse of inner regions. Thus, the information that inner layers have fallen in is communicated to the outer layers only by a rarefaction wave, which moves outward at the sound speed c_s. The amount of mass per unit time that loses its pressure support and begins to fall is constant, because the rarefaction wave radius is $r = c_s t$ and the isothermal sphere mass grows linearly with r (Figure 3.2).

As a specific example, we follow Shu (1977) in envisaging a cloud core of pure molecular hydrogen at 10 K bounded by an external pressure of 1.1×10^5 cm^{-3} K, which implies an initial mass $0.96\,M_\odot$ and $r = 1.6 \times 10^{17}$ cm. For a sound speed of

$0.2\,\mathrm{km\,s^{-1}}$, the expansion wave takes 2.5×10^5 yr to reach the outer boundary of the cloud, at which point 49% of the total mass has fallen in. Beyond this point in time the similarity solution is clearly not applicable because of the effects of the boundary. In particular, one expects a compression wave to form at the outer boundary and steepen into a shock as it propagates into the interior, so the details of the infall will change from the similarity solution.

3.3 Generalized models of protostellar collapse

While the similarity solution for the singular isothermal sphere presented by Shu (1977) provides a particularly clear and elegant way of conceptualizing the basic physics of the gravitational collapse problem, real protostellar clouds will probably depart from this similarity behavior in detail. For example, the assumed hydrostatic singular isothermal sphere does not represent a realizable state because it is intrinsically unstable (e.g., Whitworth *et al.* 1996).

The collapse of an initially stable Bonnor–Ebert sphere, with its flattened inner density distribution, cannot be represented precisely by a similarity solution. Starting from the *critical* isothermal Bonnor–Ebert sphere (§2.5), Foster & Chevalier (1993) followed gravitational collapse numerically and found differences from the singular isothermal sphere similarity solution. In particular, Foster & Chevalier found that the mass infall rate in the inner envelope decays substantially with time; the infall rate is generally higher than predicted by the singular isothermal sphere similarity solution, particularly in the earliest stages (where it more nearly resembles the Larson-Penston solution); the overall collapse occurs more quickly, with the mass infall rate falling to zero in a time $\sim (2/3)R_{crit}/c_s$.

The higher initial mass infall rates can be explained in the context of the structure of the critical Bonnor–Ebert sphere, which has a nearly constant density in its inner regions (Figure 2.4). The Larson–Penston (Larson 1969a; Penston 1969) similarity solutions, which were developed for the collapse of an initial uniform density sphere, also show higher mass infall rates than the Shu similarity solution. Some insight into this behavior is gained by reference to the case of free-fall collapse of the constant density sphere (§3.1); in that calculation, all shells reach the origin at the same time, requiring that outer regions fall in faster than inner regions. Even with retarding pressure forces, the result is a higher instantaneous central infall rate than in the inside-out collapse of the singular isothermal sphere, in which outer regions take much longer to reach the origin (Henriksen, André, & Bontemps 1997).

The supporting effects of magnetic fields against self-gravity may also be included as a perturbation of the singular isothermal sphere solution. Galli & Shu (1993a,b) considered this problem, and found little difference in mass infall rates for modest magnetic support. The qualitatively different feature of the Galli & Shu model is the development of what they called a 'pseudodisk' in the equatorial plane. The pseudodisk arises because the magnetic field threading the spherical outer cloud tends to deflect infalling material away from the radial direction. Assuming axisymmetry, the deflected material from one hemisphere shocks with material coming in the opposite direction from the other hemisphere, resulting in a pile-up of material in the equatorial plane some distance from the central mass. Since this pseudodisk is not rotationally supported (see §3.6), it must fall in toward the star. Decoupling of the interstellar magnetic field from the inner infalling envelope might cause a 'pile-up' of magnetic flux, which also could have important dynamical effects (Li & McKee 1996).

As discussed in Chapter 2, if gravitational fragmentation makes cloud cores, the initial conditions for cloud collapse will differ from both the singular isothermal sphere and the Bonnor–Ebert sphere. One simple situation which might be envisioned is fragmentation from a non-rotating isothermal sheet initially in hydrostatic equilibrium (Hartmann *et al.* 1994b). It is worth discussing the overall results for this case, as it represents one of the simplest cases of non-spherical initial cloud structure, which may be essential in the fragmentation of gas into multiple systems (Bonnell *et al.* 1991; Nelson & Papaloizou 1993; Boss 1993). Flattened cloud structure is likely to be the natural consequence of (non-isotropic) magnetic field support (Mouschovias 1976; Nakano 1984); here we consider the non-magnetic case for simplicity and for comparison with the non-magnetic, singular isothermal sphere case.

In these calculations, the equilibrium sheet of gas initially begins contracting slowly (subsonically) toward the central axis (upper left panel of Figure 3.3). However, once a central mass begins to build up (upper right), the infall begins to become supersonic, and eventually develops into radial contraction at near free-fall velocities (bottom right panel). Note that unlike the singular isothermal sphere collapse (Figure 3.2), this is initially an *outside-in* collapse, because the infall velocities initially are largest on the fragmentation scale (§2.2). As the central mass grows, the added gravitational attraction overwhelms pressure support, and the collapse velocities become largest on small scales, just as in the singular isothermal sphere similarity solution. (Sometimes reference is made in the literature to 'inside-out' collapse when all that is meant is that the infall velocities are highest in the inner regions. The sheet collapse calculations show that even with an *initially* outside-in collapse, inner regions will reach the center first; the velocity field will exhibit very similar properties to the inside-out collapse model, once the acceleration is dominated by a central gravitating mass.)

A qualitatively different feature of the collapse of a sheet (or any non-spherical distribution) compared with the sphere is the development of highly-flattened structure. Once a significant central mass is formed, evacuated cavities appear along the axis of symmetry (bottom panels of Figure 3.3). The reason is simply that the material at the shortest initial distances from the central gravitating mass falls in first. This structure is also seen in simulations of fragmentation in which magnetic fields play an important dynamical role (Nakamura, Hanawa, & Nakano 1995); collapse occurs first along magnetic field lines, producing a nearly hydrostatic disk-like or toroidal structure like the sheet of in the upper left panel of Figure 3.3, which then collapses very similarly to the results shown in the bottom panels. (In effect, the fragmentation produces a large-scale 'pseudodisk' which is almost in hydrostatic equilbrium across the plane of flattening.) The importance of this large-scale flattened collapse structure is that there might be lines of sight to central protostars over a significant range of solid angle along which the protostar is not heavily extincted (see §4.5).

The central mass infall rate in this simulation is never precisely constant, but exhibits a 'plateau' phase with an infall rate approximately three times that predicted by the Shu (1975) similarity solution for the singular isothermal sphere. It can be shown from dimensional arguments that the mass infall rate for this simulation scales as c_s^3, just as in the Shu similarity solution (Hartmann, Calvet, & Boss 1996).

To summarize, collapse models which include additional effects – stable initial

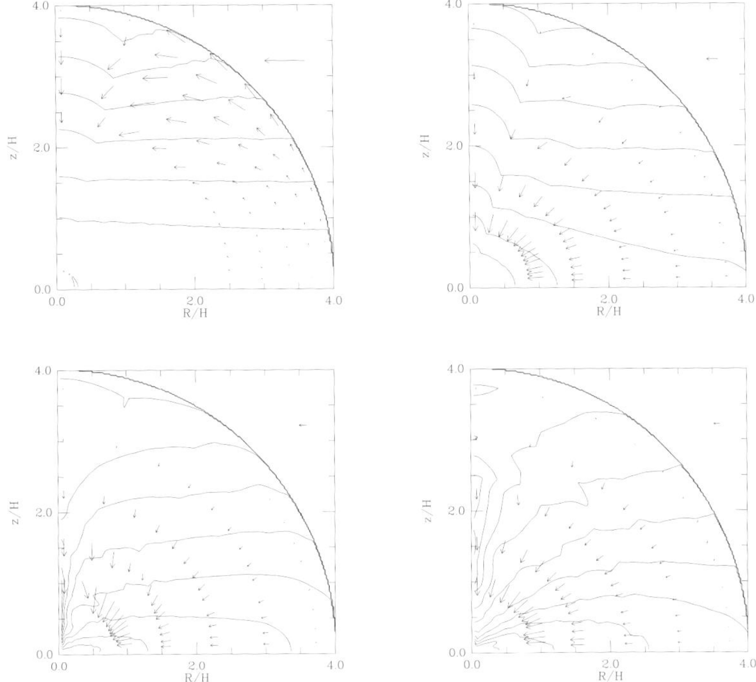

Fig. 3.3. Numerical solution for flat sheet collapse. The figures show a meridional cut through the density distribution, with velocity vectors superimposed (gray-scale representations of the density distributions are illustrated in Figure 4.10). The collapse is assumed to be axisymmetric, and a spherical outer boundary is imposed; only one quadrant of the flow is shown because of symmetry. The length of the arrow in the upper right of each figure indicates the magnitude of sound velocity; the velocities are rescaled from panel to panel to follow the development of supersonic flow. The density contours are spaced by factors of $10^{1/2}$. If the properties of this isothermal solution are scaled to a temperature of $T = 10$ K, and an initial cloud mass of $1\,M_\odot$, then the outer radius is 5750 AU, the scale height H of the initial cloud is 1440 AU, and the four panels represent elapsed times of 1.24×10^5, 2.76×10^5, 3×10^5, and 3.2×10^5 yr. See text for discussion. From Hartmann *et al.* (1994b).

equilibrium conditions, magnetic fields, non-spherical initial structure – exhibit many features in common with the Shu (1977) similarity solution for the singular isothermal sphere, but there can be quantitative differences in properties such as the mass infall rate(s), which are no longer constant in time. The principal qualitative differences produced by magnetic forces and non-spherical initial structure are in the geometry of the collapse; magnetic forces may produce a large-scale collapsing pseudodisk, while initial flattening of the parent cloud core produces relatively evacuated cavities in the infalling envelope, increasing the observability of protostars at short wavelengths (Chapter 4).

3.4   Rotating collapse

As discussed in §2.7, if collapse occurs on a dynamical timescale, it is unlikely that the angular momentum remaining in the cloud core at the start of rapid collapse can be transferred efficiently to the external medium. The large size of the initial cloud core implies that even modest initial rotational velocities will cause the infalling material to land first on a rotationally-supported disk rather than a pressure-supported star. In general, the disk may be sufficiently massive to break up or fragment into stars (Yorke, Bodenheimer, & Laughlin 1993; Laughlin & Bodenheimer 1994; Bonnell & Bate 1994; see review in Boss 1995). Here we concentrate on the simpler case where the disk mass does not affect the infall pattern. The simplest analysis of rotating collapse assumes that pressure forces are negligible and so the problem can be analyzed using ballistic trajectories. The results for the case of the collapse of a spherically-symmetric cloud in uniform (solid-body) rotation were initially worked out by Ulrich (1976), with subsequent extension by Cassen and Moosman (1981) to disk formation and by Terebey, Shu & Cassen (1984; = TSC) to the collapsing singular isothermal sphere.

Before developing the mathematical solution it is again worth making an initial estimate of its overall properties. For simplicity assume a fixed central mass of M. If material with specific angular momentum h falls in and ends up in a circular orbit while maintaining its angular momentum, then the radius R of the circular orbit is

$$R = h^2/GM.$$

$$(3.22)$$

In the similarity solution for 'inside-out' collapse, all the material which arrives at the center at a given instant of time started from the same initial cloud radius r_o. If the protostellar cloud core is initially in uniform rotation, with angular velocity Ω, then the specific angular momentum at r_o varies with the angle θ from the rotation axis as $h = \Omega r_o^2 \sin\theta$. Thus, material falling in from different directions will have different angular momenta and arrive at the midplane (the plane perpendicular to the rotation axis) at differing radii. Material near the rotation axis will fall in close to the central star because it has low angular momentum, while mass falling in from regions near $\theta \sim \pi/2$ will fall in to a maximum 'centrifugal radius'

$$r_c = r_o^4 \Omega^2/GM.$$

$$(3.23)$$

This material arriving at $\theta \sim \pi/2$ from 'above' will collide with material arriving at the same position from 'below'; this infall produces a flat structure with rotation, i.e. a disk (see §3.6).

Now we proceed to the detailed solution. For simplicity we assume that essentially all of the mass is contained in the center, that the gas falls in from from a very large distance with essentially zero total energy, so that the motion is approximately parabolic. This free-fall approximation neglects pressure forces and the mass of the inner envelope, which can be justified from the inner limits of the similarity solution (3.18).

For two-body parabolic motion around a central gravitating mass, the equation of the particle orbit in its own plane is

$$r = \frac{h^2/GM}{1 - \cos\alpha},$$

$$(3.24)$$

where α is the direction angle of the particle measured from the origin to apastron, and h is specific angular momentum. We define θ_o as the angle between the orbital plane and the rotation axis of the system (see Figure 3.4).

The transformations to convert from (r, α) in the orbital plane to (r, θ, ϕ) are

$$\cos\theta = \cos\theta_o \cos\alpha$$

$$(3.25)$$

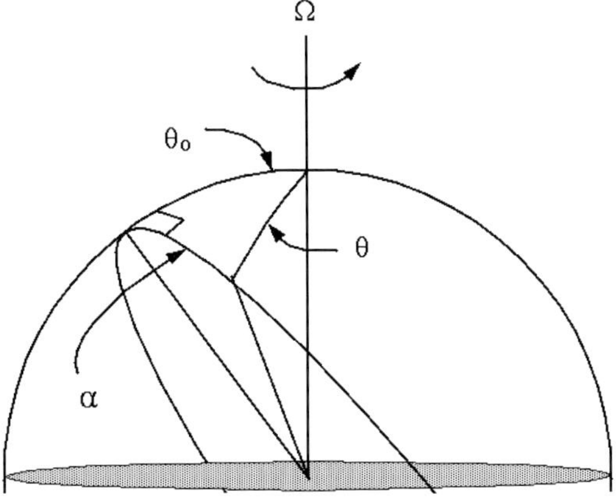

Fig. 3.4. Geometry of rotating collapse solution (see text).

and

$$\tan\phi \;=\; \frac{\tan\alpha}{\sin\theta_\circ}\,.$$

(3.26)

From the standard solution for the two-body problem,

$$u \;\equiv\; \frac{1}{r} \;=\; \frac{GM}{h^2}\,(1 - \cos\alpha),$$

(3.27)

$$\dot\alpha \;=\; hu^2,$$

(3.28)

$$\dot r \;=\; -\frac{\dot u}{u^2} \;=\; -\frac{du}{d\alpha}\frac{\dot\alpha}{u^2} \;=\; -\frac{GM}{h}\sin\alpha,$$

(3.29)

so that

$$h \;=\; [GMr\,(1 - \cos\alpha)]^{1/2}\,.$$

(3.30)

The radial velocity is

$$v_r \;=\; \dot r \;=\; -\left(\frac{GM}{r}\right)^{1/2}\left(1 + \frac{\cos\theta}{\cos\theta_\circ}\right)^{1/2}.$$

(3.31)

The meridional velocity is given by

$$v_\theta \;=\; r\dot\theta \;=\; r\frac{d\theta}{d\alpha}\dot\alpha \;=\; \frac{h}{r}\frac{d\theta}{d\alpha},$$

(3.32)

which can be rearranged to the form

$$v_\theta \;=\; \left(\frac{GM}{r}\right)^{1/2}(\cos\theta_\circ - \cos\theta)\left(\frac{\cos\theta_\circ + \cos\theta}{\cos\theta_\circ \sin^2\theta}\right)^{1/2}.$$

(3.33)

Finally, the azimuthal velocity is

$$v_\phi \;=\; r\sin\theta\,\dot\phi \;=\; r\sin\theta\,(d\phi/d\alpha)\dot\alpha \;=\; \frac{h}{r}\sin\theta\,d\phi/d\alpha,$$

(3.34)

which after some manipulation can be written as

$$v_\phi = \left(\frac{GM}{r}\right)^{1/2} \left(1 - \frac{\cos\theta}{\cos\theta_o}\right)^{1/2} \frac{\sin\theta_o}{\sin\theta}. \tag{3.35}$$

It is straightforward to verify that $v^2 = v_r^2 + v_\theta^2 + v_\phi^2 = 2GM/r$.

The quantity h is the specific angular momentum measured around the axis perpendicular to the orbital plane. It is convenient to introduce the specific angular momentum of the particle relative to the overall cloud rotation axis,

$$H_l = h\sin\theta_o. \tag{3.36}$$

Then the trajectory of the particle is given by

$$r = \frac{H_l^2}{\sin^2\theta_o}\frac{1}{GM(1 - \cos\alpha)} = \frac{H_l^2}{\sin^2\theta_o}\frac{1}{GM(1 - \cos\theta/\cos\theta_o)}. \tag{3.37}$$

The particle lands on the (thin) disk at $\theta = \pi/2$, at a radial distance

$$r(\pi/2) = \frac{H_l^2}{\sin^2\theta_o\, GM}. \tag{3.38}$$

For the ballistic solution to be valid, one requires that the streamlines of particles with different θ_o do not intersect; otherwise, shocks would result. It is clear from equation (3.38) that $H^2/\sin^2\theta_o$ must be a monotonically increasing function over $0 \leq \theta_o \leq \pi/2$ to avoid such intersections.

One simple rotation law which satisfies this condition is

$$H_l^2 = r_o^4\Omega_o^2\sin^4\theta_o, \tag{3.39}$$

which corresponds to uniform rotation of a sphere of radius r_o at an angular velocity Ω_o. Defining $r_c = r_o^4\Omega_o^2/GM$, the particle trajectories in the meridional plane can be written as

$$\frac{r}{r_c} = \frac{\sin^2\theta_o}{1 - \cos\theta/\cos\theta_o}. \tag{3.40}$$

Figure 3.5 shows the streamlines of particles in the meridional plane, which are labeled by θ_o. With the assumption of solid-body rotation, particles falling near the rotational axis have little angular momentum, and thus fall nearly radially. Particles with larger θ_o have larger angular momentum and thus fall to the disk at larger radii. All of the streamlines intersect the disk plane interior to r_c.

The density can be evaluated by assuming that the mass infall rate is steady. If $r_o \gg r_c$, $\cos\theta \approx \cos\theta_o$ (equation (3.40)) and the flow is radial. Assuming that the cloud at r_o is nearly spherical, the mass flow in a flow tube spanned by $d\theta_o$ is

$$d\dot{M} = \frac{2\pi r^2\sin\theta_o\, d\theta_o\, \dot{M}}{4\pi r^2} = \frac{1}{2}\sin\theta_o\, d\theta_o\, \dot{M}. \tag{3.41}$$

The density at r, θ can be found by following the streamlines corresponding to θ_o and $\theta_o + d\theta$,

$$\rho = \frac{d\dot{M}}{2\pi r^2\sin\theta\, d\theta\, |v_r|} = \frac{1}{2\pi r^2\sin\theta\, |v_r|}\left(\frac{d\dot{M}}{d\theta_o}\right)\left(\frac{d\theta_o}{d\theta}\right). \tag{3.42}$$

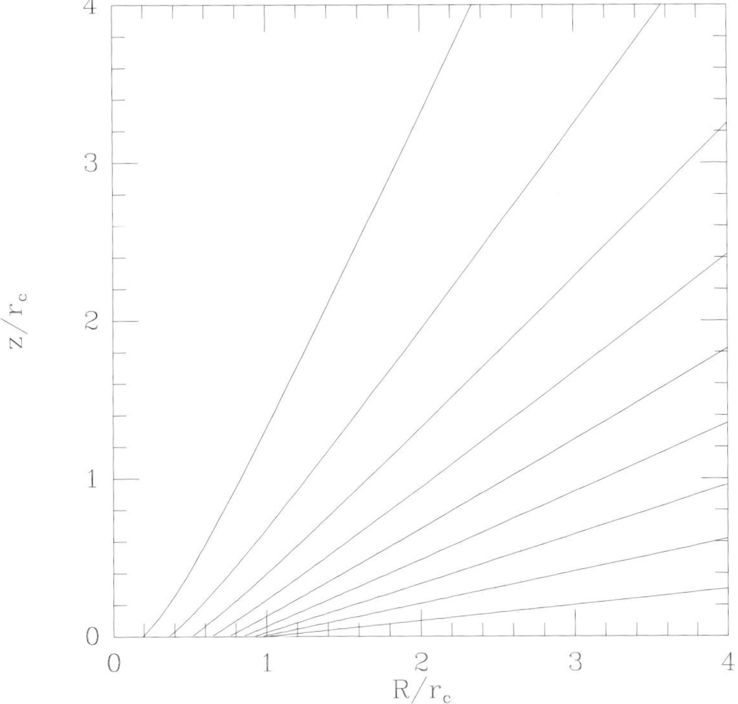

Fig. 3.5. Streamlines for the rotating collapse solution described in the text. Distance scales for the polar axis z and the cylindrical radius R are given in units of the centrifugal radius r_c. The streamlines shown are in steps of 0.1 in $\cos\theta_\circ$, with the lowest streamline for $\cos\theta_\circ = 0.9$. Since equal intervals in $\cos\theta_\circ$ correspond to equal intervals of mass in the outer cloud, the tendency of the material to pile up at the outer edge of the initial disk $(R \sim r_c)$ is evident.

Using equations (3.31) and (3.40), this becomes

$$\rho = \frac{\dot{M}}{4\pi(GMr^3)^{1/2}} \left(1 + \frac{\cos\theta}{\cos\theta_\circ} \right)^{-1/2} \left(\frac{\cos\theta}{\cos\theta_\circ} + \frac{2\cos^2\theta_\circ}{r/r_c} \right)^{-1}. \tag{3.43}$$

In Figure 3.6 we show contours of constant density in a meridional plane for this infall solution. The density is nearly spherically symmetric at large distances, where the effects of rotation are small. At distances $\lesssim r_c$, the density distribution becomes quite flattened, as material falls non-radially onto the disk. This behavior can be seen directly from equation (3.43) in its limits. For $r \gg r_c$, $\theta \to \theta_\circ$, and

$$\rho \sim \frac{\dot{M}}{4\pi(2GM)^{1/2}} r^{-3/2}, \tag{3.44}$$

which is precisely the density distribution for free-fall at a constant mass infall rate toward a point mass M. For $r \ll r_c$, the streamlines are nearly vertical and so $\cos\theta_\circ \approx 1$; therefore,

$$\rho \sim \frac{\dot{M}}{8\pi r_c(GM)^{1/2}} (1 + \cos\theta)^{-1/2} r^{-1/2}. \tag{3.45}$$

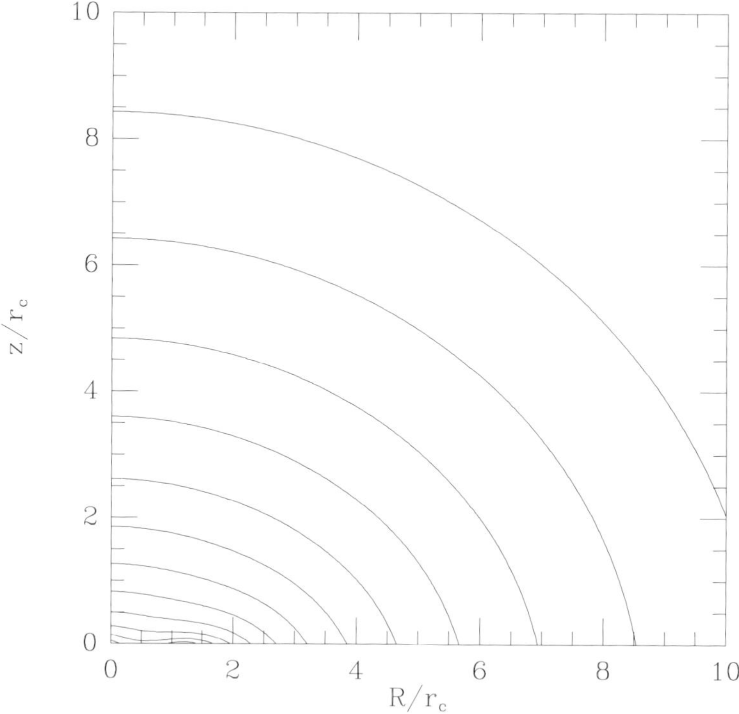

Fig. 3.6. Contours of constant density for the rotating collapse solution. Distance scales for the polar axis z and the cylindrical radius R are given in units of the centrifugal radius r_c. Each contour represents a factor of $2^{1/2}$ difference in density, with the outer contours representing the lowest densities. The flattening of the density distribution near the disk is evident.

This limit of equation (3.43) illustrates an important effect of rotation on the infall density distribution; inside r_c, the density increases less rapidly with decreasing radius than the $\rho \propto r^{-3/2}$ expected for spherically symmetric collapse. To emphasize the point, we consider the spherical average of the density distribution, i.e. the average density at a given radius r, which can be written as (Adams & Shu 1986)

$$< \rho(r) > = \int_0^{\pi/2} \rho(r,\theta) \sin\theta d\theta = Cr^{-3/2}A(r/r_c), \qquad (3.46)$$

where

$$A(u) = (2u)^{1/2} \ln \left[\frac{1 + (2u)^{1/2}}{(1 - u)^{1/2} + u^{1/2}} \right], \quad u \le 1 ; \qquad (3.47)$$

$$A(u) = (2u)^{1/2} \ln \left[\frac{1 + (2u)^{1/2}}{(2u - 1)^{1/2}} \right], \quad u \ge 1, \qquad (3.48)$$

with $u = r/r_c$ and $C = \dot{M}/[(4\pi(2GM)^{1/2})]$.

Figure 3.7 shows this angle-averaged density distribution. There is a break at $r \sim r_c$ between the $\rho \propto r^{-1/2}$ and the $\rho \propto r^{-3/2}$ regimes, because the angular momentum of the infalling material causes it to fall onto a disk at $\theta = \pi/2$, $r \le r_c$, and material

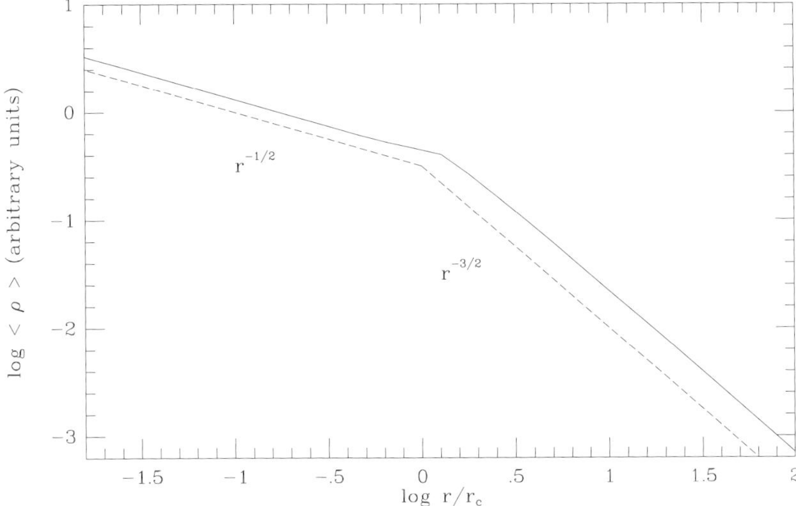

Fig. 3.7. The spherically-averaged density distribution of the rotating collapse solution as a function of radial distance in units of the centrifugal radius r_c. The dashed lines denote pure power-law distributions of $r^{-1/2}$ and $r^{-3/2}$ for comparison. The density distribution follows the spherical free-fall result outside of r_c, but departs at smaller radii as material falls onto the disk.

'disappears' from the infall solution. In effect, the mass infall rate across any sphere of radius $r < r_c$ decreases with decreasing radius because of the deposition of material into the disk. This behavior has an important effect on the observable emission from infalling protostellar envelopes, which we discuss in Chapter 4.

3.5   Time evolution of rotating collapse

In an important and influential paper, TSC used a perturbative analysis to include the effects of rotation (§3.4) as a small perturbation to the dynamics of the initial singular isothermal sphere (§3.2). The TSC solution can be used to follow the evolution of the collapse during the period when the approximation of self-similarity is valid. While the details of the mathematics are complicated, the basic evolution can be understood qualitatively as follows. The protostellar cloud initially approximates a singular isothermal sphere in structure. The cloud has a small uniform rotation at large scales at angular frequency Ω_o, so that departures from spherical symmetry on large scales are small. Collapse can proceed along the lines of the similarity solution presented in §3.2 on large scales; at small radii, rotation becomes important, and the results of §3.4 may be used. At any instant of time, equation (3.43) or its angle-averaged form (3.46) may be used to estimate the density distribution interior in the collapsing region. The estimate (3.44) for large scales is not exact, because it has been derived for the case where all of the mass is in the central star, whereas the mass in the infalling envelope is generally not negligible (cf. §3.2). Nevertheless, since most of the protostellar emission is concentrated toward the central regions (Chapter 4), equation (3.43) provides a reasonable approximation to the more detailed TSC result.

In the similarity solution the central mass varies with the elapsed time after the beginning of collapse as $M = m_{\circ} c_s^3 t / G$ (3.19), where $m_{\circ} = 0.975$. This mass was contained within a radius $r_{\circ} = (m_{\circ}/2) c_s t$ in the original singular isothermal sphere configuration. The specific angular momentum of the material relative to the axis of symmetry at this radius was

$$H^2 = r_{\circ}^4 \Omega_{\circ}^2 \sin^4 \theta_{\circ} = m_{\circ}^4 c_s^4 t^4 \Omega_{\circ}^2 \sin^4 \theta_{\circ} / 16. \tag{3.49}$$

This is the angular momentum of the material that is arriving at (near) the origin at the present time t. The material along the streamline $\theta_{\circ} = \pi/2$ lands at the largest disk radius, given by

$$r_c(t) = \frac{m_{\circ}^4}{16} \frac{c_s^4 t^4 \Omega_{\circ}^2}{GM} = \frac{m_{\circ}^3}{16} c_s t^3 \Omega_{\circ}^2. \tag{3.50}$$

Thus, the centrifugal radius increases rapidly with time, $r_c \propto t^3$. Initially most of the mass falls close to the center, because the material that falls in first has small angular momentum. As collapse proceeds, and material from larger radii is added to the central core, r_c increases rapidly with time, and material is added to the disk rather than to a central star (see Figure 3.5).

3.6   Disk formation

In the simple infall model discussed above, where the initial cloud is spherically symmetric and exhibits axisymmetric rotation, the infall solution has complete symmetry above and below the equatorial plane $\theta = \pi/2$. With this assumption, the momentum fluxes of infalling material on either side of the disk (equatorial) plane perpendicular to the disk are equal in magnitude and opposite in direction. The result is that the infalling gas must pass through a shock at the equator, which dissipates the kinetic energy of motion perpendicular to the equatorial plane. If the shocked gas cools rapidly, as expected for many plausible (though not necessarily all) conditions (Neufeld & Hollenbach 1994), the result is that material accumulates in a thin structure in the equatorial plane, i.e. a disk.

After the energy dissipation in the shock removes most of the velocity component normal to the disk plane, the shocked gas initially retains its velocity parallel to the disk plane. In general, this parallel velocity is not consistent with circular motion at the radius of entry into the disk. The result is that the shocked material must mix with existing material; further dissipation of energy and angular momentum transport must occur before ending up with a disk in circular rotation (Cantó, D'Alessio, & Lizano 1995; Yorke *et al.* 1993). If the initial magnetic field structure of the cloud produces a 'pseudodisk' (§3.3), there will be additional material near the equatorial plane falling in from larger radii. Thus, the point at which envelope material enters the disk cannot represent its final orbital radius (Cassen & Moosman 1981). Nevertheless, one would imagine that the 'characteristic' disk radius should be dependent upon the average angular momentum of the infalling material at any time, and this provides some insight into what can be expected for the disk mass distribution.

If the initial cloud can be represented roughly as a singular isothermal sphere of constant angular rotational velocity, then the centrifugal radius of the infalling material will vary with time as $r_c \propto t^3$; and since the total mass that has landed on the disk increases linearly with time, one might expect that the disk mass initially would grow with cylindrical radius R as $M_d \propto R^{1/3}$. This is very approximate, and ignores angular momentum transfer within the disk. However, it suggests that the disk might be regarded schematically as having a relatively concentrated, high-surface-density inner region, with an extended, low-surface-density outer region.

Ultimately disk material will be redistributed by the processes of angular momentum transport and energy loss which drive disk accretion (Chapter 5). The rapidity with which initial conditions are altered depends upon the rate of angular momentum transfer (e.g., Cassen & Moosman 1981), which may be sufficiently fast to eliminate 'memory' of the initial state of the disk on relatively short timescales (Chapter 9).

4

Protostellar collapse: observations vs theory

As described in the previous chapters, low-mass star formation is thought to occur when a molecular cloud core becomes unstable and collapses at nearly free-fall velocities to produce a stellar core and a rotating disk. Many attempts have been made to detect protostellar cloud collapse by detecting the infall motions directly. This has proved to be difficult for several reasons. The velocities of collapse on large scales are modest in comparison with random velocities of (external) molecular gas; the ubiquitous bipolar outflows from YSOs (Chapter 8) can push out molecular gas at substantial velocities, masking or overwhelming small infall velocity shifts; and the central regions of a collapsing cloud, which exhibit the highest infall velocities, can be hard to detect because of the small amount of mass in the region, and may require the use of particular molecular diagnostics which trace high-density gas. Nevertheless, there is an increasing body of evidence which generally supports the rapid collapse model of protostar formation.

The most general and extensive evidence for protostellar collapse is indirect: the detection of heavily-extincted infrared sources of low-luminosity in star-forming regions. The radiation from a central protostellar core is absorbed in a dust 'cocoon', which is thereby heated; the cocoon reradiates the absorbed energy at longer wavelengths from an extended dust envelope 'photosphere'. Simple arguments can be made which relate the characteristic wavelengths of this radiation to the rate of infall, indicating that the dusty photospheres should emit primarily longward of $10\,\mu$m. For this reason, the study of protostellar sources received dramatic impetus from the launch of the *IRAS* satellite, which made possible studies of the far-infrared radiation from many low-luminosity sources. These studies have led to the development of a statistical picture of the protostellar collapse phase.

One of the first tests of collapse theory for low-mass protostars was to reproduce the infrared emission of Class I sources. Although infrared spectra were calculated for many numerical simulations, such as Larson's (1969b), the subject took a major step forward with the application of the TSC collapse solution to the calculation of protostellar SEDs by Adams & Shu (1986) and Adams, Lada, & Shu (1987 = ALS). These pioneering investigations made the compelling identification of *IRAS*-detected Class I sources (§1.4) as protostars.

To see why protostars should generally have large dust extinction (for dust abundances characteristic of the interstellar material in the solar neighborhood), consider a $1\,M_\odot$ molecular cloud core with the radius $\sim 1.5 \times 10^{17}\,$cm of a critical Bonnor–Ebert sphere at 10 K (equation (2.23)). If the density in this sphere were uniform,

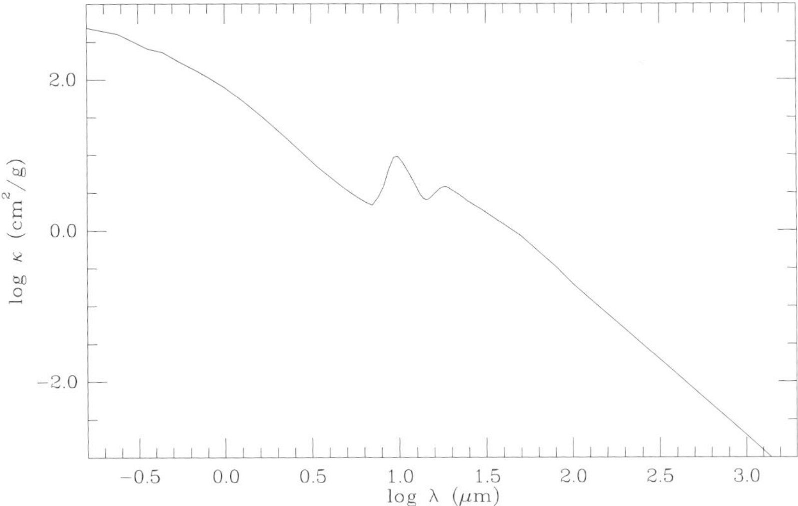

Fig. 4.1. Dust opacity for typical interstellar material, given per gram as a function of wavelength. From Draine & Lee (1984).

the column density to the center would be $\sim 2.1 \times 10^{-2}\,\mathrm{g\,cm^{-2}}$. Using the 'standard' dust opacity (extinction) distribution of Draine & Lee (1984; Figure 4.1), this column density would correspond to a radial optical depth at a wavelength of $1\,\mu$m of about $\tau \sim 2$, or an attenuation of any light emerging from a central source by a factor $\exp(-2)$. The actual column density of the critical Bonnor–Ebert sphere is about a factor of two larger because of the increase of density in the interior. Thus, any central low-mass protostar, which is likely to emit most of its energy at wavelengths $\lesssim 1\,\mu$m (Chapter 9), will have most of its radiation absorbed in this dusty envelope (and the energy emitted at longer wavelengths). A *collapsing* envelope should generally exhibit very much larger optical depths than the static cloud core, because the infalling material dramatically increases densities in the inner envelope and thus increases the extinction to the central object (protostar).

The similarity solution for thermally-supported clouds at a typical molecular cloud temperature of 10 K (equation (3.20)) predicts a mass infall rate of

$$\dot{M} \sim c_s^3/G \sim 1.6 \times 10^{-6}\,\mathrm{M_\odot\,yr^{-1}}. \tag{4.1}$$

This suggests that a $0.5\,\mathrm{M_\odot}$ cloud will take approximately $\sim 3 \times 10^5$ yr to collapse. The ages of the pre-main-sequence stars in Taurus are mostly ~ 1–2 Myr; therefore, the collapse theory predicts that about 10–20% of the pre-main-sequence population in Taurus should be protostars (i.e., young stars still surrounded by their dusty infalling envelopes). So far, about 25 far-infrared Class I sources have been found in Taurus out of a total population of around 200 pre-main-sequence objects, supporting the identification of these objects as protostars. The study of protostellar sources in dense regions of star formation has proved to be much more difficult, principally because the *distributed* dust pervading these regions can have visual extinctions of 25–50 magnitudes; the effects of this distributed dust on the SEDs of the YSOs embedded in the region are not yet fully understood. It has been suggested that the so-called Class 0 sources are the 'youngest' or even 'true' protostars (e.g., André, Ward-Thompson, & Barsony 1993), but this issue requires further exploration.

As discussed in §3.4, much of the infalling matter of a collapsing core should land on a circumstellar disk. The size of the disk depends sensitively upon the angular momentum of the parent cloud, and also on the age of the source ($r_c \propto t^3$ in the TSC model; equation (3.50)). As an order of magnitude estimate, using typical measured velocity gradients of molecular cloud cores in the Taurus complex $\sim 1\,\mathrm{km\,s^{-1}pc^{-1}}$, or $\Omega \sim 3 \times 10^{-14}\,\mathrm{s^{-1}}$ (Goodman *et al.* 1993), and adopting a critical Bonnor–Ebert sphere for a $0.5\,\mathrm{M_\odot}$ molecular cloud at 10 K with an outer radius of roughly 0.02 pc, the outermost cloud material would arrive at the disk at a centrifugal radius

$$r_c \sim r_o^4 \Omega_o^2 / GM \sim 15\ \mathrm{AU}.\qquad(4.2)$$

While this result is sensitive to the assumed initial conditions, it clearly demonstrates that solar-system-sized disks can result from protostellar collapse.

The luminosity of a protostellar system can be written as

$$L = L_* + L_{acc}\qquad(4.3)$$

where L_* is the luminosity of the central stellar core produced by gravitational contraction and deuterium fusion, and L_{acc} is the luminosity produced by accretion, which includes the dissipation of kinetic energy as the infalling material lands on the disk, the energy lost as material accretes through the disk, and finally the energy released as the accreting material lands on the central stellar core. In Taurus, the Class I sources exhibit the same luminosities ($\sim 1\,\mathrm{L_\odot}$) as the (slowly-accreting) CTTS or even the (non-accreting) WTTS (§4.1); there is little evidence for a large additional accretion luminosity component. This is surprising because the implied accretion luminosities,

$$L_{acc} \sim GM_* \dot{M}_{acc}/R_*\qquad(4.4)$$

are $L_{acc} \sim 8\,\mathrm{L_\odot}$ for $\dot{M}_{acc} \sim \dot{M} \sim 1.6 \times 10^{-6}\,\mathrm{M_\odot\,yr^{-1}}$, using reasonable estimates for protostellar masses and radii. One possible explanation of this discrepancy is that the disk accretion rate $\dot{M}_{acc}$ onto the star is much smaller than the infall rate $\dot{M}$, so that mass is stored in the disk at $R >> R_*$. However, this mass storage cannot go on indefinitely. The disk must eventually dump most of the accumulated mass onto the central star; conceivably this could occur during rapid bursts of FU Ori disk accretion (Chapter 7).

Observational surveys of infrared objects have turned up increasing numbers of protostar candidates with properties roughly similar to those described above. In addition, attempts to solve the difficult problem of detecting infall motions have provided very suggestive results. In this chapter some of the progress that has been made in identifying protostars is indicated, along with some outstanding problems.

4.1   SEDs of low-mass 'protostars'

The study of heavily-extincted pre-main-sequence stars was greatly advanced by the launch of the *IRAS* satellite, which provided unprecedented sensitivity in the far-infrared spectral regions where protostellar objects radiate most of their energy. *IRAS* observations demonstrated the existence of low-luminosity objects with SEDs similar to those predicted by infall models (Beichman *et al.* 1986; Myers *et al.* 1987; ALS; Butner *et al.* 1991, 1994; Kenyon *et al.* 1993).

Any discussion of protostellar candidates must begin with surveys of the Taurus molecular cloud region, because it is nearby, relatively sparse, and exhibits modest extinction ($A_V \lesssim 5$). This last property means that the optically-detectable pre-main-sequence population is well studied, permitting detailed comparisons of the infrared-bright populations with optical stars of known ages and masses. The sparseness of Taurus is also important because the *IRAS* beam sizes are relatively large; ~ 1.5–2 arcmin at 60 μm and $\sim$ 3–4 arcmin at 100 μm in the narrowest dimension. At the distance of Taurus, $d \sim 140$ pc, 2 arcmin corresponds to 0.08 pc, comparable to or smaller than the expected radii of the parent molecular cloud cores. Unfortunately, even at these scales source confusion is not negligible; about 25% of the T Tauri stars observed with *IRAS* have two or more well-resolved pre-main-sequence stars within the same long-wavelength beam (Rucinski 1985), not including close binaries with separations $\lesssim 100$ AU.

IRAS surveys of the Taurus region (Beichman *et al.* 1986; Myers *et al.* 1987; Kenyon *et al.* 1990, 1994; Beichman, Boulanger, & Moshir 1992) have identified about 25–30 infrared sources as potential protostars. These objects are extremely faint or unobservable at optical wavelengths, exhibit very red near-infrared spectra suggestive of heavy extinction, and have spectral energy distributions peaking at wavelengths ~ 10–50 μm. A typical low-luminosity Class I SED is shown in Figure 1.3 (see also Figure 4.7). The observed 1–2 μm colors of Class I sources typically suggest a 1 μm extinction of $\gtrsim 10$ mag, much higher than the general extinction in the distributed molecular gas in Taurus, consistent with enhanced local extinction due to infall (§4.2).

The distribution of observed luminosities of the Class I sources in Taurus is shown in Figure 4.2. The median luminosity of all the classes is $\lesssim 1\, L_\odot$, suggesting that the central objects of the (generally) optically-invisible Class I sources are similar to those in the optically-visible pre-main-sequence (Class II and Class III) population. As discussed in the introduction to this chapter, and more fully in Chapter 9, such luminosities are substantially lower than would be expected for accretion onto a central star with T Tauri-like properties. The decrease in source numbers at the lower luminosities does not appear to be an observational selection effect, suggesting that the known population of Class I objects is reasonably complete in Taurus (Kenyon *et al.* 1994). This cloud complex is known to contain $\gtrsim 150$ pre-main-sequence stars, and the median age is $\sim 1.5 \times 10^6$ yr using the D'Antona & Mazzitelli pre-main-sequence evolutionary tracks (Figure 1.2; Chapter 9). Thus, the detection of ~ 25 Class I objects implies a lifetime in this stage of $\sim 3 \times 10^5$ yr, which is the timescale required for a $0.5\, M_\odot$ molecular cloud core to collapse at $\dot{M} = 1.6 \times 10^{-6}\, M_\odot\, \mathrm{yr}^{-1}$, the latter being the predicted mass infall rate for a thermally-supported 10 K molecular cloud core (§3.2).

Many studies of dense or clustered star-forming regions have been undertaken. The results are difficult to summarize at present because the situation is rapidly evolving. In addition, it is more difficult to interpret studies of these regions than those in Taurus because of the large amounts of distributed dust they contain. For example, the ρ Ophiuchus molecular cloud complex provides an interesting contrast to the Taurus region because the star-forming gas is significantly denser and hotter than in Taurus ($T \sim 25$ K). Many of the Class I objects lie in regions of ~ 50 mag of

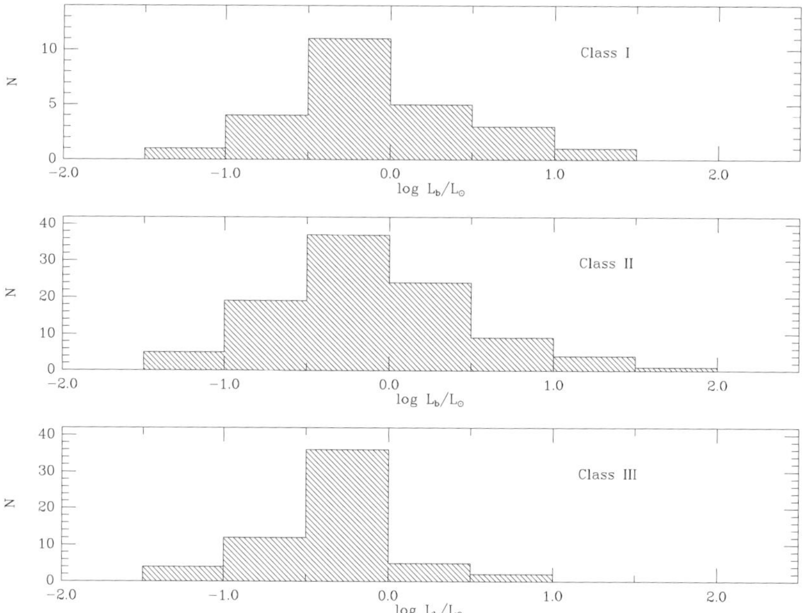

Fig. 4.2. Luminosity distributions of Class I, II, and III sources in Taurus (see Figure 1.3). The Class I sources, heavily extincted and in general protostar candidates, exhibit roughly the same median luminosities as the Class II sources (CTTS or stars with inner dusty disks) and the Class III (WTTS or stars without inner dusty disks). Adapted from Kenyon & Hartmann (1995).

visual extinction (1 μm extinction $\gtrsim$ 25 mag (Wilking & Lada 1983; Loren *et al.* 1990)). This *distributed* extinction is comparable to or larger than the *local* extinction produced by a collapsing protostellar cloud (§4.2). Objects might be identified as Class I objects and thus protostars when they are intrinsically Class II or III sources but with very large foreground extinctions that redden their SEDs enough to change their classification. The effect of large extinction on surveys has been discussed by André & Montmerle (1994) and Greene *et al.* (1994). Obviously, this large extinction will make it difficult to find *low-luminosity* protostars.

The distributed dust is also a strong source of *thermal* radiation at long wavelengths, and thus provides a bright background which makes it difficult to use the long-wavelength (60–100 μm) *IRAS* data to study any but the most luminous sources. This is especially true since only a few of the most luminous sources dominate the dust heating of sources in the region. Thus, the SEDs of many heavily-extincted, infrared-bright YSOs in Ophiuchus are uncertain at far-infrared wavelengths, and the use of ground-based observations at 10 μm and 20 μm, which have much smaller beam sizes than the corresponding *IRAS* measurements, are crucial in studying infrared excess emission in the region.

The simple theory for thermally-supported protostellar cloud cores predicts that mass infall rates should be larger if the gas temperatures are higher (Chapter 3). Surveys suggest that the luminosities of the Class I sources in Ophiuchus may indeed be larger than those in Taurus (Greene *et al.* 1994); however, the *median* luminosity of the Class I sources $\sim 4\,L_\odot$ is not very much larger than the median luminosity of the Class II + Class III sources in Ophiuchus ($\sim 2\,L_\odot$). Though Ophiuchus seems to contain a few Class I sources that are much more luminous than their counterparts in Taurus, it also has a few much more luminous Class III *stars* as well (see also Greene & Meyer 1995). In other words, it is not clear whether the higher Class I luminosities in Ophiuchus are due to accretion, or whether the pre-main-sequence stars just have generally higher luminosities. In any event, as in the case of Taurus, the median Class I luminosity in Ophiuchus is lower than the $\sim 30\,L_\odot$ resulting from the predicted infall rate of $\sim 6 \times 10^{-6}\,M_\odot\,\mathrm{yr}^{-1}$ which would result from collapse of a $T = 25$ K thermally-supported molecular cloud core (§3.2) (again assuming a M_*/R_* ratio typical of T Tauri stars; see Chapter 9).

Ophiuchus forms stars with much greater efficiency than Taurus ($\sim 10\%$ of the total gas mass in stars, vs. a few per cent in Taurus (Cohen & Kuhi 1979)). It also harbors a few of the 'cold' Class 0 sources (Chapter 1; André *et al.* 1993; André & Montmerle 1994). André & Montmerle have argued that the Class 0 sources are the 'true protostars', in the sense that these sources have yet to accrete most of their final mass, in contrast with the Class I sources, based on estimates of the envelope mass from measurements of mm-wave dust emission (see §4.4).

Near-infrared surveys have turned up many new clusters of young stars in highly-extincted regions (e.g., Lada *et al.* 1991). Currently, sufficient wavelength coverage is not available to determine the SEDs of most of these sources. The situation will be improved in important ways by results from the *ISO* observatory and by the hoped-for launch of the Space Infrared Telescope Facility *(SIRTF)*.

4.2   SEDs of spherical infalling envelopes

The identification of Class I sources in Taurus and Ophiuchus as protostars is basically due to the matching of the SEDs with the dust emission predicted for infalling dusty envelopes (Adams & Shu 1986; ALS; Butner *et al.* 1991). Although many hydrodynamic simulations have been used to predict dust envelope emission (e.g., Yorke *et al.* 1993; Boss & Yorke 1995), the advantage of the ALS method lies in its use of simple analytic models of the collapsing envelope at a snapshot in time. Because the thermal gas pressure (and radiation pressure) is generally negligible in the envelopes of low-mass protostars, one can decouple the hydrodynamic problem from the radiative transfer problem. The use of an analytic model allows greater freedom in handling the radiation transfer and thermal equilibrium, and makes it relatively easy to explore the effects of different envelope parameters on the emergent spectrum.

We begin with an exploration of dust envelope emission for spherical collapse. In the following discussion we assume a familiarity with the basic results of radiative transfer. Some essential relations are summarized in Appendix 3; others can be found in Mihalas (1978).

As discussed in §3.2, the density distribution of a spherical, collapsing, protostellar cloud can be reasonably approximated by 'free-fall' behavior, where the density of the infalling matter is

$$\rho \sim \frac{\dot{M}}{4\pi r^2 v_{ff}} = \frac{\dot{M}}{4\pi (2GM)^{1/2}} r^{-3/2} \tag{4.5}$$

(equation (3.21)). The radial optical depth integrated inward to radius r from the center, assuming a large outer radius, is

$$\tau_\lambda \;=\; \frac{\kappa_\lambda \dot{M}}{2\pi(2GM)^{1/2}} r^{-1/2}, \qquad\qquad (4.6)$$

where κ_λ is the opacity at wavelength λ.

To make a quantitative estimate of the amount of dust extinction through the envelope, we again adopt the standard curve for interstellar dust opacity from Draine & Lee (1984), as illustrated in Figure 4.1. This extinction curve is not unique; for example, it is known that the ultraviolet dust extinction varies strongly in regions with massive star formation (e.g., Mathis 1990). Far-infrared studies also indicate larger dust opacities near 200 μm (Hildebrand 1983). More recent studies suggest that the opacities in the sub-mm and mm range in dense protostellar cores might be a factor of 4–5 larger (Ossenkopf & Henning 1994). However, Figure 4.1 provides a starting point for exploration.

Most low-mass (T Tauri) stars have peaks in their photospheric SEDs at wavelengths $\lesssim 1\,\mu$m. The optical depth at 1 μm through the infalling envelope to a radius r_{in} is

$$\tau_1 \;\sim\; 10 \left(\frac{\dot{M}}{2\times 10^{-6}\,\mathrm{M_\odot\,yr^{-1}}} \right) \left(\frac{M}{0.5\,\mathrm{M_\odot}} \right)^{-1/2} \left(\frac{r_{in}}{15\,\mathrm{AU}} \right)^{-1/2}. \qquad (4.7)$$

From equation (A3.7), the light from the central star escaping along the radial direction is extinguished by a factor of $\exp(-\tau)$.[*] Thus, as long as the envelope extends into radii of tens of AU or less, for typical mass infall rates one expects the central star to be extinguished by a factor of $\sim 10^4$, or around 10 mag ($\Delta mag = 1.086\tau$). This simple estimate suggests that the central star will be essentially invisible at optical wavelengths unless the dust envelope is somehow removed at radii $\lesssim 100$ AU.

Because protostellar envelopes are very optically thick at the characteristic wavelengths of radiation from the inner region, they will absorb essentially all of the light from the central source (as long as the envelope is nearly spherically-symmetric). This means that, to a first approximation, the details of the central star's spectrum are not important, and the emergent spectrum seen at the Earth depends mainly upon the properties of the dust envelope. It is straightforward to show that the radiative cooling times of the grains are shorter than infall timescales, and therefore the envelope emission can be computed in the limit of radiative equilibrium.

The observed protostellar radiation at a given wavelength is a weighted sum of the radiation from several layers in the envelope. However, only a limited range of the envelope can contribute significantly, since optically-thin regions contribute little flux and very few photons escape from layers at large optical depths. Crudely, one may say that the observed emission arises from a 'characteristic' layer at $\tau_\lambda \sim 2/3$ (see Appendix 3). In this approximation, the dust envelope will have a 'photospheric'

---

[*] The amount of light from the protostar escaping through the envelope is not simply the stellar emission which is not absorbed by the envelope along the line between the star and observer. In addition, the envelope dust *scatters* protostellar emission emitted in other directions into the line of sight of the observer. In general, one cannot neglect dust scattering at optical and near-infrared wavelengths, and so equation (4.7) greatly underestimates the amount of protostellar emission that can be observed directly, especially since envelopes are not spherically symmetric (see §4.6; Figure 4.9). However, the extremely large extinctions of protostellar envelopes do make short-wavelength detections difficult.

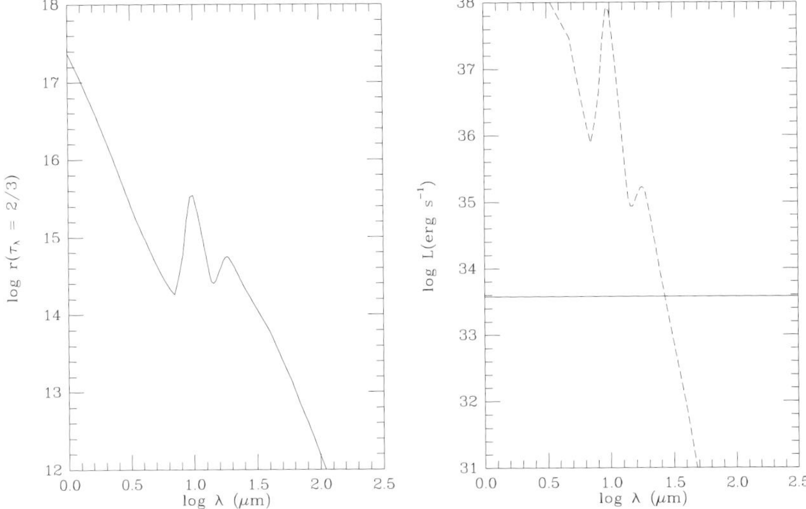

Fig. 4.3. Approximate solution for the dust 'photosphere' of a protostar, as described in the text. The left-hand panel shows the photospheric radius of the spherical collapsing envelope as a function of wavelength (equation (4.8)). The dashed curve in the right-hand panel shows the estimated envelope luminosity as a function of its characteristic wavelength λ_m. The horizontal line corresponds to a typical protostellar luminosity of 1 $L_\odot$. The intersection of the two provides an estimate of the approximate λ_m.

radius r_λ, defined as the radius where $\tau_\lambda = 2/3$,

$$r_\lambda = \left(\frac{9}{4}\right) \frac{\kappa_\lambda^2 \dot{M}^2}{8\pi^2 GM}.$$ (4.8)

In the case of most stellar atmospheres, the density distribution varies so rapidly with radius that the photospheric radius is essentially fixed over a wide range of wavelengths and opacities. However, the extended density distribution of the infalling envelope produces different photospheric radii at differing wavelengths of observation. The left-hand panel in Figure 4.3 shows r_λ as a function of wavelength using the opacity in Figure 4.1 and $\dot{M} = 2 \times 10^{-6} M_\odot \, \text{yr}^{-1}$.

To make an initial guess at the protostellar spectrum, we adopt the approximation that the dusty envelope emits like a blackbody with a peak wavelength λ_m. Equivalently, we assume that the dust emission has a characteristic effective temperature $T = 0.3/\lambda_m$. Thus,

$$L = 4\pi r_{\lambda_m}^2 \sigma \, T(\lambda_m)^4,$$ (4.9)

where r_{λ_m} is the radius at which $T(\lambda_m)$ occurs. If we then assume that r_{λ_m} is the 'photospheric' radius and so occurs at a radial optical depth $\tau_{\lambda_m} = 2/3$, then for a given dust opacity (Figure 4.1), a mass infall rate $\dot{M}$, and total central luminosity L, one can solve for λ_m, $T(\lambda_m)$, and r_{λ_m}.

The right-hand panel in Figure 4.3 shows the results of equation (4.9) applied to the infall case for $\dot{M} = 2 \times 10^{-6} M_\odot \, \text{yr}^{-1}$ as the dashed curve running from the upper left to the lower right. The intersection of this curve with the appropriate source luminosity

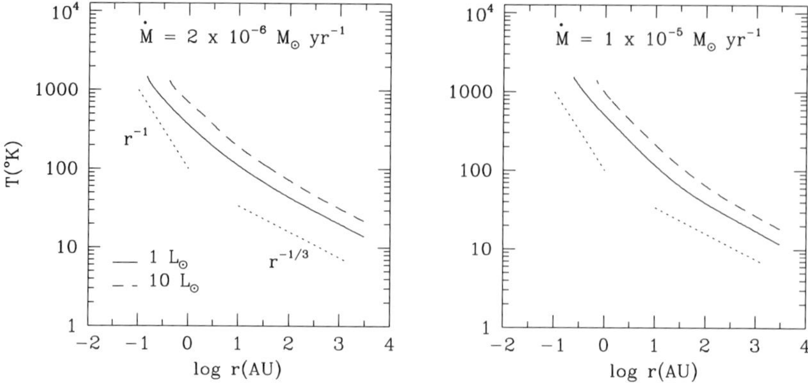

Fig. 4.4. Temperature distributions for spherical envelope models, shown for two different mass infall rates and two different source luminosities. The dotted lines exhibit the asymptotic behaviors predicted for optically-thin and optically-thick radiative equilibrium (see text). Courtesy N. Calvet.

provides an approximation to the photospheric radius of the dust envelope and the wavelength of peak SED emission. For an assumed luminosity of $1\,L_\odot$ (horizontal line), the intersection occurs at a wavelength $\sim 30\,\mu$m, or a characteristic temperature $T(\lambda_m) \sim 100$ K. Thus, the model predicts that typical protostars should mainly emit in the far-infrared.

An analytic estimate of λ_m can be made if we approximate the opacity as $\kappa_\lambda = \kappa_\circ(\lambda/\lambda_\circ)^{-\beta}$. Substituting,

$$\frac{\lambda_m}{\lambda_\circ} = \left(\frac{0.3}{\lambda_\circ}\right)^{1/(1+\beta)} \left(\frac{4\pi\sigma}{L}\right)^{1/(4+4\beta)} \left(\frac{9\,\dot{M}^2\kappa_\circ^2}{32\pi^2 GM}\right)^{1/(2+2\beta)}. \tag{4.10}$$

If we take the far-infrared opacity to be $\kappa_\lambda = 0.3\,(\lambda/100\,\mu\text{m})^{-2}$, then the peak of the spectrum occurs at

$$\lambda_m \approx 30\left(\frac{L}{L_\odot}\right)^{-1/12} \left(\frac{\dot{M}}{2\times10^{-6}\,M_\odot\,\text{yr}^{-1}}\right)^{1/3} \left(\frac{M}{0.5\,M_\odot}\right)^{-1/6} \mu\text{m}. \tag{4.11}$$

Thus, the free-fall collapse picture, coupled with predicted infall rates for simple models of cloud cores, indicates that protostars will emit most of their energy in the mid- to far-infrared spectral region. This conclusion is almost independent of the assumed central luminosity, and is only moderately dependent upon the mass infall rate.

To proceed beyond these simple estimates to more detailed calculation of a protostellar spectrum requires the application of detailed radiative transfer techniques to compute the radiative equilibrium temperature structure of the dust envelope and the emergent spectrum. Detailed radiative equilibrium temperature distributions for spherically-symmetric dusty envelope models are shown in Figure 4.4. Results are presented for two mass infall rates, $\dot{M} = 2\times10^{-6}\,M_\odot\,\text{yr}^{-1}$, and $\dot{M} = 10^{-5}\,M_\odot\,\text{yr}^{-1}$, for central source luminosities of 1 and 10 $L_\odot$.

The overall features of these temperature distributions can be understood from

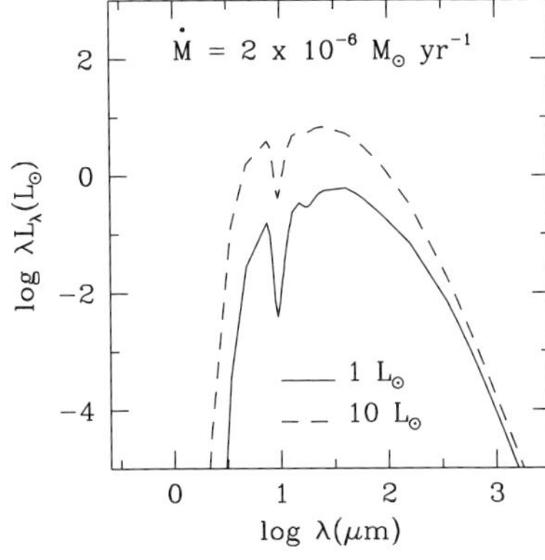
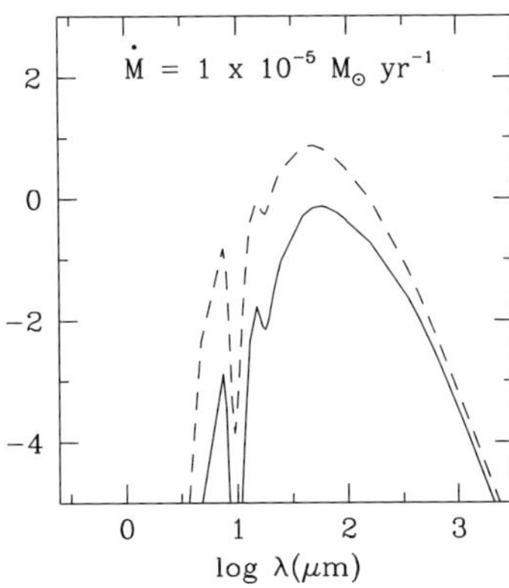

Fig. 4.5. SEDs for the spherical infall models whose temperature distributions are shown in Figure 4.4. The wavelength λ_m of the peak emission moves to longer wavelengths for higher mass infall rates, but is not very sensitive to the central source luminosity, as described in the text. Courtesy N. Calvet.

simple arguments. Compare, for example, the results in Figure 4.4 for $L = 1\,\mathrm{L_\odot}$ and $\dot{M} = 2 \times 10^{-6}\,\mathrm{M_\odot\,yr^{-1}}$ with the results of Figure 4.3 or the approximate relations (4.11) and (4.8). The predicted value of $\log \lambda_m$ is ~ 1.5, corresponding to a 'photospheric' radius of about 1.5×10^{14} cm ~ 10 AU. Roughly speaking, the detailed temperature distribution in Figure 4.4 changes slope near this radius, marking the transition between inner 'optically-thick' and outer 'optically-thin' behavior.

At small optical depths, the radiative equilibrium equation

$$\int \kappa_\nu (J_\nu - B_\nu)\, d\nu = 0 \tag{4.12}$$

(see Appendix 3) simplifies because $J_\nu \to$ constant $\times\ r^{-2}$. It is straightforward to show that the appropriate (Planck) mean opacity varies as $\kappa_P \propto T^\beta$ for $\kappa_\lambda \propto \lambda^{-\beta}$. Therefore,

$$T \propto r^{-2/(4 + \beta)} \quad \text{(optically thin)}. \tag{4.13}$$

For $\beta = 2$, characteristic of the far-infrared opacity adopted (Figure 4.1), one can use equation (4.13) to show that $T \propto r^{-1/3}$. The temperature distributions in Figure 4.5 show this property at large distances. (If one adopts $\beta = 1.5$ for the long-wavelength opacity, as suggested by observational and theoretical treatments (e.g., André *et al.* 1993; Ossenkopf & Henning 1994), the results are only slightly different, $T \propto r^{-0.36}$.)

The temperature distribution changes its character inside the 'photospheric' radius, where the envelope is very optically thick; trapping of radiation causes the temperature gradient to become steeper than in the optically-thin case. In the limit of large optical depth, one can use the usual diffusion approximation,

$$L = -\frac{64\pi\sigma r^2 T^3}{3\kappa_R \rho}\frac{dT}{dr} \tag{4.14}$$

(Appendix 3), where κ_R is the Rosseland mean opacity. Again, for a power-law dependence of opacity $\kappa_\lambda \propto \lambda^{-\beta}$, $\kappa_R \propto T^\beta$. Taking a power-law distribution of density $\rho = \rho_{\circ}(r/r_{\circ})^{-n}$, equation (4.14) can be integrated with the boundary condition that $T \to 0$ as $r \to \infty$, with the result that

$$T \propto (L\rho_{\circ}r_{\circ}^{n})^{1/(4-\beta)}r^{-(1+n)/(4-\beta)}. \qquad (4.15)$$

The optically-thick portions of the solutions shown in Figure 4.4 pass through temperatures ~ 300 K, where the peak of the blackbody radiation is at $10\,\mu m$; in this region, $\beta \sim 1.5$, and so we find $T \propto r^{-1}$, in reasonable agreement with the detailed calculations.

Figure 4.5 shows SEDs for spherical infalling envelopes calculated from detailed radiative transfer solutions for two values of both central source luminosities and mass infall rates. The simple scaling law described above provides a reasonable first approximation to the results, although there are obviously important spectral features, such as the silicate features near 10 and 20 μm, which require detailed calculations. The resulting SEDs are slightly broader than a single-temperature blackbody distribution, because a range of temperatures is present in layers that contribute to the emission.

4.3   SEDs for rotating collapse models

The observed SEDs of Taurus Class I sources (Figure 1.3) exhibit far-infrared emission similar to that predicted by the spherical collapse models (Figure 4.5). However, the models predict far too little near- to mid-infrared emission to explain the observations, indicating that the extinction of the central hot regions must be reduced. As equation (4.6) indicates, most of the optical depth in the infalling envelope accumulates at small radii. For this reason, Myers *et al.* (1987) originally suggested that there were inner 'holes' in Class I envelopes to reduce the extinction and increase the amount of short-wavelength light that escapes.

The inclusion of rotation in the envelope models of ALS produced SEDs in better agreement with the observations. Because rotation causes the infalling material to land on the disk (§3.4), it also reduces the central envelope extinction. To see this in broad terms, consider the spherical average of the TSC density distribution (Figure 3.7; §3.4). In this approximation, the density distribution inside r_c has the form $\rho \propto r^{-1/2}$ (equation (3.46)). Rotation therefore limits the column density or optical depth ($\propto \int \rho dr$) through the envelope; material drops onto the disk and is therefore removed as an opacity source.

In a radial infall model, the inner radius of the dust envelope is limited by the dust destruction radius, i.e. where temperatures become too high for dust to survive. Typical estimates suggest that the dust destruction temperature is ~ 1500 K (Larson 1972; Stahler, Shu, & Taam 1980a,b; Wolfire & Cassinelli 1987). For a central star of a few solar luminosities, the dust destruction radius is of order 10^{-1} AU. Now most of the optical depth of a TSC model is accumulated near r_c; in terms of the spherical average, the total optical depth through the envelope is roughly given by equation (4.7) if r_{in} is replaced by r_c and the scaling constant is doubled. Thus, a TSC infall model with a centrifugal radius of 100 AU has a total column density roughly 30 times smaller than the corresponding radial collapse model with the same total mass infall rate.

This reduction of envelope optical depth due to rotation has a profound effect on the amount of near- and mid-infrared emission emitted by a protostellar envelope

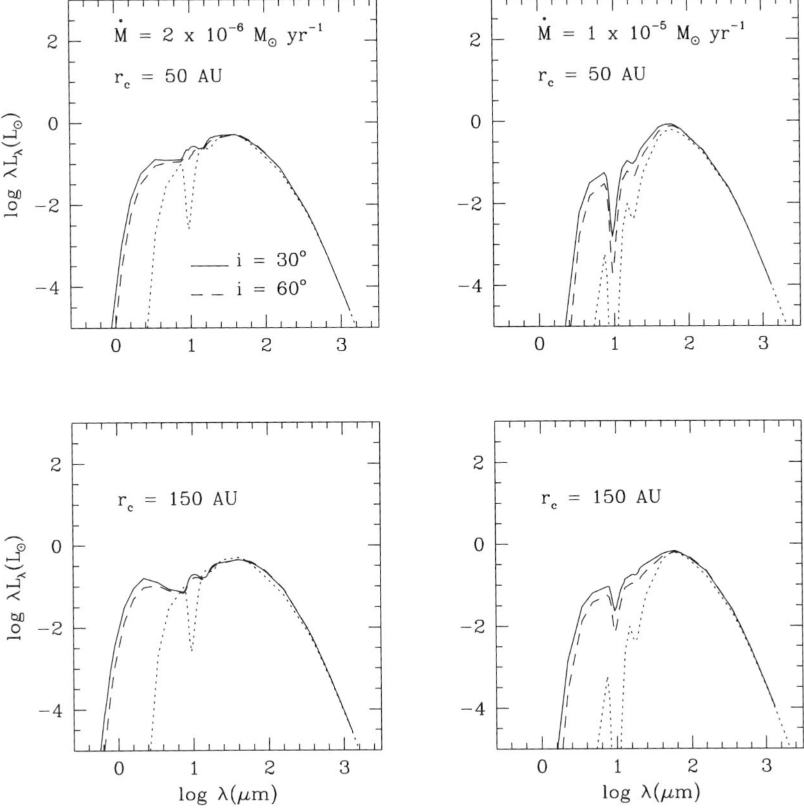

Fig. 4.6. Effects of rotation on the SEDs of infalling envelopes. The dotted lines repeat the non-rotating SEDs with $L = 1\,L_\odot$ from Figure 4.5 for comparison. The models including rotation exhibit vastly more mid- and near-infrared emission because of the reduction in inner envelope optical depth as infalling dusty material falls onto the disk. The silicate feature at $\lambda \sim 10\,\mu$m can change from the deep absorption seen in spherical models to emission for plausible infall rates and centrifugal radii r_c (§3.4). The SED and the depth of the silicate feature also depend upon the viewing angle i between the rotation axis and the line of sight in addition to the mass infall rate and centrifugal radius. Courtesy N. Calvet.

(see Figure 4.6). Adams & Shu (1986) and ALS made the initial applications of the TSC infall models to Class I sources. Because the density variation from pole to equator in the TSC model at a constant radius is generally no more than $\sim$ a factor of two (see Figure 3.3), ALS estimated the radiative equilibrium temperature using the spherical average of the opacity distribution in the envelope (cf. Efstathiou & Rowan-Robinson 1991). The resulting calculations showed that modest amounts of rotation in the initial protostellar cloud, consistent with the velocity gradients observed in molecular cloud cores (Goodman *et al.* 1993), could explain much of the observed mid-infrared emission of these infrared sources.

ALS found reasonable agreement of source luminosities with reasonable theoretical parameters for the assumed accretion rate. In their treatment, the central luminosity comes from both accretion onto the star and through the disk, and the accretion luminosity is generally

$$L_{acc} \sim \frac{3}{4}\frac{GM_* \dot{M}}{R_*}.$$

(4.16)

The factor of 3/4 arises from the assumption that some of the accretion energy is not radiated at the star but goes into ejecting a wind, or spinning up the star, etc. (see Chapter 9 for further dicussion). Encouragingly, ALS found that they could reproduce their Taurus source luminosities for $\dot{M} = 2 \times 10^{-6}\,\mathrm{M_\odot\,yr^{-1}}$ with central masses of $\sim 0.5\,\mathrm{M_\odot}$. However, later investigators (Kenyon *et al.* 1990, 1994) found disagreement between predicted and observed Class I luminosities, mainly because ALS only modelled some of the brightest Class I sources in Taurus and Ophiuchus; their Taurus sources have luminosities almost an order of magnitude larger than the median value.

This result emphasizes the need to obtain systematic samples of the protostellar populations. Observations of any star-forming region yield only a snapshot of proto-stars of differing masses in various stages of evolution. For these reasons, results for as complete a sample as possible are required to make further progress. Currently the most systematic study of the SEDs of protostellar sources in a star-forming region is the survey by Kenyon *et al.* (1993) of Taurus Class I objects. Kenyon *et al.* also adopted the TSC model to compute dusty envelope emission, and also used the spherical average of the TSC density distribution to calculate the dust equilibrium temperature, but departed from ALS in taking the correct density distribution into account when calculating the emergent spectrum. As shown in Figure 4.6, the emer-gent spectrum is strongly affected by the envelope rotation; at short wavelengths, the extinction of the outer layers is much larger in the equatorial direction than along po-lar regions, even though the differences in column densities are modest (cf. Figure 3.6). Kenyon *et al.* (1993) also included the effects of dust scattering in a systematic way, and, unlike ALS, assumed that the source luminosity is a free parameter. With these assumptions, the TSC SED models essentially have four fitting parameters: the source luminosity, the centrifugal radius, the viewing angle relative to the rotation axis, and a parameter setting the density of the infalling envelope. This last parameter can be related to the mass infall rate for a given assumed central mass.

Kenyon *et al.* modelled the SEDs of 21 Class I sources and found a median infall rate $\sim 4 \times 10^{-6}(M_*/0.5\,\mathrm{M_\odot})^{1/2}\,\mathrm{M_\odot\,yr^{-1}}$ (cf. Figure 4.7). All but two of the objects had infall rates within a factor of three of this median value. Given all the uncertainties in the calculation – departures from spherical symmetry, uncertainties in dust properties, possible role of magnetic fields, approximations in the radiative transfer, etc. – this result is reasonably close to the $\dot{M} \sim 1.6 \times 10^{-6}\,\mathrm{M_\odot\,yr^{-1}}$ predicted by the simplest singular isothermal sphere collapse theory for the $T = 10$ K temperatures in Taurus. However, this infall rate is unlikely to be the accretion rate onto the central star, because for the expected stellar mass to radius ratio (Chapter 9), the accretion luminosity should be more like $8\,\mathrm{L_\odot}$ than the median luminosity $\sim 0.6\,\mathrm{L_\odot}$ (Figure 4.2).

One possible resolution of the luminosity problem suggested by Kenyon *et al.* (1990) and Kenyon *et al.* (1993) is that matter is piling up in the disk. This cannot continue indefinitely, since a massive disk is probably gravitationally unstable. The accumulation of matter in the disk can only explain the observations if rapid accretion occurs during short-lived outbursts, such as in the FU Ori objects (§ 1.6, Figure 1.7;

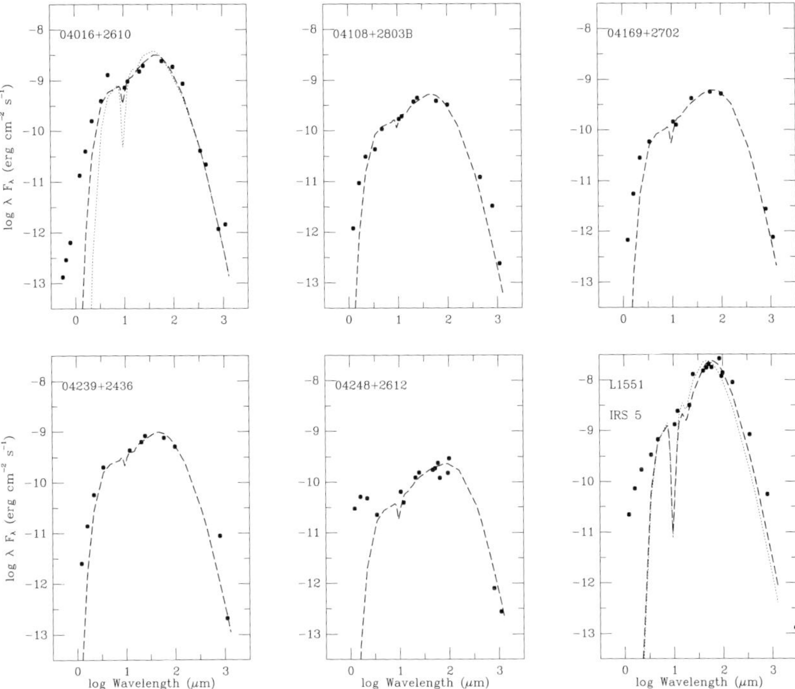

Fig. 4.7. Rotating infall model SEDs fitted to observations of Taurus Class I sources. Two fits are shown for IRAS 04016+2610 and L1551 IRS 5 to illustrate the sensitivity to parameters. In the case of IRAS 04016+2610, the best-fit model (dashed line) corresponds to $\log \dot{M} = -5.4$ (in solar masses per year, assuming a central mass of $0.5\,M_\odot$), $r_c = 70$ AU, and an inclination angle $i = 60°$; the dotted curve shows a model with the same infall density but $r_c = 10$ AU and $i = 30°$. In either case the models cannot explain the observed radiation at wavelengths $\lesssim 1\,\mu$m. The two models shown for L1551 IRS 5 are of roughly comparable goodness of fit; the dashed curve corresponds to $\log \dot{M} = -4.9$ (again assuming a central mass of $0.5\,M_\odot$), $r_c = 70$ AU, and $i = 60°$; the dotted curve corresponds to a model with $\log \dot{M} = -4.4$, $r_c = 300$ AU, and $i = 30°$. In general the fit parameters are interdependent; for example, to a (limited) extent the larger extinction produced by smaller r_c can be compensated for by making i smaller, or $\dot{M}$ smaller, although this last parameter is constrained by the wavelength of peak emission (§4.2). Adapted from Kenyon *et al.* (1993).

Chapter 7). Indeed, observations suggest that the most luminous Class I source in Taurus, L1551 IRS 5, is a (relatively) low-luminosity FU Ori object (Chapter 7), with current rapid disk accretion.

The TSC model fitting also produces some questionable patterns. To produce enough near-infrared emission, almost all of the sources must have $r_c \gtrsim 70$ AU. This seems unlikely, since the centrifugal radius grows with time as $r_c \propto t^3$ (equation (3.50)), and it is difficult to imagine why a common instantaneous centrifugal radius should appear among objects of differing evolutionary status. The sensitivity of r_c to the initial cloud rotation makes this result seem even more unlikely. Similarly, the median inclination derived from the SED modelling is $i \sim 30°$, rather than the average value $i \sim 60°$ expected for a random distribution of inclinations. The small values of inclination are needed to produce enough short-wavelength emission, similarly

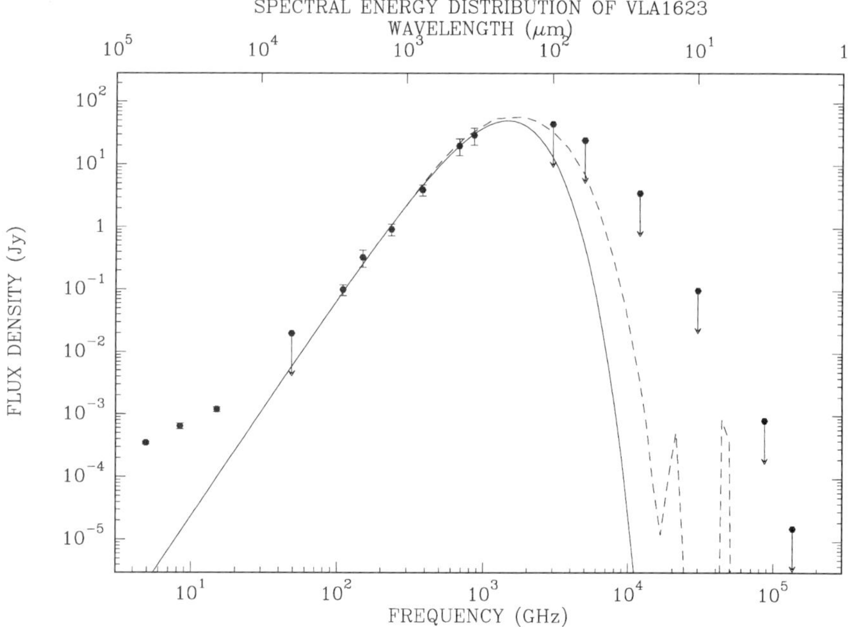

Fig. 4.8. SED of the Class 0 source VLA 1623 (see text). From André *et al.* (1993).

implying less extinction in the line of sight than present in a typical TSC model at average inclination.

4.4   The Class 0 sources

At present the Class 0 sources pose somewhat of a problem for the standard collapse model. The SEDs of these objects peak near $\lambda_m \sim$ 150–200 μm (Figure 4.8), which according to equation (4.11) would require a mass infall rate $\gtrsim 4 \times 10^{-4} \, M_\odot \, \mathrm{yr}^{-1}$ for typical luminosities $L \sim 10 \, L_\odot$. (For comparison, the accretion rate predicted at $T = 25$ K in singular isothermal sphere similarity solution is $\dot{M} \sim 6 \times 10^{-6} \, M_\odot \, \mathrm{yr}^{-1}$.) Using alternative forms for the long-wavelength opacity than Draine & Lee (1984), such as a smaller wavelength dependence of the opacity $\beta = 1.5$ (e.g., André *et al.* 1993), as might be suggested by theoretical calculations including ice mantles (Ossenkopf & Henning 1994) reduces the required infall rate only by a modest amount. One possibility is that the Class 0 sources represent the non-similar collapse of the protostellar cloud in its initial phase (Henriksen *et al.* 1997). As discussed in §3.3, the earliest stages of the collapse of a Bonnor–Ebert sphere or other cloud with a flat innermost density distribution will exhibit much higher mass infall rates than predicted by the singular isothermal sphere similarity solution.

Detailed observational studies suggest additional difficulties for the standard picture of protostellar collapse. In a careful analysis of the Class 0 source VLA 1623.4-2418, André *et al.* (1993) argued that the spatial distribution of the sub-mm continuum emission was not consistent with infall, but instead suggested a density distribution $\rho \propto r^{-1/2}$, with an outer radius of $r_{out} \sim 3.5 \times 10^{16}$ cm and a density at this radius $\rho(r_{out}) \sim 4 \times 10^{-17} \, \mathrm{g \, cm^{-3}}$. A cloud with a uniform density of $4 \times 10^{-17} \, \mathrm{g \, cm^{-3}}$ and

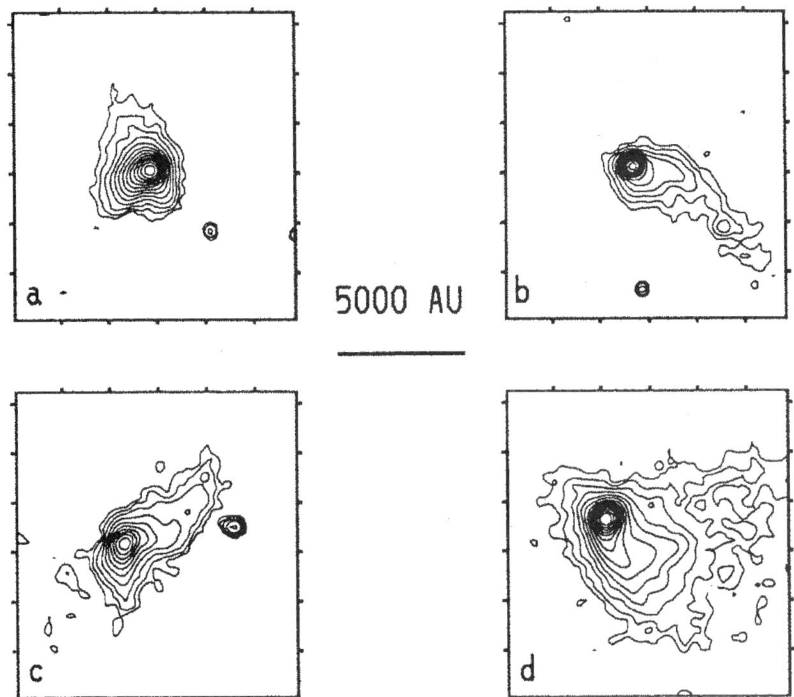

Fig. 4.9. Images of Class I sources in Taurus taken at $\lambda = 2.2\,\mu$m. The images have been scaled to a resolution of 2.4 arcsec. (a) 04016+2610 (cf. Figures 1.3, 4.7); (b) 04169+2702; (c) 04248+2612; (d) 04287+1801. From Tamura *et al.* (1991).

this outer radius would have a total mass of $\sim 3.6\,M_\odot$ and a free-fall collapse time of $t_{ff} \sim 10^4$ yr, suggesting appropriate conditions for producing infall at $\sim 10^{-4}\,M_\odot\,\mathrm{yr}^{-1}$. However, it is difficult to understand what (if anything) is supporting this material against gravity. (For comparison, a $1\,M_\odot$ critical Bonnor–Ebert sphere at $T = 25$ K has a radius of $\sim 6 \times 10^{16}$ cm and a mean density of only $\sim 2 \times 10^{-18}\,\mathrm{g\,cm}^{-3}$.) Models have been constructed for cores in hydrostatic equilibrium, but supported mostly by supersonic turbulent pressure (Lizano & Shu 1989; Myers & Fuller 1992; Caselli & Myers 1995), but the applicability of such models is not clear.

The large circumstellar masses of Class 0 objects led André *et al.* (1993) to suggest that the Class 0 sources are the 'true', 'youngest' protostars (see also André & Montmerle (1994)). It may be that there is no conflict with results in Taurus, where source statistics are consistent with infall rates determined by SEDs; perhaps star formation in dense regions like Ophiuchus proceeds by extremely rapid infall at higher densities. (Note that Taurus contains at least one borderline Class 0 source, 04368 + 2557 (L1527); André *et al.* 1993). Very high infall rates, however, may involve turbulent pressure support, collapse from cores not in hydrostatic equilibrium, or other departures from the standard model of star formation discussed in Chapters 2 and 3.

4.5 Envelope 'holes', outflows, and flattened collapse

The rotating (e.g., TSC) collapse models generally predict far too little dust emission near $\lambda \lesssim 1\,\mu$m in Taurus Class I sources, even if systematically large r_c and small

i values are adopted. The main reason for this discrepancy is probably that real protostellar envelopes have 'holes' or low-extinction paths through which short-wavelength light preferentially escapes. Figure 4.9 shows near-infrared images of several Taurus Class I sources. The observed spatially-extended emission is not locally produced; the temperatures on these distance scales ~ 100 AU are far too low to produce appreciable emission at such short wavelengths (cf. Figure 4.4). Instead, this light must originate in relatively hot regions near or at the central (proto)star, and be reflected from dust on much larger scales. The geometry of the envelope must include some sort of cavity to allow enough short wavelength radiation to escape.

Scattered light cavities are commonly observed near highly-extincted young stellar objects (e.g., Bastien & Ménard 1988). The classic example of extended scattered light is the luminous Class I source L1551 IRS 5 in Taurus (see, e.g., Stocke *et al.* (1988) and references therein). A large ~ 4 arcmin ~ 0.15 pc optical reflection nebula extends to the southwest of the central source. The identification of the infrared source IRS 5 as the illuminating source of this nebula is also demonstrated by the pattern of the polarization of the nebular light. Estimates of the visual extinction along the line of sight to the central source are $A_v \sim 90$, while the reflection nebula is so bright that the total extinction along the path from the central source to the point of reflection and then on to Earth cannot be much larger than $A_v \sim 2 - 3$. Clearly this huge variation in column density is inconsistent with a simple TSC model. Because the reflection nebula is spatially coincident with the blueshifted lobe of a powerful bipolar molecular outflow seen in CO emission (Chapter 8), it is likely that mass ejection is responsible (at least in part) for evacuating the infalling/remnant dusty material along the axis of ejection (Snell *et al.* 1980; cf. Figure 1.6).

In addition to outflows which may blow away the enshrouding protostellar and circumstellar material of Class I sources, departures from initial non-sphericity in protostellar clouds may help produce envelope holes as well. To see why, consider again the simple calculation of sheet collapse discussed in §3.3. Figure 4.10 illustrates the sheet density distribution initially shown in Figure 3.3, displayed in gray scale to emphasize the regions of high density. As collapse proceeds, most of the material becomes concentrated toward the equatorial plane, producing evacuated cavities in the polar directions. The resulting envelope asymmetry is very much greater than in a TSC model.

To explore the implications of this density structure for reflection nebulae, Hartmann *et al.* (1996) developed a modification of the TSC collapse model which matches the numerical simulation on large scales. Figure 4.11 illustrates the resulting scattered light nebula at $\lambda = 1.25\,\mu$m (J-band) predicted for typical Taurus Class I parameters $\dot{M} = 4 \times 10^{-6}(M_*/0.5\,\mathrm{M_\odot})^{1/2}\,\mathrm{M_\odot}\,\mathrm{yr}^{-1}$, $r_c = 50\,\mathrm{AU}$, and $i = 60°$. The four panels show the results for increasing flatness (increasing parameter η) of the collapsing envelope, which in turn suggests how the reflection nebula might evolve with time. The results for $\eta = 2$ to $\eta = 3$ correspond roughly to the two bottom snapshots in Figure 4.10 (also Figure 3.3). At early times, the flattening of the envelope is not large, and so the enhancement of scattered light is small; in these stages one would expect that the observed scattered light would be dominated by the hole punched into the envelope by the bipolar outflow. At later times, the envelope 'cavity' becomes wider, and a bright fan-shaped nebula can be observed (cf. Figure 4.9). This occurs even if the outflow is highly collimated (Chapter 8) and therefore unable to produce a wide-angle cavity.

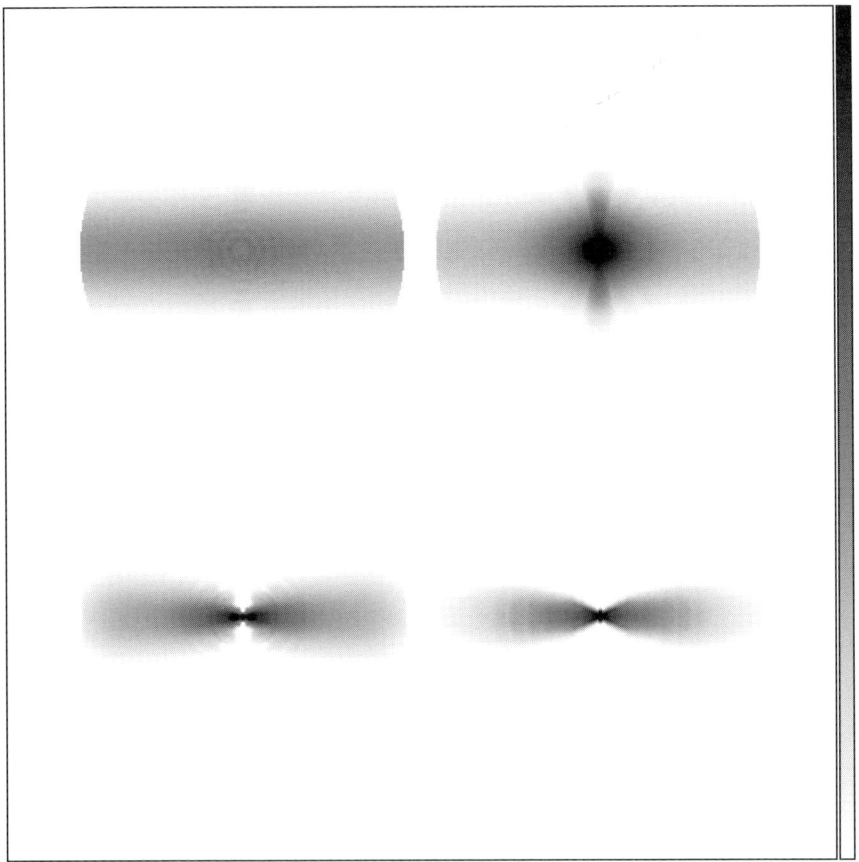

Fig. 4.10. The density distribution of the sheet collapse model discussed in §3.3, shown as a logarithmic gray scale in a meridional plane (see Figure 3.3). From Hartmann *et al.* (1996).

Protostellar cloud cores are not observed to be spherical (§2.1). Indeed, if magnetic fields play a role in supporting the cloud or in modifying its dynamics (Mouschovias & Spitzer 1976; Nakamura *et al.* 1995; Li & Shu 1996), the collapsing cloud cannot be spherical on large scales. Therefore, it seems inevitable that low-extinction cavities will be present in the envelopes of protostars, even in the absence of outflows, since some material must have already fallen in to produce the protostellar core.

One implication of cavities in infalling envelopes is that analyses of short-wavelength emission in SEDs do not provide strong constraints on r_c and i, since holes can drastically affect the results. An even more important result is that protostellar envelopes with wide-angle cavities allow one to observe the central protostar at short wavelengths with low extinction over a substantial range of viewing angles. We return to this possibility in Chapter 6.

4.6   Spatial distribution of emission

If the thermal dust emission from the envelope can be resolved, one can test the consistency of the density distribution with infall, especially when observed at long

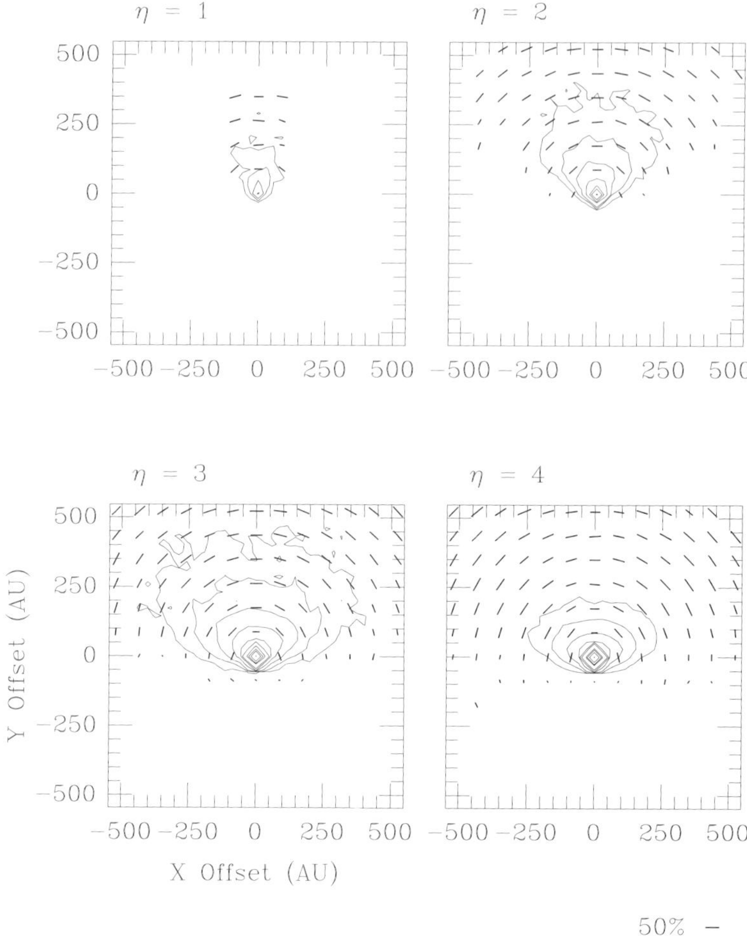

Fig. 4.11. Calculations of light scattering from a collapsing flattened sheet. An analytic approximation to the numerical calculations presented in Figure 4.10 (also Figure 3.3) is used to improve resolution of the density structure near the central mass. The parameter η is used to indicate the amount of flattening in the envelope; roughly speaking, it corresponds to the original radial distance from the center of material currently falling onto the central star and disk, measured in units of the initial sheet scale height H; e.g., when $\eta = 3$, material that started out at $r \sim 3H$ is falling into the central regions. The bottom two panels ($\eta = 3, 4$) approximately correspond to the two bottom density distributions shown in Figure 4.10. Polarization vectors are also shown as lines, with the length of the line indicating the magnitude of linear polarization. From Hartmann *et al.* (1996).

wavelengths where the dust is optically thin. The observed intensity on the sky is (cf. Appendix 3)

$$I_\nu(p) = \int_0^\tau S_\nu \exp(-\tau_\nu) d\tau_\nu, \qquad (4.17)$$

where the impact parameter p is the distance measured perpendicular to the line of sight from the center. The optical depth at frequency v along the line of sight at p is

$$\tau_v = \int dz\, \rho \kappa_v \,, \qquad (4.18)$$

where κ_v is the dust opacity per unit mass at frequency v and z is the coordinate along the line of sight at constant p.

One simple limit is when the envelope is optically thin. This applies at long wavelengths and large radii, where the envelope will be more spherical, and the source function will be entirely thermal. Then

$$I_v(p) = \int_0^\tau B_v d\tau_v = 2 \int_0^\infty B_v k_v \rho\, dz \,, \qquad (4.19)$$

where B_v is the Planck function and the factor of two takes the symmetry of the envelope into account.

If the frequency of observation is sufficiently low that the Planck function is in the Rayleigh–Jeans limit $hv \ll kT$, and if we take power-law distributions $T = T_\circ (r/r_\circ)^{-m}$, $\rho = \rho_\circ (r/r_\circ)^{-n}$, then

$$I_v(p) = \frac{4v^2 k r_\circ T_\circ \rho_\circ \kappa_v}{c^2} \left(\frac{p}{r_\circ} \right)^{1-(m+n)} \int_1^\infty \frac{x^{1-(m+n)} dx}{(x^2 - 1)^{1/2}} \,. \qquad (4.20)$$

Thus, the surface brightness depends upon both the density and the temperature gradient. For the standard case of infall in the optically-thin temperature distribution, $m = 1/3$ and $n = 3/2$, $I_v(p) \propto p^{-5/6}$.

There are several problems with carrying out this analysis in practice. The intensity distribution is very sharply peaked at small radii (where very little of the mass resides), while the extended wings of the distribution produced at large radii (where most of the mass resides) may be difficult to distinguish from background emission. This makes it generally difficult to follow the surface brightness of the cloud out to several beam sizes; usually, comparisons are made between the main peak of the intensity distribution and the point spread function (e.g., Butner *et al.* 1991). A pure power-law distribution is scale free, so it does not have a 'size'; inner and outer cutoffs play an important role in determining the apparent source 'size' when convolved with a (roughly Gaussian) observational point spread function (see discussion in Terebey, Chandler, & André (1993) and Ladd *et al.* (1991)). Notwithstanding these difficulties, some careful investigations of the spatial distribution of long-wavelength dust emission suggest density distributions reasonably consistent with infall (Butner *et al.* 1991; Ladd *et al.* 1991; Terebey *et al.* 1993; Chandler & Sargent 1993).

The infalling envelope should have most of its mass on large scales, with little mass near the star. A few Class I sources appear to have large amounts of dust $\sim 0.1\,M_\odot$ on size scales comparable to 100 AU (Keene & Masson 1990; Lay *et al.* 1994). These structures are probably circumstellar disks produced by the infall (Terebey *et al.* 1993; Chapter 6).

4.7   Detection of infall Doppler shifts

Although direct detection of infall motions is the clearest indication of collapse, as mentioned in the introduction to this chapter, this has been a very difficult measurement to make. One reason is the confusing effects of outflows which may accelerate molecular gas (cf. Zhou & Evans (1994) and references therein). Another reason is that there is very little mass at high velocities. One can see this by noting that, in spherical infall, the mass interior to r is $M_r \propto r^{3/2}$; but since the infall velocity $v_{ff} \propto r^{-1/2}$, $M_r \propto v^{-3}$. In addition, if material does drop out onto a disk of size $r_c \sim 100\,\mathrm{AU}$, the amount of mass at velocities greater than a few $\mathrm{km\,s^{-1}}$ will be even smaller.

Radio-frequency spectral lines are the most promising diagnostic of protostellar collapse, since such lines will not suffer attenuation from dust. To illustrate the nature of the line profile in a very approximate way, we first make some simplifying approximations. Consider an optically-thin emission line formed in a spherically-symmetric envelope undergoing free-fall collapse. The total emission per unit volume (cf. Appendix 3) can be approximated by

$$4\pi j_v = 4\pi\kappa_\circ S(r)\rho\,\delta(v - v_\circ), \tag{4.21}$$

where $\kappa_\circ$ and S are the line opacity and source function integrated over the line profile, and it is assumed that the thermal line width is much smaller than the flow velocities so that the profile can be taken as a Dirac delta function.

The projected radial velocity of an element with velocity v_z along the line of sight z is

$$v_z = v(r)\cos\theta, \tag{4.22}$$

where θ is the angle between the radius vector to the point in question and the line of sight. By approximating the line profile as a delta function, we neglect the intrinsic line width; then the emission from a volume element $2\pi r^2 dr\sin\theta d\theta$ will be spread over a corresponding velocity interval

$$d|v_z| = v(r)\sin\theta d\theta. \tag{4.23}$$

Consider a thin shell with negligible thickness dr. The emission from this shell in a projected velocity interval $d|v_z|$ is

$$dL(v_z) = 4\pi\kappa_\circ S_\circ\rho 2\pi r^2\sin\theta d\theta\, dr = 8\pi^2\kappa_\circ S_\circ\rho r^2\frac{d|v_z|}{v(r)}\,dr. \tag{4.24}$$

This emission is not a function of v_z, i.e. the emission per unit projected velocity interval is a constant, i.e. the velocity profile of this optically-thin, spherically-symmetric shell is rectangular.

The total emission profile is a sum over all contributing shells. To see how this is added up in a simple way, note that the emission from an individual shell at velocities $v > v_0$ is a fraction $(1 - v_0/v)$ of the total shell emission (because the profile is rectangular). The total emission from the entire envelope is given by the sum of this fraction over all shells. Since only radii less than r_0 can contribute to the emission at $v(r) > v_0 = v(r_0)$, this integrated emission is

$$L(v > v_0) = \int_0^{r_o} 4\pi r^2\kappa_\circ S(r)\rho(r)[1 - v_0/v(r)]dr. \tag{4.25}$$

If we assume that the opacity $k_\circ$ is constant, $\rho = \rho_o(r/r_o)^{-3/2}$, $v = v_o(r/r_o)^{-1/2}$, and

$S = S_\circ(r/r_0)^{-q}$, then

$$L(v > v_0) = 4\pi r_o^3 \kappa_\circ \rho_0 S_\circ \int_0^{r_o} \left(\frac{r}{r_o}\right)^{1/2-q} \left(\frac{dr}{r_o}\right) \left[1 - \frac{v_o}{v(r)}\right] \tag{4.26}$$

$$= 8\pi r_o^3 \kappa_\circ \rho_0 S_\circ \int_1^\infty \frac{dw}{w^{4-2q}} \left(1 - \frac{1}{w}\right), \tag{4.27}$$

where $w = v(r)/v_0$. The integral provides only a numerical factor; the dependence of the factors outside of the integral is

$$L(v > v_0) \propto r_o^{3/2-q} \propto v_o^{-(3-2q)}. \tag{4.28}$$

This is the integrated luminosity at $v > v_0$, so the line profile is the derivative of this quantity,

$$L(v) \propto v^{-2(2-q)} \tag{4.29}$$

(Kuiper, Rodriguez-Kuiper, & Zuckerman 1978). Thus, the emission line profile of a collapsing cloud should be centrally peaked, with the high-velocity infalling material observable only in faint wings for typical source function gradients. As a simple example, consider a radio-frequency line in LTE (local thermodynamic equilibrium) in the long-wavelength or Rayleigh–Jeans limit, so that the source function is equal to the Planck function $S = 2v^2 kT/c^2 \propto T$. Since $0 \leq q \leq 1$ (see §4.2), $L(v) \propto v^{-4} - v^{-6}$.

This simple analysis is of limited practical use in seaching for infall, in part because of complications such as non-LTE excitation, molecular chemistry, etc., but mostly because optically-thin emission line profiles cannot be used to distinguish between infall, outflow, and turbulent motions. In any of these situations, the line profiles will generally appear to be symmetric. In the case of optically-thick line emission, however, it is possible to make a distinction between net expansion or collapse of the medium. This analysis utilizes the line optical depth to distinguish between the near side and the far side of the envelope, along with the general tendency of line source functions to decrease with increasing distance from the central source.

A more complicated analysis than presented above is needed to model the optically-thick line emission from moving envelopes (Kuiper *et al.* 1978; Anglada *et al.* 1987). Again consider a spherical envelope in radial free-fall. Simplifying the situation once more, assume that only material with exactly the appropriate Doppler shift can contribute to the observed emission at a frequency shift Δv from the line center frequency $v_\circ$. In other words, only gas with a Doppler shift in the observer's direction of $\Delta V_n = c\Delta v/v_\circ$ contributes to the flux at Δv. This simplifying assumption is part of the so-called Sobolev approximation, which is used to make the treatment of line transfer in moving media much more tractable (in the radio astronomical literature, this is often called the Large Velocity Gradient or LVG approximation). The points in space which satisfy this requirement lie on a surface given by the equation

$$\Delta V_n = v_\circ(r_\circ/r)^{1/2} z/r, \tag{4.30}$$

where z is the distance in the observer's direction measured from the plane perpendicular to the observer's direction passing through the center.

Figure 4.12 shows typical surfaces of constant velocity (in the observer's plane). It is evident from the figure that a ray passing through the envelope at a given impact parameter p will intersect the constant velocity surface twice, and so two

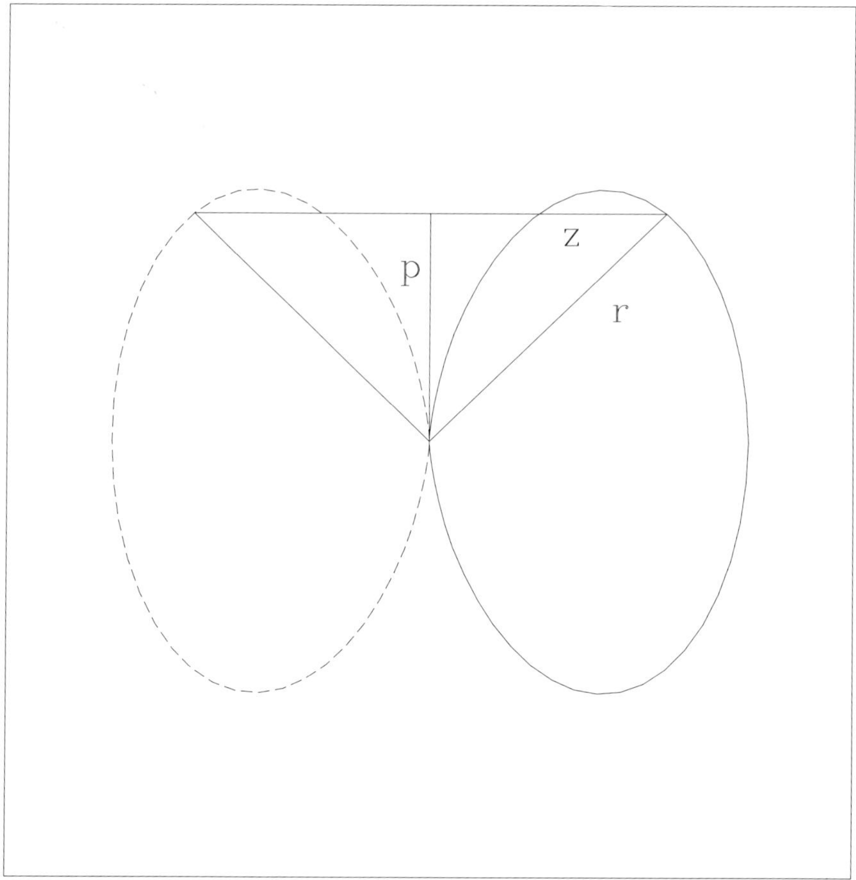

Fig. 4.12. Constant velocity surfaces for radial infall in a (meridional) plane. If the observer is on the right-hand side, the solid curve corresponds to a redshifted surface of constant projected velocity, while the dashed curve is the corresponding surface for the same velocity blueshifted. The line of sight to the observer, taken along coordinate z at impact parameter p from the center, generally intersects the constant velocity surface at two points. If the source function of the spectral line varies radially, then the differing order of radial distance at which the line of sight intersects the constant velocity surface on the blueshifted and redshifted sides produces a line asymmetry (see text).

distinct regions will contribute to the flux at a given velocity shift. Denote these intersections as r_1, r_2, where the former corresponds to the intersection closest to the observer. We make the further assumptions of the Sobolev approximation that the source functions are slowly-varying across the region near the constant velocity surface. This requirement is most likely to be met in the case where the relevant path length $dl \ll r$, which is most easily satisified if the medium has large velocity gradients. Then the formal solution of the equation of transfer (equation (4.17)) at impact parameter p in the direction z toward the observer is

$$I^+(\Delta v) \;=\; S(r_1)\{1 \,-\, \exp[-\tau(r_1)]\} \;+\; S(r_2)\{1 \,-\, \exp[-\tau(r_2)]\}\exp[-\tau(r_1)]. \quad (4.31)$$

That is, the specific intensity in the line of sight is given by the emission from the outer layer plus the emission from the inner layer reduced by the extinction of the overlying part of the constant velocity surface.

In the case of a static medium, the line width is $\delta v^2 = v_{th}^2 + v_t^2$, where $v_{th} = (2kT/m)^{1/2}$ is the thermal broadening for a species of mass m and v_t is the turbulent broadening velocity. For a static medium, this line broadening helps determine the line optical depth, but in the Sobolev limit, the line optical depths τ depend instead on the envelope velocity field. To see this, note that regions of the envelope that have the same line-of-sight velocity within δv of v are able to contribute to the absorption at v. The path length over which the medium is in radiative contact is then

$$dl \approx v_{th}/(dv/dr)_s, \tag{4.32}$$

where $(dv/dr)_s$ is the velocity gradient along the line of sight at the resonant position s where the line-of-sight velocity in the medium is equal to the desired velocity shift. If we approximate the line profile as square with a width given by $\delta v = v_{th}$, and take the line absorption cross-section per molecule for spectral line m to be α_m, then the optical depth at velocity v through the resonant region becomes

$$\tau_m \approx \frac{\alpha_m N_s}{v_{th}} dl = \frac{\alpha_m N_s}{v_{th}} \frac{v_{th}}{(dv/dr)_s} = \frac{\alpha_m N_s}{(dv/dr)_s}, \tag{4.33}$$

where N_s is the number density at the resonant position s. The velocity width of the line δv thus drops out of the optical depth determination.

Because of the symmetry of the problem, the radii of intersection and optical depths are the same at $\pm \Delta v$. However, the ordering of the intersections with the constant velocity surface differs, and this is responsible for producing a profile asymmetry for non-constant line source functions S. In many situations the source function will decrease with increasing radius, either because of outwardly-decreasing temperatures, or densities, or both. Consider the limiting case of an outwardly-decreasing source function, namely $S(r_1) = 0$. Then the emission from the material on the side of the envelope nearest the observer, which is redshifted and thus contributes to the long-wavelength portion of the line profile, is reduced by a factor of $\exp[-\tau(r_1)]$ relative to the emission on the correspondingly blueshifted portion of the line profile. This asymmetry holds as one sums over all impact parameters to obtain the total observed flux. Therefore the blue wing of the line is brighter than the red wing (Anglada *et al.* 1987). The centroid of the emission profile will exhibit a net blueshift. A similar effect is discussed in §8.12, when infall line profiles are revisited in the context of accretion onto the surfaces of T Tauri stars.

The principal difficulties in detecting this infall asymmetry are defining the central velocity shift of the object and avoiding the effects of outflows; the latter can produce confusing effects, especially because they do not have the quasi-spherical symmetry assumed implicitly in the above discussion (Figure 4.12). The question of the central velocity is especially important, because even small errors in this parameter can affect the derived asymmetry dramatically. Initial studies (e.g., Walker *et al.* 1986) utilized absorption reversals in optically-thick lines to define the central velocity (cf. Figure 4.13). The problem with this method is that it can be difficult to distinguish absorption in the outer infalling region from foreground absorption by an independent cloud with an unrelated motion. The currently-favored approach (cf. discussion by Zhou & Evans (1994)) is to use the emission from another optically-thin spectral line

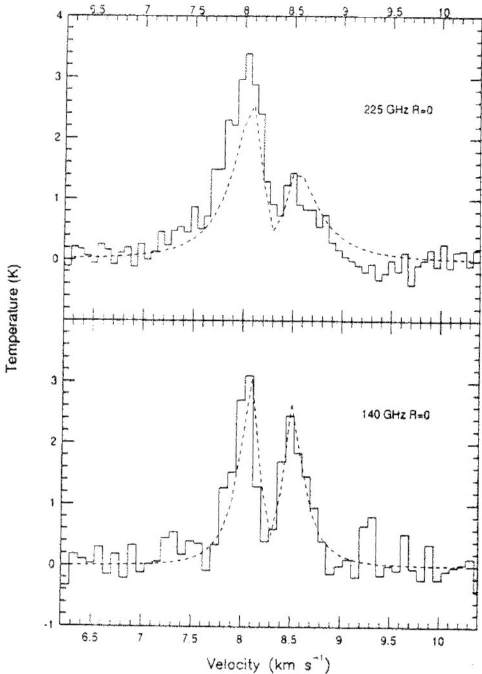

Fig. 4.13. Emission line profiles of H_2CO in the Bok Globule B335. The central reversal is presumed to be absorption from the outer, nearly stationary, regions, while the emission wings are asymmetric in a sense consistent with infall. The dashed lines are predicted line profiles for a simple model based on the singular isothermal sphere similarity solution for collapse (§3.2). From Zhou *et al.* (1993).

which is too optically-thin to show the effects of outflow and foreground low-density clouds, and therefore is more likely to define the system velocity for asymmetry analysis (Figure 4.13). Using this approach, several promising candidates for protostellar infall have been found, although uncertainties remain (Zhou 1994, and references therein; also Choi *et al.* 1995, Wang *et al.* 1995, Myers *et al.* 1995; Mardones *et al.* 1997). In addition, mm-wave interferometric mapping provides evidence for spatially-resolved collapse motions (e.g., Hayashi *et al.* 1993; Ohashi *et al.* 1996). This situation is evolving rapidly and larger statistical samples which will become available in the near future should produce more robust results.

5

Disk accretion

As described in Chapters 3 and 4, the collapse of protostellar clouds with plausible amounts of angular momentum generally should result in the formation of disks as well as protostars. Disk formation during the collapse phase is then followed by a longer phase of *disk accretion* during which angular momentum is transferred to a small fraction of disk particles at large radial distances, permitting the accretion of most of the disk mass onto the central star (with some fraction possibly forming planets). The subsequent evolution of a star-disk system will be controlled by the rate at which angular momentum is transported in the disk. Unfortunately, the mechanisms of angular momentum transfer are poorly understood, making it impossible at present to predict disk evolution from first principles. It may be possible to use observations to constrain disk evolution and by implication, the mechanisms of angular momentum transport.

Fairly general predictions about the emission of circumstellar disks can be made without knowing the specifics of the angular momentum transfer, as long as accretion is relatively steady and the disk is optically thick, as shown many years ago for T Tauri stars by Lynden-Bell & Pringle (1974). The essence of this result can be derived from energy conservation and blackbody radiation. If material accretes at a rate $\dot{M}$ through an annulus ΔR in a Keplerian disk at radius R from a star of mass M_*, the gravitational potential energy released by accretion must be radiated away by the disk surface; thus,

$$\frac{GM_* \dot{M}}{2R} \frac{\Delta R}{R} \sim 2 \times 2\pi R \, \Delta R \, \sigma T_d^4, \tag{5.1}$$

and so the disk surface temperature T_d is predicted to vary as

$$T_d \sim \left(\frac{GM_* \dot{M}}{8\pi\sigma R^3}\right)^{1/4}. \tag{5.2}$$

While this result reproduces the basic dependence of disk temperature on radial distance, one must also consider energy fluxes carried by viscous stresses to obtain the exact relation (§5.2).

To understand the temperature distributions of typical pre-main-sequence disks it is also necessary to account for the radiation from the central star intercepted by the disk, which can often exceed the heating resulting from accretion. In the limit of no accretion energy release, the absorption of light from the star by an optically-thick disk balances the energy loss by disk radiation, so that

$$\frac{L_*}{4\pi R^2} < \cos\gamma > \ \sim \ \sigma T_d^4 \,, \tag{5.3}$$

where $< \cos\gamma >$ is an average angle with respect to the disk normal at which rays from the star enter the disk. With $< \cos\gamma > \sim R_*/R$, where R_* is the stellar radius, the temperature distribution for the case of disk *irradiation* is

$$T_d \ \sim \ \left(\frac{L_* R_*}{4\pi\sigma R^3}\right)^{1/4} . \tag{5.4}$$

Thus, whenever $L_* > GM_* \dot{M}/R_*$, which is often the case for T Tauri stars (Chapter 6), disk irradiation can dominate the disk temperature distribution. Irradiation can be especially important in the outer regions of disks, where the finite vertical thickness of the disk may increase $\cos\gamma$ substantially.

Since the disks of pre-main-sequence stars may have very large radial extensions, they can radiate over a wide range of wavelengths, extending from the near-infrared to mm wavelengths and beyond. The success of the predictions of simple disk models in accounting for a wide variety of observations provides the basis for identifying circumstellar disks around T Tauri stars.

In this chapter a brief sketch of disk physics is presented. Specific applications of disk theory to observations of T Tauri stars are discussed in Chapter 6. Fuller treatments of disk physics can be found in several other sources: a basic and particularly clear elaboration is given by Frank, King, & Raine (1992); a useful outline is presented by Pringle (1981); and a recent review is given by Papaloizou & Lin (1995).

5.1   Energy minimization and angular momentum conservation

To illustrate some basic issues involved in disk accretion, we follow Lynden-Bell & Pringle (1974) in considering an idealized situation involving only two bodies orbiting around a central mass. Suppose that the bodies of (small) mass m_1 and m_2 are in circular Keplerian orbits about the central mass M. Then the energy and angular momentum of this system are

$$E \ = \ -\frac{GM}{2}\left(\frac{m_1}{r_1} + \frac{m_2}{r_2}\right), \tag{5.5}$$

and

$$J \ = \ (GM)^{1/2}(m_1 r_1^{1/2} + m_2 r_2^{1/2}), \tag{5.6}$$

where r_1 and r_2 are the corresponding radial coordinates of the bodies. Now suppose that the orbits are perturbed by small amounts while conserving the overall angular momentum J. Then the relation between the perturbations is

$$m_1 r_1^{-1/2}\Delta r_1 \ = \ -m_2 r_2^{-1/2}\Delta r_2 \,, \tag{5.7}$$

and the corresponding change in the system energy, written in terms of the first body, is

$$\Delta E \ = \ -\frac{GM m_1 \Delta r_1}{2r_1^2}\left[\left(\frac{r_1}{r_2}\right)^{3/2} - 1\right]. \tag{5.8}$$

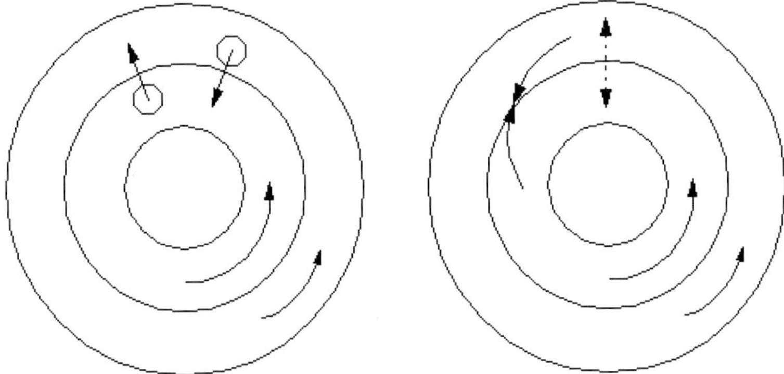

Fig. 5.1. Schematic treatment of angular momentum transfer between two annuli in a shearing disk. The angular velocity is assumed to decrease outwards. If there is friction or communication between two neighboring annuli, the resulting torques will attempt to bring the two annuli into corotation. This means that a net torque will be exterted on the outer annulus in the sense of spinning it up, i.e. gaining angular momentum. In the left-hand panel, the torque or transport of angular momentum is indicated by the interchange of parcels of material from each annulus, which bring differing angular momenta to the annuli to which they are transferred. The right-hand panel illustrates the potential effects of magnetic fields or gravitational instabilities schematically, shown as if adjacent disk annuli are tied together by an (elastic) string tethered at each end. Starting with an initially radial configuration (dashed line), the differing angular velocities of the annuli will cause the string to be stretched. The restoring force of the string (tension forces in the case of the magnetic field; gravitational attraction by excess mass in the case of gravitational instability) work in the direction of trying to spin up the outer annulus and spin down the inner annulus, i.e. to transfer angular momentum outward.

Suppose we wish to reduce the energy E of the system. If body 1 is at a larger radial distance from the center than body 2, we can decrease the energy of the system by a positive displacement Δr_1, i.e. moving body 1 further away from the center. If body 1 is originally closer to the center, then the energy can be reduced by a negative displacement Δr_1, moving body 1 closer in. Thus, the energy can be reduced while conserving orbital angular momentum by moving the initially closer body in and moving the initially outer body further away. This is the basic action of the accretion disk; energy is released as material both accretes and spreads to larger distances. If mass can be transferred between bodies, one can show (e.g., Lynden-Bell & Pringle 1974) that the system energy can be minimized by transferring mass from the outer body to the inner body. With many bodies, the energy is minimized by moving most of the mass inwards as far as possible; this can be accommodated with conservation of angular momentum if some (small) mass is moved outward to very large distances.

This process requires some way of connecting different particles in the disk. One schematic way of picturing this is to imagine rings or annuli in a disk which rub against each other (Figure 5.1). In all situations considered in this book, the angular velocity of the disk system decreases outward (in the specific case of the Keplerian disk, $\Omega = (GM/r^3)^{1/2}$). This means that an inner annulus will rotate faster than its neigboring outer annulus. Friction between the two will try to spin up the outer annulus and spin down the inner annulus, i.e. angular momentum tends to be removed from the inner annulus and transferred outward. Because the gravitational potential constrains the orbital motion, this transfer of angular momentum results in moving material from the inner annulus inwards while outer material gaining angular momentum will move to larger radii. Energy is lost due to frictional dissipation, and this must cause the net gravitational potential energy of the system to decrease, i.e. the net motion of disk mass must be inward.

The situation is more complicated in a gaseous disk because material diffuses in both directions from all radii. One physical picture for the torque g between neighboring annuli supposes that the gas exhibits (small) turbulent, random motions which cause mixing in the radial direction. As indicated in Figure 5.1, this means that material between adjacent annuli will be exchanged. Since the material originating in the two annuli will have different specific angular momenta, this will cause a transfer of angular momentum between annuli.

Following Frank *et al.* (1992), we can calculate the magnitude of the angular momentum transfer in terms of a *kinematic viscosity*. The basic picture is one in which turbulent elements of the gas moving at a typical random velocity w travel a mean free path λ before mixing with other material. Thus, the net torques or angular momentum transfer at cylindrical radius R (Figure 5.1) will be produced (schematically) by the differing angular momenta of two streams of material; one from material originating at $R - \lambda/2$ and moving outward across R to mix with annular material centered at $R + \lambda/2$; and the other starting at $R + \lambda/2$ and moving inward across R to mix with the inner annulus at $R - \lambda/2$.

In this kinematic viscosity model, no net angular momentum is transported unless there is shearing orbital motion, $d\Omega/dR \equiv \Omega' \neq 0$. The net angular momentum fluxes can be calculated in the following way. As seen by an observer at R rotating with the disk, material originating at $R - \lambda/2$ has an angular momentum difference of

$$\Delta J_{in} = (R - \lambda/2)^2 \left[\Omega(R - \lambda/2) - \Omega(R)\right] = (R - \lambda/2)^2 \left[- (\lambda/2)(d\Omega/dR)\right], \quad (5.9)$$

where we have approximated the difference in the angular velocities in terms of the first deriviative with respect to R. A similar expression with the negative values changed to positive applies to the material at $R + \lambda/2$. The inner material diffuses outwards at velocity w and the outer material diffuses inward at w across R. (The *net* inward motion of material in the accretion disk is assumed to be small in comparison with the turbulent velocity w.) For simplicity we integrate or average the disk structure in the z direction perpendicular to the disk plane; then the net outward transfer of angular momentum across R per unit length for a disk with mass density per unit area Σ is

$$\Sigma w \left[(R - \lambda/2)^2(-\lambda/2)d\Omega/dR - (R + \lambda/2)^2(\lambda/2)d\Omega/dR\right] = -\Sigma w \lambda R^2 d\Omega/dR, \quad (5.10)$$

where we have assumed that λ is a short distance compared with the scale over which Ω varies significantly. With this result, the total angular momentum flux outward across R, i.e. the torque of the inner annulus on the outer annulus, can be written as

$$g = -2\pi R \Sigma v_v R^2 d\Omega/dR, \quad (5.11)$$

where the viscosity v_v is

$$v_v = \lambda w. \quad (5.12)$$

Note that a negative gradient of angular velocity (i.e. Ω decreasing outward) leads to a positive outward flux of angular momentum, as predicted by the qualitative picture of friction between neighboring annuli discussed above.

In the case in which the viscosity is *molecular*, i.e. due to random gaseous diffusion, a kinematic viscosity prescription is clearly appropriate. Unfortunately, it is easily shown (e.g., Frank *et al.* 1992) that ordinary molecular viscosity is vastly too small to have any practical effect in astrophysical disks. However, there may be other processes which produce an effective 'turbulent' viscosity (§5.3), and these situations may be modeled with appropriate choices for the 'velocity' w and 'scale length' λ.

5.2 The thin accretion disk

Consider a thin disk composed of particles moving essentially on circular orbits in a single plane. We suppose further that any radial motions are small and that radial pressure forces are negligible, so that the orbital motion of the disk is due entirely to equating the centripetal acceleration with gravity. Then, adopting cylindrical polar coordinates (R, ϕ, z), the circular velocity resulting from a gravitational potential $\Phi(R)$ is

$$\frac{v_\phi^2}{R} = \frac{d\Phi}{dR}. \tag{5.13}$$

The equation of mass conservation for an annulus of width ΔR at R, denoting the surface density of the disk again by Σ, is

$$\frac{\partial}{\partial t}(2\pi R \Delta R \Sigma) = v_R(R, t)\, 2\pi R \Sigma(R, t)$$
$$- v_R(R + \Delta R, t)\, 2\pi(R + \Delta R)\Sigma(R + \Delta R, t), \tag{5.14}$$

where v_R is the *net* radial velocity of the material. (In a turbulent viscosity model, the turbulent velocity w may be much larger than v_R, but the turbulent motions do not represent a net mass flux; §5.1.) The first term on the right-hand side of equation (5.14) is the flow of material into the annulus and the second term is the flow out. Taking the limit for small ΔR, one obtains

$$R\frac{\partial \Sigma}{\partial t} + \frac{\partial}{\partial R}(R\Sigma v_R) = 0. \tag{5.15}$$

Similarly, the equation for conservation of angular momentum can be written as

$$R\frac{\partial}{\partial t}(\Sigma R^2 \Omega) + \frac{\partial}{\partial R}(R\Sigma v_R R^2 \Omega) = -\frac{1}{2\pi}\frac{\partial g}{\partial R}, \tag{5.16}$$

or, using equation (5.11),

$$\frac{\partial}{\partial t}(\Sigma R^2 \Omega) + \frac{1}{R}\frac{\partial}{\partial R}(\Sigma R^3 \Omega v_R) = \frac{1}{R}\frac{\partial}{\partial R}\left(v_v \Sigma R^3 \frac{d\Omega}{dR}\right). \tag{5.17}$$

The mass conservation equation can be used to eliminate v_R, with the result that

$$\frac{\partial \Sigma}{\partial t} = \frac{1}{R}\frac{\partial}{\partial R}\left[\left(\frac{d\Omega R^2}{dR}\right)^{-1}\frac{\partial}{\partial R}\left(-v_v \Sigma R^3 \frac{d\Omega}{dR}\right)\right]. \tag{5.18}$$

Since the viscosity can be a function of local physical conditions in the disk, this is a non-linear diffusion equation for Σ.

In many cases of interest most of the mass is contained in the central spherical star, so that the gravitational potential is that of a central point mass, and the angular velocity takes on its Keplerian value, $\Omega = (GM/R^3)^{1/2}$. Then the diffusion equation (5.18) becomes

$$\frac{\partial \Sigma}{\partial t} = \frac{3}{R}\frac{\partial}{\partial R}\left[R^{1/2}\frac{\partial}{\partial R}(v_v \Sigma R^{1/2})\right]. \tag{5.19}$$

By assuming circular motion and a thin disk we have in effect imposed constraints on the conservation of energy in this disk system. An annulus of width ΔR is subject to torques on its inner and outer surfaces; the net torque is

$$g(R + \Delta R) - g(R) = \frac{\partial g}{\partial R}\Delta R. \tag{5.20}$$

The rate of working by this torque, i.e. the energy added, is

$$\Omega\frac{\partial g}{\partial R}\Delta R = \left[\frac{\partial}{\partial R}(g\Omega) - g\frac{d\Omega}{dR}\right]\Delta R. \tag{5.21}$$

The first term on the right-hand side is a rate of 'convected' rotational energy; integrated over the radial extent of the disk, this is just

$$g\Omega|_{outer\ edge} - g\Omega|_{inner\ edge} \tag{5.22}$$

and therefore is determined by the disk boundary conditions. The second term represents a local rate of mechanical energy dissipation. In these equations this energy is assumed to be lost from the system (i.e. radiated away). This energy loss per unit area for each of the two sides of the disk is

$$\dot{E}(R) = -\frac{g\,d\Omega/dR}{4\pi R} = (1/2)\,v_v\Sigma(R\,d\Omega/dR)^2. \tag{5.23}$$

The radiation of this dissipated accretion energy makes the disk self-luminous.

The evolution of the viscous accretion disk depends upon the behavior of the viscosity v_v, which in turn can depend in complicated and unknown ways on disk properties. It is instructive to explore some simple solutions of the diffusion equation (5.19) which illustrate some general properties of accreting disks. To begin, we examine the case in which the viscosity is constant. One may proceed with generality by considering only one initial (thin) annulus of material because the assumption of constant viscosity makes the equation linear in Σ, and therefore one can construct a general solution by adding up solutions for individual annuli.

Starting with an initial density distribution representing an annulus of material at radius R_1,

$$\Sigma(R, t = 0) = \frac{\delta(R - R_1)}{2\pi R_1}, \tag{5.24}$$

Lynden-Bell & Pringle (1974) showed that the solution to (5.19) for a constant viscosity v_v is

$$\Sigma(x, t_d) = \frac{x^{-1/4}t_d^{-1}}{2\pi R_1^2}\exp\left[\frac{-(1 + x^2)}{2t_d}\right]I_{1/4}(x/t_d), \tag{5.25}$$

where $I_{1/4}$ is the modified Bessel function of fractional order and the dimensionless distance and time are $x = R/R_1$ and $t_d = 6v_v t/R_1^2$, respectively. Plots of this solution at various times are given in Figure 5.2. One observes that the net effect of the

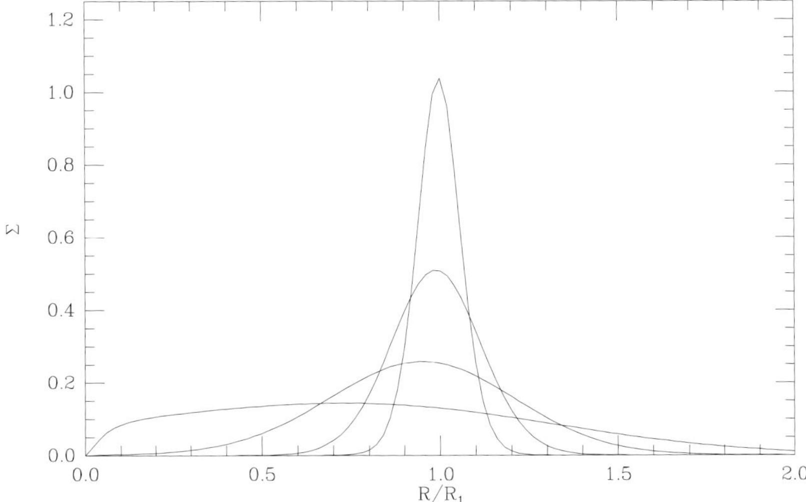

Fig. 5.2. Calculation of the diffusion of a ring of material of unit mass initially at $R = R_1$ for the case $v_v = $ constant. As time proceeds, the ring spreads, and distributes more and more mass to smaller radii; results are shown for multiples of 0.004, 0.016, 0.064, and 0.256 of the time measured in units of $R_1^2/(6v_v)$. The turn-down of the surface density at small radii is due to the inner boundary condition employed, as discussed further in §5.4. After Lynden-Bell & Pringle (1974).

viscosity is to spread the ring of material, ultimately concentrating mass at smaller radii, while small amounts of matter are pushed to large distances to conserve angular momentum.

The evolutionary timescale of this ring is of order R_1^2/v_v, a result which is expected on dimensional grounds. This can also be seen as the relevant timescale based on an argument using the turbulent diffusion picture. Individual elements travel a distance λ before merging with the background. The turbulent motion of a particle inward or outward then consists of a random walk process with a mean mean free path λ at each step. It requires $\sim N^2 \sim (R_1/\lambda)^2$ random walk 'steps' for the particle to move a distance $\sim R_1$; each 'step' takes a time $t_s \sim \lambda/w$; thus the time taken, on average, for a particle to diffuse a distance R_1 is $\sim R_1^2/(\lambda w) \sim R_1^2/v_v$.

Another illustrative result can be obtained with the assumptions that the viscosity is independent of time and is a power-law function of radius. Recall that in the case of free-fall collapse of an initial power-law density distribution (the singular isothermal sphere; Chapter 3), it was possible to find a similarity solution provided inner and outer boundary conditions did not become important. It is also possible to find similarity solutions to the disk evolution case for the above conditions, as shown by Lynden-Bell & Pringle (1974), and we develop one of these solutions to explore some general properties of viscous accretion disks.

It is useful to recast the equations in terms of the specific angular momentum $h = \Omega R^2$ as a variable. Expanding equation (5.16) in terms of h, and using the mass continuity equation,

$$\frac{\partial g}{\partial R} = -\dot{M}\frac{\partial h}{\partial R}, \tag{5.26}$$

where the (outward) mass flux is $\dot{M} = 2\pi R \Sigma v_R$. This equation can be written as

$$\frac{\partial g}{\partial h} = -\dot{M}. \tag{5.27}$$

The continuity equation (5.15) becomes

$$\frac{\partial \Sigma}{\partial t} + \frac{1}{2\pi R}\frac{\partial \dot{M}}{\partial R} = 0. \tag{5.28}$$

Substituting for Σ and $\dot{M}$ with g and h, we have

$$-\frac{\partial}{\partial t}\left(\frac{g}{v_v R^2 d\Omega/dR}\right) - \frac{\partial}{\partial R}\left(\frac{\partial g}{\partial h}\right) = 0, \tag{5.29}$$

or

$$\frac{\partial^2 g}{\partial h^2} = -\frac{\partial}{\partial t}\left[\frac{g}{v_v R^2 (d\Omega/dR)dh/dR}\right]. \tag{5.30}$$

For a central point mass dominating the gravitational potential,

$$\Omega = (GM/R^3)^{1/2}, \quad h = (GMR)^{1/2}. \tag{5.31}$$

so that

$$\frac{\partial^2 g}{\partial h^2} = \frac{4h^2}{3v_v(GM)^2}\frac{\partial g}{\partial t}. \tag{5.32}$$

This equation simplifies greatly if we assume that the viscosity depends only upon R as a power law. For simplicity we take the special case $v_v = v_o(R/R_o)$, which may be appropriate in some circumstances (Chapter 9). Then we may write

$$\frac{\partial^2 g}{\partial h^2} = \frac{4R_o}{3v_o GM}\frac{\partial g}{\partial t} \equiv \kappa_g^2 \frac{\partial g}{\partial t}, \tag{5.33}$$

where κ_g^2 is a constant. Note that because

$$\Sigma = -\frac{g}{2\pi R^3 v_v d\Omega/dR}, \tag{5.34}$$

finding a solution for g is equivalent to solving for Σ.

Following Lynden-Bell & Pringle (1974), suppose we look for modes of the form

$$g \propto \exp(-st). \tag{5.35}$$

If we set $k^2 = \kappa_g^2 s$, the equation for a single mode is simply

$$\frac{\partial^2 g_k}{\partial h^2} = -k^2\frac{\partial g}{\partial t}. \tag{5.36}$$

For simplicity we assume that g vanishes at the origin $h = 0$, i.e. there is no central torque. Then the solutions to this equation are of the form

$$g(t,h) = \int_0^\infty dk\, e^{-tk^2/\kappa_g^2} A_k \sin(kh). \tag{5.37}$$

From the Fourier theorem, we have

$$A_{k'} = \frac{2}{\pi}\int_0^\infty \int_0^\infty A_k \sin(kh)dk \,\sin(k'h)dh = \int_0^\infty dh\, g(0,h)\sin(k'h). \tag{5.38}$$

This equation can be used to find the coefficients A_k with an initial form for the torque.

For reasons that will become apparent, we try the initial solution

$$g(0, h) = C_g h e^{-a^2 h^2},$$
(5.39)

where C_g and a are constants. Substituting this into the equation (5.38),

$$A_k = \frac{2}{\pi} \int_0^\infty C_g h e^{-a^2 h^2} \sin(kh) dh \, ;$$
(5.40)

integrating by parts, we find

$$A_k = C_g \frac{k}{2a^2} \frac{1}{\pi^{1/2} a} e^{-k^2/(4a^2)} \, .$$
(5.41)

Then substituting back in to (5.37),

$$g(t, h) = \frac{C_g}{2a^3 \pi^{1/2}} \int_0^\infty dk \, k \, e^{-[(t/\kappa_g^2) + (1/4a^2)]k^2} \sin(kh) \, .$$
(5.42)

This integral is exactly of the same form as the one needed to determine the A_k. After some manipulation we finally arrive at the result

$$g(t, h) = \frac{C_g h}{\left[(4a^2 t/\kappa_g^2) + 1\right]^{3/2}} \exp\left[-\frac{ah^2}{(4a^2 t/\kappa_g^2) + 1}\right] \, .$$
(5.43)

This solution has the same form at all times, i.e. it represents a similarity solution. To see this more clearly, define

$$T_g = \frac{4a^2}{\kappa_g^2} t + 1 \, .$$
(5.44)

Then we can write the solution (5.43) as

$$g = C_g h T^{-3/2} e^{-a^2 h^2/T} \, .$$
(5.45)

If we make the identification

$$a^2 = (GMR_1)^{-1},$$
(5.46)

where R_1 is the appropriate length scale for the initial density distribution, we can solve for the mass flux and the surface density of the disk in terms of physical variables,

$$\dot{M} = C_g T_g^{-3/2} \exp\left(-\frac{R/R_1}{T_g}\right) \left[1 - \frac{2R/R_1}{T_g}\right],$$
(5.47)

and

$$\Sigma = \frac{C_g T_g^{-3/2}}{3\pi v_\circ (R/R_\circ)} \exp\left(-\frac{R/R_1}{T_g}\right),$$
(5.48)

where

$$T_g = \frac{3v_\circ t}{R_\circ R_1} + 1 = \frac{t}{R_1^2/3v_v(R_1)} + 1 \, .$$
(5.49)

Solutions for this case are plotted in Figure 5.3. The disk expands to take up the angular momentum as accretion proceeds. The disk mass decreases with increasing

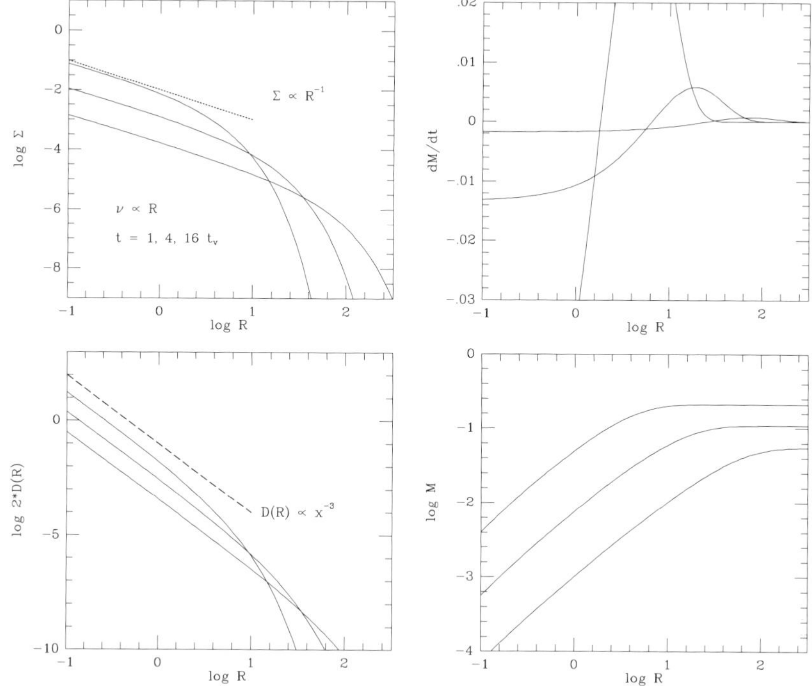

Fig. 5.3. Similarity solution for $v_v \propto R$. If R is measured in units of R_1, then the solutions shown correspond to $T_g = 1, 4$, and 16 (see text). The surface density (upper left panel) exhibits the same shape at all times when scaled as discussed in the text. The mass flux (upper right) is negative at small radii (accretion) and positive at large radii (expansion). The viscous dissipation rate is shown in the lower left panel, while the disk mass shown in the lower right panel exhibits a $T_g^{-1/2}$ scaling. The dashed lines show the $\Sigma \propto R^{-1}$ and $D(R) \propto R^{-3/4}$ scaling expected for steady accretion (see §5.4).

time, because material is flowing into the origin (onto the central star; we have ignored the change in central stellar mass in this calculation). The inner disk regions approach a constant mass accretion rate (as a function of radius); overall, the inner mass accretion rates decrease with time as the disk empties out.

Equation (5.47) shows that the mass flux changes sign at a radius $R_{tr} = R_1 T_g/2$. At larger radii, the *net* flow of matter is outward; at smaller radii, the net flow of mass is inward. Eventually most of the disk mass is accreted onto the origin; however, one may show by integrating the equation for Σ with radius that most of the mass in the disk at any given instant of time is actually moving *outward*.

This *self-similar* solution scales with the scaled non-dimensional time T_g as $R \propto T_g$. This solution thus expands linearly in size with time (for $T_g \gg 1$), like the singular isothermal sphere similarity solution discussed in §3.2. A qualitative difference in the nature of these similarity solutions is that, in the case of the expansion-wave solution for infall, the initial condition is important, and at late times there is no guarantee that the similarity solution will hold (unless the initial singular isothermal sphere were truly infinite). In contrast, for the disk accretion case, the surface density approaches the similarity solution as the disk expands far beyond its initial (characteristic) radius R_1, no matter what the initial surface density distribution was. This can be seen from a Green's function analysis, which shows that the Green's function for this problem asymptotically approaches the similarity form given above with increasing time.

This similarity solution obviously is not unique; Lynden-Bell & Pringle (1974) give the general solutions for all cases where the viscosity has a power-law dependence on radius. In general one cannot expect the viscosity to scale precisely as any power-law in radius, or be independent of other disk properties. Nevertheless, this simple similarity solution illustrates some generic properties of viscous accretion disks, and provides a simple basis upon which to explore disk evolution, as outlined briefly in Chapter 9.

5.3   Sources of viscosity and/or angular momentum transport

It is evident that a complete theory of disk accretion requires a knowledge of the viscosity. Unfortunately, viscous transport processes are not well understood at present. Molecular viscosity is so small that disk evolution due to this mechanism of angular momentum transport would be far too slow to be of interest. A number of other mechanisms have been considered, but it is only recently that a promising possibility has been identified, and our understanding is still rapidly evolving (see, e.g., the review by Livio (1995)).

Turbulent *convection* in disks has been a popular mechanism for viscosity for some time (Lin & Papaloizou 1980); however, several investigations (Ryu & Goodman 1992; Cabot & Pollack 1992; Stone & Balbus 1996) suggest that convection might actually transport angular momentum *inward* more effectively than *outward*, i.e. the wrong way. Another problem for this mechanism is that many T Tauri disks might be convectively stable (e.g., Ruden & Pollack 1991), given the importance of external disk heating or irradiation by the central star (§5.6; Chapter 6). Other gas dynamic waves might be important in carrying angular momentum, particularly if they are forced by some external factor, such as the gravitational field of a companion star (e.g., Vishniac & Diamond 1989; Rozyczka & Spruit 1993). However, the presence of a companion star or massive planet does not seem to be sufficiently general to account for pre-main-sequence disk accretion in many or most cases.

The most promising possibility at present is that of the 'magnetic instability' initially discussed by Velikhov (1959) and specifically developed for accretion disks in a series of papers by Balbus & Hawley (1991) and collaborators (related discussions were also provided by Chandrasekhar (1960) & Fricke (1969), and a recent review has been presented by Balbus & Hawley (1997)). The essence of this instability is indicated in Figure 5.1. Imagine that a magnetic field line initially connects two neighboring annuli in a radial direction, as shown by the dotted line. Because these two annuli have differing angular velocities, the field line will tend to become stretched as the shear proceeds (solid curve). The magnetic field will try to oppose the shear, but if it is not initially 'strong', it cannot resist. The magnetic field line will then try to straighten out, which requires speeding up the outer annulus relative to the inner annulus, i.e. transferring angular momentum outward. As the inner annulus loses angular momentum, it will fall in deeper to the potential well, increasing its angular velocity further, while the outer annulus will tend to move outward and slow down, enhancing the stretching of the magnetic field. This is the essence of the

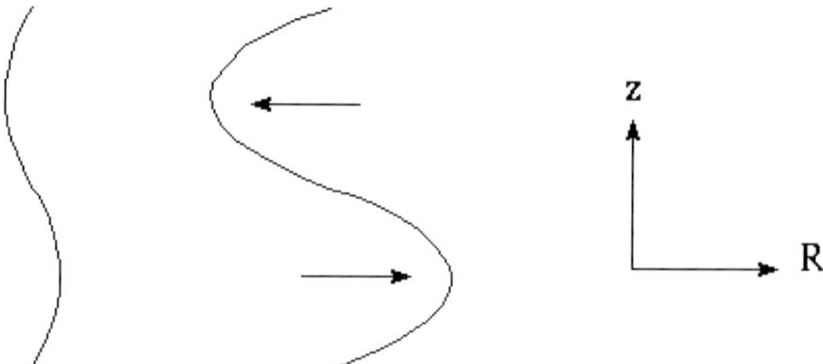

Fig. 5.4. Schematic diagram of BH instability in the meridional plane for a disturbance with a finite vertical wavenumber, as described in the text.

instability. Three-dimensional numerical simulations suggest that this 'Balbus–Hawley' or BH instability results in vigorous magnetohydrodynamic turbulence, which may in principle provide the needed viscosity for accretion disks.

To see schematically how this instability can give rise to turbulent motions, consider a magnetic field line which is initially vertical, i.e. perpendicular to the disk plane (Figure 5.4). Suppose that a small sinusoidal perturbation is applied to the field line. Then, for the reasons discussed in the previous paragraph, the outward 'bulges' will tend to be stretched further outward, while the inward bulges will tend to be stretched further inward. The result is increasingly long (in the radial direction) loops of magnetic field, allowing the interchange of material moving inward and outward at different (randomly varying) levels z in the disk. This behavior is shown dramatically in the numerical simulations of Hawley, Gammie, & Balbus (1995) and Stone *et al.* (1996).

What if there is no magnetic field to begin with? The *stretching* of the magnetic field shown in Figure 5.1 amounts to an *amplification* of the field. The instability can amplify small (perturbation) fields to larger fields, i.e. it can exhibit dynamo activity (Tout & Pringle 1992; Vishniac & Diamond 1992; Curry, Pudritz, & Sutherland 1994; Brandenburg *et al.* 1996; Hawley, Gammie & Balbus 1996), so the instability may arise even if there is a vanishingly small initial magnetic field. On the other hand, if the magnetic field is too strong, it can enforce co-rotation of the neighboring annuli and quench the instability. This places a limit on the magnetic field such that the magnetic pressure must be less than the gas pressure.

Another constraint arises from the requirement that the magnetic field couple effectively with the gas. Since the magnetic field acts directly on charged particles, the efficiency with which the BH mechanism can work depends upon whether there are enough ions in the gas; otherwise, collisions between ions and neutrals may not be frequent enough to transfer momentum effectively from the former to the latter. In the cold, dusty disks around young stars, the ionization fraction may become quite low (cf. Umebayashi & Nakano 1988; Reyes-Ruiz & Stepinski 1995; Gammie 1996b), and this can strongly reduce the effectiveness of the BH instability (Blaes & Balbus 1994).

Efforts to characterize the magnitude of the BH viscosity quantitatively are in their initial stages. Three-dimensional numerical results (Stone *et al.* 1996; Brandenburg *et al.* 1996; Heyvaerts, Priest, & Bardou 1996) suggest a viscosity

$$\nu_v \sim 0.01 c_s H, \tag{5.50}$$

where the disk sound speed and scale height are c_s and H, respectively, but this result is highly preliminary.

There are other possible mechanisms for transporting angular momentum in disks which are not directly the result of a turbulent viscosity. The possibility that most of the mass of a typical star initially resides in its disk (Chapters 3 and 4) suggests that, at least in the earliest stages of accretion, disks might be self-gravitating, and gravitational instabilities may also transfer angular momentum. The basic idea can again be understood from the schematic drawing in the right-hand panel of Figure 5.1. Suppose that instead of a magnetic field line there is a radially-extended mass concentration or perturbation along in the dotted line. If the gravitational instability can keep this mass concentration together despite the shearing effects of the disk and tidal acceleration, the excess mass can be confined to a trailing spiral arm. There are excess gravitational forces due to this mass concentration; the inner part of the spiral arm pulls on the outer part of the spiral arm, trying to accelerate the outer regions. As in the BH instability, this tendency to accelerate outer regions in the direction of motion causes a net flow of angular momentum outward, and thus in principle can produce the torque needed for mass accretion.

Under what conditions can gravitational instability arise? In Chapter 2 we discussed gravitational instabilities in a non-rotating sheet. Although we require non-axisymmetric instabilities to transfer angular momentum (cf. Figure 5.1), it turns out that the criterion for such instabilities is nearly the same as the requirement for axisymmetric instability. The dispersion relation for symmetric modes in a thin rotating disk is (Binney & Tremaine 1987)

$$\omega^2 = \kappa^2 + c_s^2 k^2 - 2\pi G\Sigma \, | \, k \, | \, . \tag{5.51}$$

This relation is similar to the result for the non-rotating sheet discussed in Appendix 2; it differs in the term involving the epicyclic frequency κ. (When the disk rotation is Keplerian, $\kappa = \Omega$, the local angular velocity.) For negative ω^2, perturbations grow exponentially and the disk is unstable. The limiting condition occurs when $\omega = 0$; the condition for axisymmetric instability is then

$$Q \equiv \frac{c_s \kappa}{\pi G\Sigma} < 1. \tag{5.52}$$

The precise numerical constant depends upon the specific conditions assumed. For example, in a finite thickness, isothermal disk, the stability criterion is $Q < 0.676$ (Goldreich & Lynden-Bell 1965).

This instability constraint on the parameter usually called the 'Toomre' Q can be understood qualitatively from a simple argument (Toomre 1964). The basic idea is to consider a small region or fragment in a self-gravitating disk. When this fragment is compressed, its self-gravity is increased; but it also tends to spin more rapidly, producing a centrifugal force to oppose gravity. The balance of these two effects leads to the instability criterion, as described in the following paragraphs.

Consider a region in the disk of size ΔR which is initially in equilibrium. Suppose the mass in this region is compressed so that it is now confined within a radius

$\Delta R - \delta R$. The force of gravity per unit mass is then changed to

$$F_G = \frac{GM}{(\Delta R - \delta R)^2} \approx \frac{GM}{(\Delta R)^2}\left(1 - 2\frac{\delta R}{\Delta R}\right). \tag{5.53}$$

Next, one must consider the effects of angular momentum conservation in spinning up this region or fragment around its own center. The inital specific angular momentum of the region or fragment is

$$l \sim \Omega \Delta R^2. \tag{5.54}$$

After contraction, if angular momentum is conserved, the new angular velocity is

$$\Omega' \sim \frac{l}{(\Delta R - \delta R)^2} \sim \Omega\left(1 + 2\frac{\delta R}{\Delta R}\right). \tag{5.55}$$

The centrifugal acceleration of this fragment or region around its center is

$$\frac{v'^2}{(\Delta R - \delta R)} \sim \Omega'^2(\Delta R - \delta R) \sim \Omega^2 \Delta R\left(1 + 3\frac{\delta R}{\Delta R}\right). \tag{5.56}$$

For stability the change in centrifugal acceleration must more than balance the increased gravity, so with $M = \pi \Delta R^2 \Sigma$,

$$3\Omega^2 \delta R > 2\pi G \Sigma \delta R / \Delta R, \tag{5.57}$$

so that

$$\Delta R > \frac{2\pi G \Sigma}{3\Omega^2}. \tag{5.58}$$

A similar analysis for small length scales provides a constraint for gas pressure to stabilize the perturbation against gravity.

$$\Delta R < \frac{c_s^2}{\pi G \Sigma}. \tag{5.59}$$

This is basically just a Jeans analysis (Chapter 2); because pressure stabilizes on small scales, there is a minimum length scale (a Jeans length) for gravitational instabilities. If the length scale over which pressure stabilizes the gravitational perturbation is *larger* than the length scale on which rotation provides stability, then

$$\frac{2\pi G \Sigma}{3\Omega^2} < \Delta R < \frac{c_s^2}{\pi G \Sigma}, \tag{5.60}$$

or

$$\frac{3\Omega^2 c_s^2}{2(\pi G \Sigma)^2} > 1. \tag{5.61}$$

This reproduces the basic result for the Q parameter if the epicyclic frequency κ is comparable to Ω.

The above analysis applies to axisymmetric gravitational instabilities, whereas non-axisymmetric modes are required to transport angular momentum. Although disks are formally stable to all local non-axisymmetric disturbances (Goldreich & Lynden-Bell 1965; Julian & Toomre 1966), global instabilities may occur (i.e. loosely-wound spiral arms) which may be either linear or non-linear, depending upon the so-called swing amplification and boundary conditions (see §6.3 in Binney & Tremaine (1987) for further discussion). Roughly speaking, when Q approaches unity, nonaxisymmetric waves may appear which can transfer angular momentum efficiently (Larson 1984; Shu *et al.* 1990).

It is instructive to consider the requirement the gravitational instability criterion places on the disk mass. To do this we first need to develop the relation between the disk thickness and the sound speed. Force balance in the vertical z direction, perpendicular to the disk plane, is given by the equation of hydrostatic equilibrium,

$$\frac{dP}{dz} = -\frac{GM\rho}{(R^2 + z^2)^{3/2}} z, \tag{5.62}$$

where we assume that the gas pressure P balances the vertical gravity (due to the central stellar mass alone; this implies that the disk is not far from Keplerian). Assuming an ideal gas law $P = \rho c_s^2$, we take the sound speed c_s to be constant in z for simplicity. Finally, we assume that the disk is thin, i.e. $R >> z$. Then the density structure is given by

$$\rho = \rho_0 \exp\left(-\frac{z^2}{2H^2}\right), \quad H = (R^3 c_s^2/GM_*)^{1/2}, \tag{5.63}$$

where the scale height H can be written in terms of the sound speed and the Keplerian velocity v_ϕ,

$$\frac{H}{R} = \frac{c_s}{v_\phi} = \frac{c_s}{\Omega R}. \tag{5.64}$$

Returning to the Toomre criterion, we multiply the numerator and denominator of (5.52) by the outer disk radius R_d^2, and make the approximation that the disk mass is $M_d \sim \pi R_d^2 \Sigma$; then the condition for gravitational instability is

$$c_s \kappa R_d^2 < GM_d. \tag{5.65}$$

Making the further approximation that the disk motion is nearly Keplerian, the requirement for instability is roughly (Pringle 1981)

$$M_d > \frac{H}{R} M_*. \tag{5.66}$$

Thus, gravitational instability occurs only when the disk has an appreciable mass compared with the central star (Larson 1984), as might be expected intuitively. The central stellar mass enters because it is responsible for producing the initial angular velocity of a contracting fragment; for larger central masses, Ω becomes larger, and therefore the disk mass must be larger to overcome the centrifugal resistance to contraction. The disk temperature is also important. For a given ratio of disk mass to central mass, cooler disks have smaller scale heights H and are thus more susceptible to gravitational instability.

Numerical simulations (see Tomley, Steiman-Cameron, & Cassen (1994), Laughlin & Bodenheimer (1994), and references therein; also discussion in Lin & Pringle (1990)) indicate that gravitational instabilities can generate turbulence which appears as unsteady, wave-like, spiral density structures, and can transfer angular momentum as suggested, provided that there is some mechanism for 'cooling' the disk, i.e. a way of reducing the random motions. If the disk is not cooled effectively, it tends to heat up until Q increases to values near the stability limit (see above). (Similar results were found in the case of stellar disks of galaxies, and the need for cooling for gaseous disks is related to the development of 'hot' galactic stellar disks in the absence of a massive halo; see the discussion in Binney & Tremaine (1987).) Quantitative estimates of the angular momentum transport rates are still difficult to obtain, and it is not clear whether this mechanism can be characterized by some sort of viscous prescription (Tomley *et al.* 1994).

The single-armed spiral mode is another variant of gravitational instability, which might be very effective because it can propagate long distances in the disk (Adams, Ruden, & Shu 1989; Shu *et al.* 1990). The effectiveness of this mechanism is under debate because it appears to be sensitive to the outer boundary condition in the disk.

Yet another possibility is angular momentum transport by magnetically-coupled disk winds. As discussed in Chapter 8, strong winds are associated with rapidly-accreting disks, and the most plausible origin for these winds is that they are 'slung off' the disk surface by magnetic fields rooted in the disk. The magnetic fields couple the disk to ejected material which carries away angular momentum. If the accelerating magnetic fields are strong enough, disk winds can take away most of the angular momentum needed for accretion. As in the case of the gravitational instability, viscous processes are not involved in this angular momentum transport. In addition, as discussed in Chapter 8, if the wind carries away all the angular momentum, it carries away all the accretion energy as well. Observations which suggest that some disks are self-luminous due to viscous dissipation (Chapters 6, 7) and other considerations (Chapter 8) suggest that winds are not the dominant mode of angular momentum transport of pre-main-sequence accretion disks.

5.4   The steady optically-thick disk

Given the uncertainties in understanding angular momentum transport in accretion disks, it is important to develop results which are not sensitive to v_v. The similarity solutions shown in Figure 5.3 suggest that disks tend to evolve toward a nearly constant accretion rate (as a function of radius at a given time) in their inner regions. Using the assumption of steady flow, the emission from the accretion disk can be predicted independently of the viscosity, provided the disk is optically thick.

For steady accretion the mass conservation equation becomes

$$\dot{M} = -2\pi R \Sigma v_R, \tag{5.67}$$

where we have assumed that the constant mass flux $\dot{M}$ is inward, and the conservation of angular momentum equation can be integrated to yield

$$\Sigma R^3 \Omega v_R = v_v \Sigma R^3 \frac{d\Omega}{dR} + C, \tag{5.68}$$

where C is a constant of integration. Alternatively, this result can be written as

$$-v_v \Sigma \frac{d\Omega}{dR} = \Sigma \Omega(-v_R) + \frac{C}{R^3}. \tag{5.69}$$

Most T Tauri stars are slowly rotating (Chapter 8), at rates much less than the Keplerian velocity. It follows that the angular velocity in the disk must eventually decrease (if the disk extends up to the stellar surface, which may not be the case in general; see §5.6). At the point where the derivative of the angular velocity goes to zero (Figure 5.5), $C = -\dot{M}\Omega R^2/2\pi$; therefore C is proportional to the angular momentum flux at the point where the radial derivative of Ω changes sign.

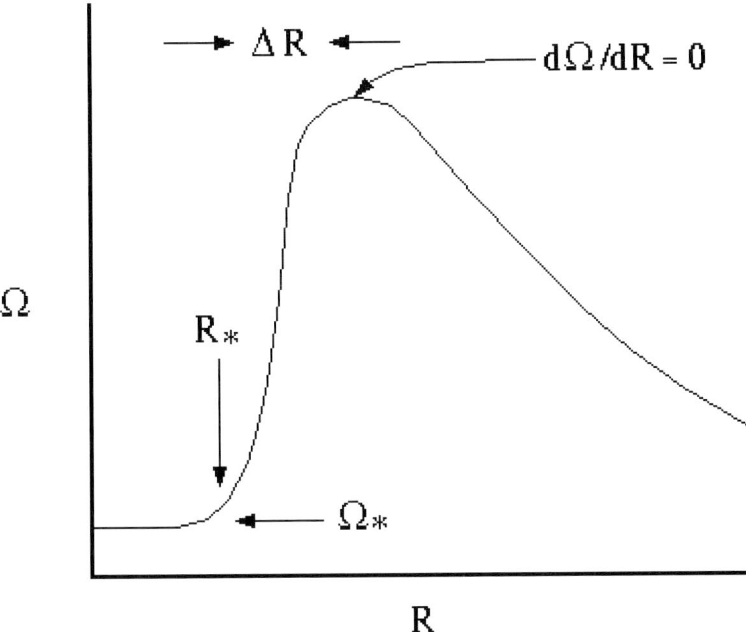

Fig. 5.5. Schematic diagram of the angular velocity in the region where the disk reaches the stellar surface. The star is assumed to be rotating at a rate Ω_* that is much less than the Keplerian velocity near the surface. Therefore, there must be a maximum in the angular velocity Ω of the disk. The point where $d\Omega/dR = 0$ is assumed to be a small distance $\Delta R \ll R_*$ exterior to the stellar surface. The narrow region where the disk material loses most of its rotational kinetic energy is called the boundary layer (see §5.6).

In simple disk theory, the point where the shear goes to zero is expected to be close to the stellar surface, for reasons outlined in §5.6. For the moment we assume this and write $C = -\dot{M}(GMR_*)^{1/2}/2\pi$, where R_* is the stellar radius, so that

$$v_v \Sigma = \frac{\dot{M}}{3\pi}\left[1 - \left(\frac{R_*}{R}\right)^{1/2}\right]. \tag{5.70}$$

Far from the inner boundary condition, the steady disk has $\Sigma \propto v_v^{-1}$; this is consistent with the results for the inner, nearly steady regions of the similarity solution discussed in §5.2.

To obtain the surface temperature distribution of the disk, and therefore its emission properties, we use the result that the viscosity generates dissipation of energy in the disk at a rate $D(R)$ per unit area per unit time (cf. equation (5.23)),

$$D(R) = \dot{E} = \frac{1}{2}v_v\Sigma\left(R\frac{d\Omega}{dR}\right)^2 = \frac{3GM\dot{M}}{8\pi R^3}\left[1 - \left(\frac{R_*}{R}\right)^{1/2}\right]. \tag{5.71}$$

Note that in the limit of constant mass accretion rate, the energy release is independent of the viscosity v_v. Integrating the dissipated energy over radius yields the total energy release by accretion,

$$L_d = \frac{1}{2} \frac{GM\dot{M}}{R_*}. \tag{5.72}$$

This is one-half the total accretion energy available, as can be seen by considering bringing in a parcel of material which starts out at infinite radius and therefore has zero energy; at the inner edge of the disk, the gravitational potential energy is twice the kinetic energy of the inner Keplerian orbit. A comparable amount of energy can be released if the kinetic energy of the material at the inner edge of the disk is completely dissipated as it comes to rest upon a slowly-rotating star (see §5.6).

A peculiarity of this solution is that the energy released in an annulus at $R \gg R_*$ is three times the energy produced locally by accretion through the annulus. The extra energy is provided by a large viscous energy flux transporting energy away from the inner boundary (equation (5.22)). There is a deficit of energy release near R_* as the viscous energy is convected outward by the action of the torque g. As discussed by Lynden-Bell & Pringle (1974), in this model the torques go to zero at the inner boundary and at infinity; thus, the torques can only redistribute the energy released by accretion in this region.

If we make the simple assumption that the disk radiates from its surface like a blackbody, then the surface or effective temperature of the disk is

$$T_d^4 = \frac{3GM\dot{M}}{8\pi R^3 \sigma}\left[1 - \left(\frac{R_*}{R}\right)^{1/2}\right]. \tag{5.73}$$

The maximum temperature of the disk,

$$T_{max} = 0.488\left(\frac{3GM\dot{M}}{8\pi R_*^3 \sigma}\right)^{1/4}, \tag{5.74}$$

occurs at $R_{max} = 49/36 = 1.36R_*$. We note that the temperature distribution given by (5.73) formally goes to zero at the inner edge of the disk. This results from the assumed boundary condition on the angular momentum, as discussed above. Although a decrease to zero temperature is unphysical, modifications require considering the disk structure and inner boundary condition in detail.

With this result the luminosity as a function of frequency emitted by the disk can be calculated as

$$L_\nu = \int_{R_{in}}^{R_{out}} \pi B_\nu \, T_d(R) \, 2\pi R \, dR, \tag{5.75}$$

where L_ν is the luminosity and R_{out} and R_{in} are the inner and outer disk radii. The results are shown in Figure 5.6.

At high frequencies or small wavelengths, the emission will be dominated by the hot inner edge of the disk, and the spectrum will have an exponential falloff to higher frequencies. At low frequencies or large wavelengths, the spectrum will depend upon on the density distribution in the outer disk regions and the opacity law. Assuming that the disk is optically thick at all wavelengths, and that it has a finite outer radius (and thus a lower limit to the temperatures at which the disk emits), at very long wavelengths the spectrum varies as $L_\nu \propto \nu^2$, because the emission from all disk annuli have this spectral form at long wavelengths regardless of their temperature. In

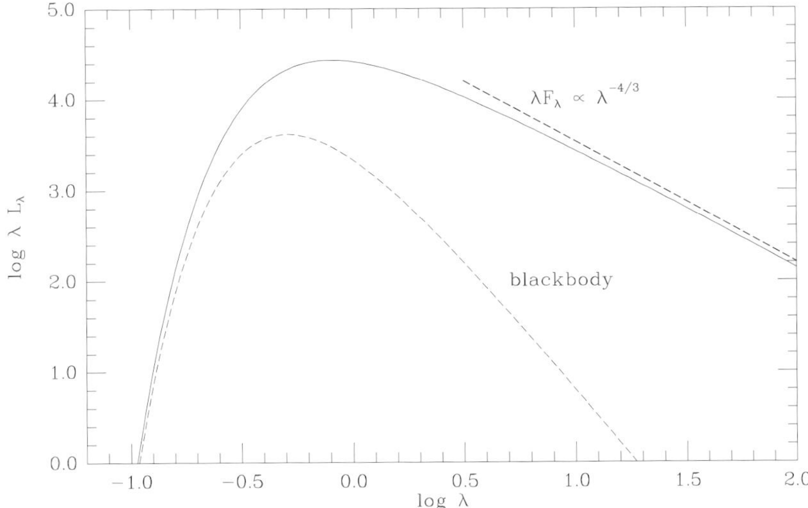

Fig. 5.6. Spectrum of the optically-thick steady disk. At short wavelengths, the SED looks like a blackbody with the temperature of the hottest disk annulus; at long wavelengths, the asymptotic $\lambda L_\lambda \propto \lambda^{-4/3}$ is observed, resulting in much more emission at long wavelengths than would result from a single-temperature blackbody (dashed line).

practice, this limit is not very important for disks around YSOs because the (dust) opacity falls off rapidly with increasing wavelengths, and so the long-wavelength emission of real disks tends to be optically thin (Chapter 6).

At intermediate wavelengths, the spectrum has a characteristic power-law shape produced by the power-law temperature distribution. This can be seen by writing the temperature in terms of a fiducial value at some specified radius, $T_d = T_\circ (R/R_\circ)^{-3/4}$, and neglecting the turn-over in the temperature near the inner edge of the disk. The integral can then be written as

$$ L_v = \frac{16\pi^2 h R_\circ^2}{3c^2} \left(\frac{kT_\circ}{h} \right)^{8/3} v^{1/3} \int_{x_{in}}^{x_{out}} \frac{x^{5/3}\,dx}{(e^x - 1)}, \tag{5.76} $$

where $x = hv/kT_d$. When the outer and inner limits are sufficiently large and small, respectively, (i.e., the inner and outer disk temperatures are sufficiently different) the integral over x is nearly constant, so the spectrum varies with frequency as $L_v \propto v^{1/3}$, or $\lambda L_\lambda \propto \lambda^{-4/3}$.

5.5   The α disk

In the absence of a detailed theory, the viscous stress may be parameterized for comparison with observations. The usual way of doing this is through the dimensionless parameter α introduced by Shakura & Sunyaev (1973). We use the modified formalism in which the viscosity is scaled in terms of a characteristic length and a turbulent velocity. It is generally assumed that the length scale of the turbulence in a disk will be less than the scale height H, and the eddy velocity will be less than the sound speed c_s. Thus, the kinematic viscosity can be written as (cf. Pringle 1981)

$$v_v = \alpha c_s H, \tag{5.77}$$

where $\alpha \le 1$. Shakura & Sunyaev also argued that this form might be expected to hold for magnetic stresses. Defining the Alfvén speed $v_A^2 = B^2/4\pi\rho$ (see Chapter 8), Shakura & Sunyaev suggested that $\alpha \sim V_A^2/c_s^2$ should similarly be be less than unity; this is consistent with the BH instability, since in that case the magnetic pressure $B^2/8\pi < \rho c_s^2/2$ (cf. §5.3).

The α parameterization can be used to illustrate some features of disk structure most easily discussed in the steady disk approximation. Substituting equation (5.77) into the steady disk angular momentum equation,

$$\alpha c_s H \Sigma = \frac{\dot{M}}{3\pi}\left[1 - \left(\frac{R_*}{R}\right)^{1/2}\right], \tag{5.78}$$

produces an equation for the surface density Σ provided that some relation is found between the total energy dissipated and the central temperature, which is most appropriate for this vertically-averaged form. (Note that the surface density $\Sigma \to 0$ as $R \to R_*$ using the temperature distribution of equation (5.73); again, the issue of conditions at the innermost edge of the disk must be settled by more detailed considerations.)

As a simple example, suppose that the disk is not very optically thick, so that it is nearly isothermal in the vertical direction (perpendicular to the disk plane). Then, neglecting the factor in brackets, $T(R) \propto R^{-3/4}$ or $c_s \propto R^{-3/8}$. Then one can show that the scale height of the disk varies as $H \propto c_s R/v_\phi \propto c_s R^{3/2} \propto R^{9/8}$, and therefore $\Sigma \propto \alpha^{-1} R^{-3/4}$.

It is also worth pointing out that for the α parameterization, the radial accretion velocity in the steady disk,

$$v_R \sim (3/2)\alpha c_s(H/R), \tag{5.79}$$

is highly subsonic, consistent with the assumptions of nearly circular motion. Note that attempts to represent the viscosities of specific mechanisms, such as the Balbus–Hawley instability, suggest values of $\alpha \lesssim 10^{-2}$, again consistent with highly subsonic radial motion.

In general, the relationship between the central temperature and the surface temperature (or the dissipated energy) is not so simple and an energy balance equation coupled with a constitutive equation must be solved to close the equations. It is, of course, also not obvious that α is constant at all radii.

5.6   Disk boundary layers

A particle which moves from an orbit far from the central star to a circular orbit just above the stellar surface must lose an amount of energy per unit mass equal to $GM_*/2R_*$. Potentially, this particle can lose almost an equal amount of energy in coming to rest on the surface of a slowly-rotating T Tauri star. Thus, the release of energy near the stellar surface is an important part of the total accretion energy, and if a substantial fraction of this energy is radiated it can have important effects on the spectral energy distribution.

Suppose that the disk rotates at Keplerian velocities right up to the stellar surface.

(Justification for this assumption will be developed below.) The angular velocity must decline as material passes from the disk to join the slowly-rotating star (Figure 5.5). The region interior to the peak in $\Omega(R)$ cannot transfer angular momentum outward to the disk, because the gradient in Ω goes the wrong way (cf. equation (5.11)). If there are no other angular momentum transport mechanisms, such as a magnetically-coupled wind (Chapter 8), the accreting material will transfer angular momentum to the star, which over a long enough time will spin the star up.

If material at $R_* + \Delta R$ is initially rotating at the Keplerian angular velocity $\Omega_K(R_* + \Delta R) \approx \Omega_K(R_*)$, and then slows down to the stellar angular velocity Ω_*, the change in kinetic energy per unit mass is $(\Omega_K^2 - \Omega_*^2)R_*^2$. However, not all of this energy is released as radiation; some of it goes into spinning up the star. The proportion between radiation and spin up depends in detail upon what is assumed about the effect of adding material to the star.

In the narrow boundary layer limit, the accretion of mass at $\dot{M}$ adds angular momentum to the star at a rate

$$\dot{J} \;=\; \dot{M}\Omega_K R_*^2 . \tag{5.80}$$

The mass added to the star carries angular momentum

$$\dot{J}_m \;=\; \dot{M}\Omega_* R_*^2 , \tag{5.81}$$

while the remaining angular momentum is added by the torque applied to the star

$$\dot{J}_t \;=\; \dot{M}(\Omega_K - \Omega_*)R_*^2 . \tag{5.82}$$

Suppose, following Narayan & Popham (1994, and references therein) that the addition of mass does not change the stellar structure, so that the change in the star's rotational energy due to the addition of mass is just the rotational energy of that mass,

$$\dot{E}_m \;=\; \frac{1}{2}\dot{M}\Omega_*^2 R_*^2 . \tag{5.83}$$

The change in the star's rotational energy due to the torque applied through the viscous stresses is then

$$\dot{E}_t \;=\; \dot{M}(\Omega_K - \Omega_*)\Omega_* R_*^2 . \tag{5.84}$$

The total rotational energy added to the star under these assumptions is

$$\dot{E}_r \;=\; \dot{E}_m + \dot{E}_t \;=\; \dot{M}\Omega_K^2 R_*^2\left(\frac{\Omega_*}{\Omega_K} - \frac{\Omega_*^2}{2\Omega_K^2}\right), \tag{5.85}$$

so that the luminosity radiated by the boundary layer can be written as

$$L_{BL} \;=\; \frac{GM\dot{M}}{2R_*} - \dot{E}_r \;=\; \frac{GM\dot{M}}{2R_*}\left(1 - \frac{\Omega_*}{\Omega_K}\right)^2 . \tag{5.86}$$

Most T Tauri stars have $\Omega_* \sim 0.1\Omega_K$ (Bouvier *et al.* 1986; Hartmann *et al.* 1986; Edwards *et al.* 1993; Bouvier *et al.* 1993), so in most cases the boundary layer luminosity is predicted to approach $L_{BL} \to GM\dot{M}/2R_*$.

Now we justify the adoption of a narrow boundary layer, implicit in the assumption that the disk exhibits Keplerian rotation almost up to the stellar surface. Basically, the boundary layer is narrow in radial dimension as long as the disk is thin, i.e. $H << R$. In the radial direction the equation of motion is

$$v_R \frac{dv_R}{dR} - \frac{v_\phi^2}{R} = -\frac{1}{\rho}\frac{dP}{dR} - \frac{GM}{R^2}. \tag{5.87}$$

Since the radial velocity is small in most cases, departures from Keplerian motion will only occur when the gas pressure forces can help balance gravity. If Ω departs significantly from Keplerian motion over a region of length R_{BL}, then the radial force balance equation requires

$$\frac{c_s^2}{R_{BL}} \sim \frac{GM}{R^2}, \tag{5.88}$$

and, using (5.64),

$$\frac{R_{BL}}{R} \sim \left(\frac{H}{R}\right)^2, \tag{5.89}$$

so that the dynamical boundary layer is thin, as assumed, as long as the disk is thin.

If the disk material in the vicinity of the boundary layer is optically thick, then even if the *dynamical* boundary layer thickness R_{BL} is small, the energy dissipated as material is slowed down must diffuse in the radial direction as well as in the z direction. In an optically-thick boundary layer, this diffusion must occur over a radial scale $\sim H$, since on much smaller scales the influence of the surface cannot be felt. A reasonable rule of thumb for the width ΔR of an optically-thick, emitting boundary layer region is $\Delta R \sim H$ (Pringle 1989).

In either optically-thin or -thick cases the boundary layer of a disk accreting onto a slowly-rotating star emits a luminosity comparable to that released in the rest of the accretion disk, but over a very small area. Therefore, the boundary layer must be much hotter than the maximum temperature in the disk proper, and so will radiate at much shorter wavelengths. For this reason Lynden-Bell & Pringle (1974) suggested that the hot optical and ultraviolet excess emission of T Tauri stars is emitted in the disk boundary layer (see Chapter 6).

However, it appears that the traditional boundary layer picture is not relevant for most T Tauri stars, because the stellar magnetic fields are generally strong enough to prevent most disks from reaching the stellar surface. The stellar magnetosphere disrupts the disk and channels the accreting material along field lines out of the disk and onto the star (§§6.4, 8.11, 8.12). The stellar magnetic field absorbs the angular momentum of the accreting material, allowing it to fall inward. It appears likely that the accreting gas falls in supersonically, at nearly free-fall velocities, and then shocks at the stellar photosphere. In this model the radiation produced as material comes to rest on the stellar surface arises from this hot shock at the stellar surface. The size of the emitting region is controlled by magnetic field strengths and geometries and not by scale-height considerations for the disk. The observational support for this picture, as well as some of the physics involved, will be discussed further in Chapters 6 and 8.

5.7   Disk irradiation

In T Tauri stars the disk accretion luminosity is often less than that of the luminosity of the central star (§6.2). In this case, the absorption of light from the central object can be the dominant mechanism heating the disk, a process called *irradiation* here.

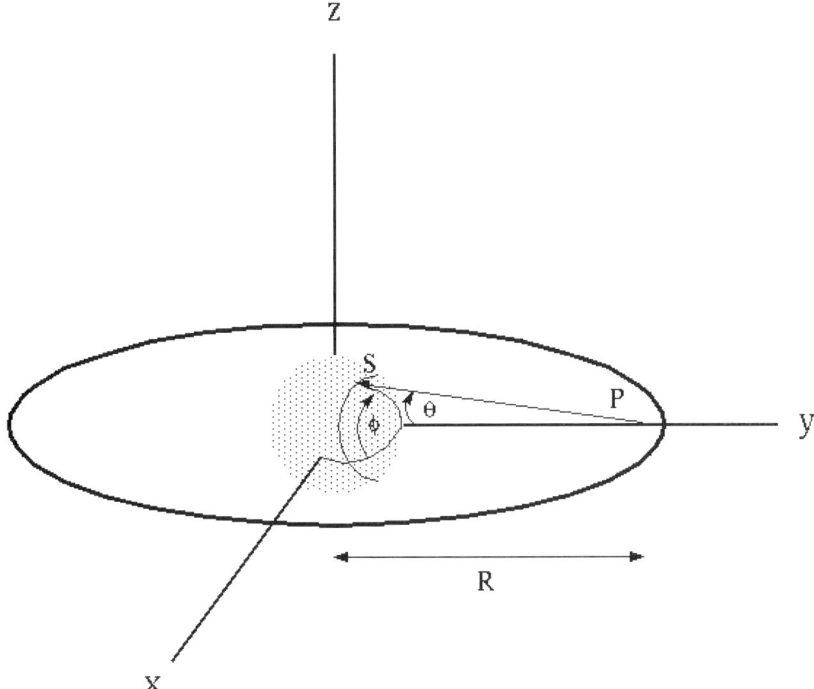

Fig. 5.7. Geometry for the calculation of the irradiation of a flat disk by the central star. The point P on the disk under consideration is along the y axis at a distance R from the star's center. The calculation is of the flux emitted along a ray from point S on the star to point P; the angle θ lies betweeen SP and the y axis; and ϕ is the angle between this projected vector and the x axis. See text.

The absorbed energy for an opaque, completely-absorbing, flat disk can be calculated simply assuming that the star radiates uniformly and isotropically from its surface. For a point on the disk at distance R from the center of the star, we take a coordinate system in which the xy plane corresponds to the disk, the z axis is perpendicular to this plane, parallel to the disk rotation axis (see Figure 5.7). Then a direction vector to a point on the star is characterized by two angles, θ measured from the y axis, and ϕ, measured between the projection of the direction vector onto the xz plane and the x axis. In this coordinate system the direction vector is then given by

$$\hat{\imath} = -\cos\theta\,\hat{\mathbf{y}} + \sin\theta\,\cos\phi\,\hat{\mathbf{x}} + \sin\theta\,\sin\phi\,\hat{\mathbf{z}}. \qquad (5.90)$$

The flux of energy from an area of the star through the disk is given by the component perpendicular to the disk of the intensity times the solid angle subtended by the stellar unit area,

$$dF_d = I\,\hat{\imath}\cdot\hat{\mathbf{z}}\,d\omega = I_\circ \sin^2\theta\,\sin\phi\,d\theta\,d\phi, \qquad (5.91)$$

where we take the stellar intensity $I_\circ$ to be constant. The total flux into the disk is

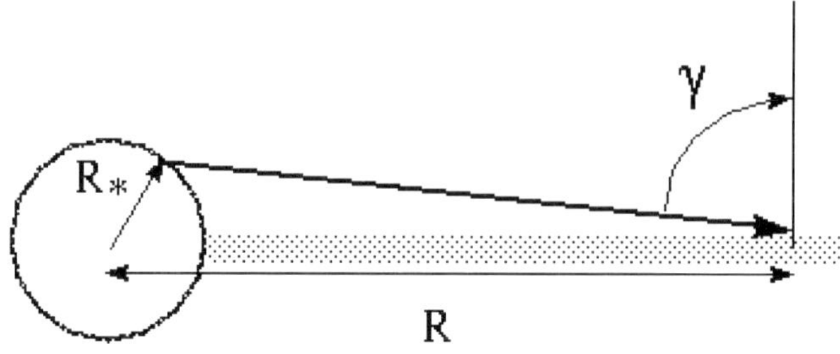

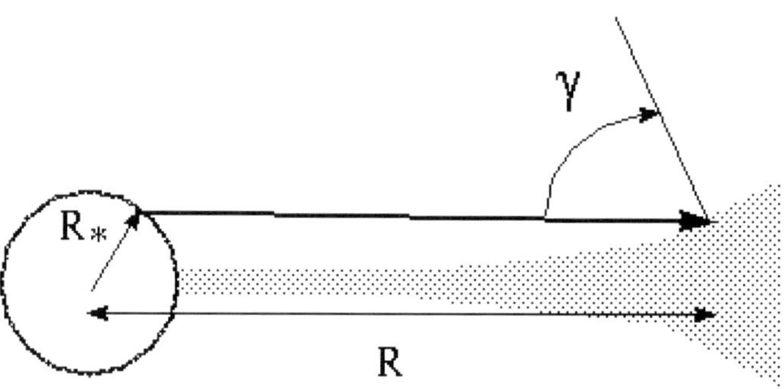

Fig. 5.8. Geometry of irradiated disks, showing how the 'flaring' of the disk affects the heating (see text).

given by the integral of dF_d over the solid angle subtended by the star,

$$F_d = \int_0^\pi d\phi \, \sin\phi \int_0^{\theta_m} d\theta \, I_\circ \sin^2\theta = 2I_\circ \int_0^{\theta_m} d\theta \, \sin^2\theta. \tag{5.92}$$

The maximum value of θ occurs when the direction vector is tangent to the stellar surface, when the radial distance is R_*; thus, $\sin\theta_m = R_*/R$ and therefore

$$F_d = I_\circ \left[\sin^{-1}\left(\frac{R_*}{R}\right) - \frac{R_*}{R}\left(1 - \frac{R_*^2}{R^2}\right)^{1/2} \right]. \; . \tag{5.93}$$

In the limit that $R \gg R_*$, $F_d \propto R^{-3}$. Assuming that the disk radiates away the incoming energy in the form of blackbody emission, so that $\sigma T_d^4 = F_d$, asymptotically $T_d \propto R^{-3/4}$. Thus, the optically-thick flat disk radiating purely as a result of irradiation by the central star will exhibit the same wavelength dependence in its SED, $\lambda F_\lambda \propto \lambda^{-4/3}$, as the steady optically-thick accretion disk (§5.4).

The total luminosity of one side of the infinite flat disk is then

$$L_d = \int_{R_*}^\infty 2\pi R \, dR \, F_d = \pi^2 I_\circ R_*^2/2. \tag{5.94}$$

By assumption, the intensity $I_\circ = B = \sigma T_*^4/\pi$, so $L_d = L_*/8$ for one side of the disk.

The total emission from both sides of the disk is therefore one-quarter of the stellar luminosity.

It is apparent from the above discussion that if the optically-thick, steady disk has a mass accretion rate producing an accretion luminosity $\lesssim L_*/4$, it will be difficult to distinguish accretion-produced energy from irradiation-induced disk emission.

One important detail of the irradiation process needs to be considered further; the possibility that the disk is sufficiently thick and curved that it absorbs stellar radiation much more efficiently than indicated by (5.93). The irradiation of outer regions of a flat disk is very inefficient because the light from the central star impinges very obliquely on the disk (Figure 5.8). As shown in Figure 5.8, the flux of stellar radiant energy entering the disk depends upon the angle γ which the impinging radiation makes with the normal to the disk surface. If we assume that the 'average' radiation from the star originates at a distance $R_*/2$ from the disk plane on the stellar surface, then the radiative flux entering the flat disk at radius R is

$$ F_d \; \sim \; \frac{L_*}{4\pi R^2} \cos\gamma \; \sim \; \frac{L_*}{4\pi R^2}\frac{R_*}{2R}, \tag{5.95}$$

reproducing the $F_d \propto R^3$ behavior derived for the exact solution (5.93).

However, disk surfaces in general will not be entirely flat, and can curve upward (e.g., Figure 1.4); this can make an enormous difference in how much stellar radiation is absorbed (Kenyon & Hartmann 1987). In most disks H/R is an increasing function of R. For example, in the steady, constant α, vertically-isothermal disk model discussed above, where $c_s \propto R^{-3/8}$, $H/R \propto R^{1/8}$. If the disk is sufficiently large, its vertical thickness can be significant, especially at large radii. Such 'flared' disks absorb much more radiation than a geometrically flat and thin disk. It is straightforward to show that, for relatively thin disks, $\cos\gamma \propto dH/dR - H/R$ (Kenyon & Hartmann 1987; Ruden & Pollack 1991). In the limit that the disk is very thick in the vertical direction and curved toward the central star, $\cos\gamma \to 1$, and $F_d \to L_*/4\pi R^2$. Disk flaring therefore can have dramatic consequences for the far-infrared SEDS and images of T Tauri disks, as discussed in the following chapter (§6.2).

6

The disks and envelopes
of T Tauri stars

The idea that the early Sun was surrounded by a rotating flattened nebula or disk out of which the planets formed has had a long history. However, a detailed application of the disk model to pre-main-sequence stars was not made until the seminal work by Lynden-Bell & Pringle (1974). These authors suggested that the excess emission of T Tauri stars could be powered by disk accretion; the extended dusty disk accounts for the excess infrared emission of T Tauri stars, while the hot gas predicted at the boundary layer between the star and disk produces the observed ultraviolet continuum emission. Lynden-Bell & Pringle further suggested that T Tauri disks could be quite massive, and might even outshine the central star in their early stages.

In retrospect, it seems clear that the field was not ready for these insights, partly due to the observational limitations of the time. The situation was further complicated by the discovery that young stars can produce substantial hot gaseous emission without disk accretion, through the action of their strong magnetic fields. Ultraviolet and X-ray observations with the IUE and Einstein satellites in the late 1970s showed that young stars exhibit high-temperature chromospheric and coronal emission at much higher levels than observed on the Sun (Gahm *et al.* 1979; Cram, Giampapa, & Imhoff 1980; Imhoff & Giampapa 1980; Gahm 1981; Giampapa *et al.* 1981; Feigelson & DeCampli 1981; Walter & Kuhi 1981; Feigelson & Kriss 1981; Mundt *et al.* 1983). It was natural to think that the excess optical and ultraviolet emission of the low-mass T Tauri stars represented the extreme youthful limit of solar magnetic activity (Herbig 1970; Cram 1979; Calvet, Basri, & Kuhi 1983). Models were constructed which attempted to use hot gaseous emission to explain the *near-infrared* emission of T Tauri stars as well (Rydgren, Strom, & Strom 1976).

Infrared studies with higher sensitivity changed the picture completely. Near-infrared observations initially indicated that dust, not hot gas, must be responsible for the excess emission (Mendoza 1966; Strom *et al.* 1972; Cohen 1973a,b,c; Cohen & Schwartz 1976; Rydgren *et al.* 1976; Cohen & Kuhi 1979; Rydgren & Vrba 1981). The case for dust emission was made very clear by far-infrared *IRAS* observations, which showed that the long-wavelength excess fluxes of T Tauri stars could not be explained by gaseous emission models. Dusty disks were the only plausible candidates for explaining the observed infrared spectra (Rucinski 1985). The disk picture provided a paradigm with which to understand the new data, eventually allowing many independent observations to fall into place in a coherent way. Further improvements in interferometry at radio wavelengths led to better understanding of disk masses and structure. The high-resolution optical imaging possible with *HST* has led to remarkable pictures of disks around young stars: some seen as shadows against the bright Orion Nebula (§1.5), others in light from the central T Tauri star scattered by the outer disk surfaces (Figure 1.4).

We currently think that Lynden-Bell & Pringle were mostly right. The angular momenta of collapsing protostellar clouds are large enough that stars are formed with substantial rotating disks around them; accretion does power the excess optical and ultraviolet emission of T Tauri stars; and the infrared excess does arise from circumstellar accretion disks. However, the disk emission of typical T Tauri stars is powered in large measure by the absorption of radiation from the central star, in addition to local accretion energy release (e.g., Adams & Shu 1986; Kenyon & Hartmann 1987). In addition, it is now thought that the hot continuum excess emission observed at optical and ultraviolet wavelengths, though powered by accretion, does not arise from a boundary layer. Instead, it is thought that accretion onto T Tauri stars is modified by a stellar magnetic field (Uchida & Shibata 1984a,b; Bertout *et al.* 1988; Camenzind 1990; Königl 1991; Shu *et al.* 1994). The stellar magnetic field responsible for highly-enhanced solar-type activity (coronal X-ray emission, flares, starspots) is strong enough to hold off the disk before it reaches the stellar surface. This magnetic field disrupts the inner disk and channels the accreting material to fall onto the star in accretion rings or spots. The observed hot continuum emission then arises from the accretion energy dissipated when the rapidly-moving magnetospheric gas shocks at the stellar surface (Figure 6.1).

The broad emission lines of T Tauri stars are thought to be produced in the magnetospheric accretion columns (Calvet & Hartmann 1992), where the gas is moving at essentially free-fall velocities. The optical emission line profiles produced by this infalling gas exhibit asymmetries (Hartmann *et al.* 1994) for the same reasons that the radio-wavelength emission lines produced in collapsing protostellar envelopes exhibit asymmetries (Chapter 4). T Tauri stars which exhibit prominent infall asymmetries were originally called YY Ori stars (Walker 1972), and were thought to be objects in the protostellar infall phase (e.g., Wolf, Appenzeller, & Bertout 1977). However, most YY Ori objects are not heavily extincted, in contrast with predictions for infalling protostellar envelopes (Chapter 4); disks can provide adequate mass reservoirs without strong dust absorption at most viewing angles. It has become clear from more recent spectral studies at high signal-to-noise that YY Ori-type asymmetries are the rule, not the exception in CTTS, indicating that most T Tauri stars gain material through stellar magnetospheres (Edwards *et al.* 1994).

Our ability to distinguish between accretion power and the energy dissipation produced by solar-type magnetic activity has been greatly aided by the existence of the low-mass, pre-main-sequence stars called WTTS. In contrast to the CTTS, the WTTS are not accreting from disks. The clear differences between CTTS and WTTS allow us to determine which properties are due to disk accretion and which to magnetic activity of the underlying star. Why some very young stars are WTTS and others are CTTS is not entirely clear, but one important and perhaps dominant reason is the presence of a binary companion, which renders disks dynamically unstable on scales comparable to the binary orbit (Jensen *et al.* 1994, 1996; Osterloh & Beckwith 1995).

Not all infrared excess emission of T Tauri stars can be explained with disks. In some cases this excess emission is almost certainly due to radiation from infalling dusty

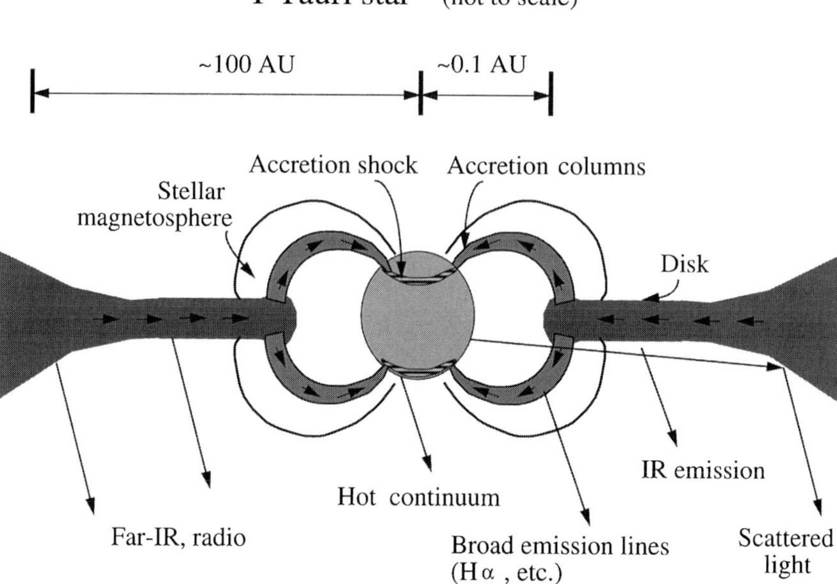

Fig. 6.1. Schematic picture of accretion in T Tauri stars. The pre-main-sequence star is surrounded by an accreting circumstellar disk which emits at infrared, sub-mm, and mm wavelengths. The inner disk is disrupted by stellar magnetic fields, which cause accreting material to be diverted out of the disk and fall rapidly onto the star. This magnetospheric material emits broad emission lines as it falls along the accretion columns, and produces a hot continuum when it crashes into the stellar surface at an accretion shock.

envelopes; the implication is that some of these objects are *optically-visible protostars* (Calvet *et al.* 1994). We now know that young stars exhibit powerful bipolar outflows which can easily blow away remnant circumstellar envelope material, at least in certain directions (Chapter 8). These outflows can carve paths through the collapsing dust through which optical light can escape. Viewing such objects along outflow holes, one can observe the protostar – which, if accreting from the circumstellar disk, should probably look much like a CTTS. Moreover, if infalling envelopes are not initially spherical, or are displaced from spherical symmetry by magnetic forces, the infalling material will become increasingly flattened as collapse proceeds, resulting in removing obscuring dust over large viewing angles (§4.5). One might therefore expect to be able to observe protostars at optical wavelengths especially well in the final stages of infall, when the envelope is the most aspheric, and the least opaque. A variety of arguments now lead fairly convincingly to the conclusion that the so-called 'flat spectrum' T Tauri stars are in fact objects in the late stages of protostellar collapse (Calvet *et al.* 1994).

Lynden-Bell & Pringle envisaged that accreting disks might be more luminous than their central objects at very early times, when disks are relatively massive and small. Most T Tauri stars outshine their disks; it appears that few T Tauri disks are more luminous than their central stars, and that for many objects, the disk heating is dominated by stellar irradiation rather than viscous heating (§5.7). However, the very young FU Ori objects exhibit highly time-variable accretion, and during outbursts the disks can become hundreds of times more luminous than the central star (Chapter 7).

Because of rapid advancements in radio-wavelength interferometric capabilities, observations at sub-mm and mm wavelengths can be used to estimate disk masses from optically-thin dust emission (e.g., Beckwith *et al.* 1990; §6.3). The interpretation of these initial observations is limited by substantial uncertainties in dust opacities, since there is evidence that dust particles in T Tauri disks may be evolving in their sizes and other properties from their interstellar values; nevertheless, reasonable guesses of opacities indicate disk masses that in many cases are similar to the minimum disk mass $\sim 0.01\,M_\odot$ needed to account for the heavy elements in the Solar System. These and other developments indicate that we are on the threshold of realizing the hope expressed by Lynden-Bell & Pringle (1974): 'If this interpretation of T Tauri stars is correct, their study will provide new and important evidence on the conditions under which the planets in the solar system were formed.'

6.1　Disks?

Most evidence for disks around T Tauri stars is circumstantial at present. In a few cases spatial images provide quite suggestive evidence for elongation (§1.5). Combining this with kinematic information showing rotation would provide the strongest case for disks. It is currently difficult to resolve disk rotation spatially (e.g., Dutrey *et al.* 1995; Koerner & Sargent 1995); current mm-wavelength interferometric observations tend to be most sensitive to larger-scale infall and rotation of circumstellar envelopes (e.g., Hayashi *et al.* 1993).

Despite limited kinematic evidence, so many independent lines of evidence point to the same conclusion that the case for circumstellar Keplerian disks around T Tauri stars is fairly secure. For instance, the only mechanism advanced so far which can explain the observed spectra of T Tauri stars from $\sim 1\,\mu\mathrm{m}$ to ~ 2 mm is dust emission. Even though the presence of emission lines suggests that gaseous free-free (bremsstrahlung) emission might be important, detailed analysis shows that such gaseous emission cannot reproduce the observed spectra. An isothermal, optically-thin volume of gas emits a spectrum $\lambda F_\lambda \propto \lambda^{-1}$; although some T Tauri SEDs agree with this, most objects have smaller spectral indices $s \gtrsim -2/3$, i.e. most observed infrared excesses decrease much more slowly with increasing wavelength; no gaseous emission models have been constructed which reproduce this behavior. Moreover, gaeseous emission models with significant near-infrared emission tend to produce far too much optical and ultraviolet emission (Cohen & Kuhi 1979; Hartmann *et al.* 1990). To make the free-free emission optically thin but yet produce a very large luminosity, it is necessary to place it in a very large emitting volume; it is hard to see how to arrange this physically, and in any event the recombination radiation of hydrogen Balmer lines would be much larger than observed. Finally, it also appears impossible to explain the steep spectral indices observed at mm wavelengths by gaseous emission, although the flatter spectra sometimes observed at cm wavelengths probably is the result of free-free emission from outflowing jets (Chapter 8).

If we conclude, as seems unavoidable, that solid grains are responsible for the infrared emission of T Tauri stars at wavelengths from $\sim 2\,\mu\mathrm{m}$ to 2 mm, then the amount of dust required is incompatible with the low extinctions to many T Tauri stars *unless* the dust is distributed non-spherically. The SEDs of many T Tauri stars

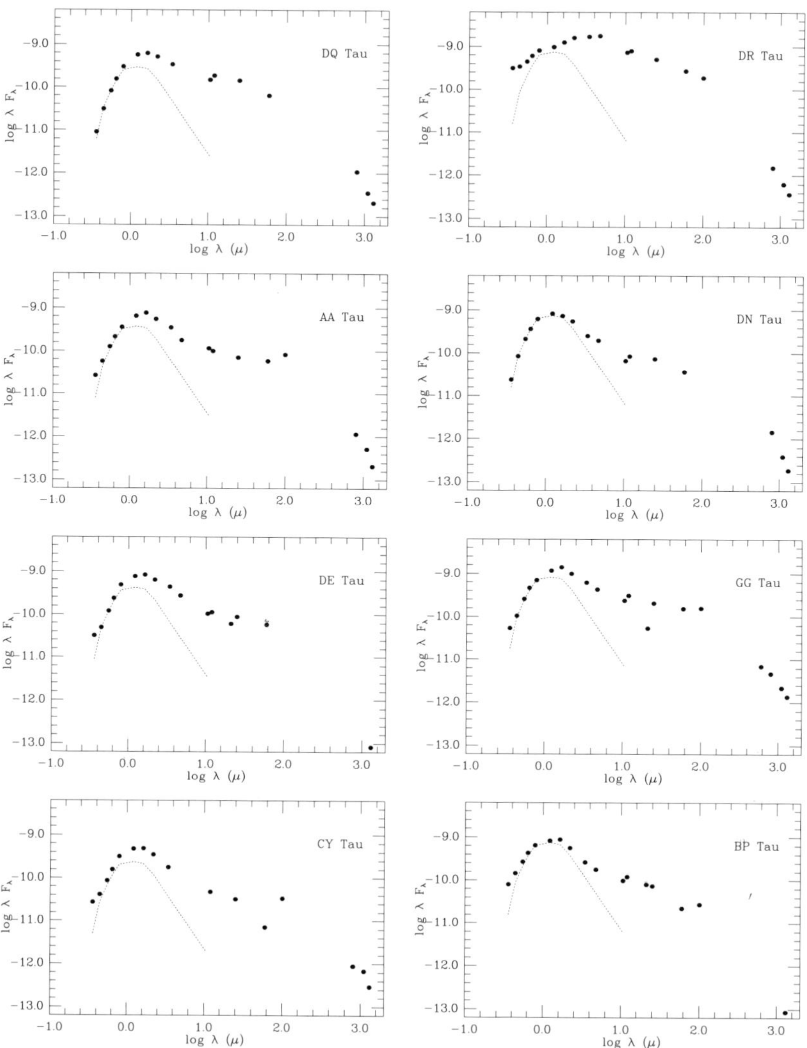

Fig. 6.2. SEDs of T Tauri stars in the Taurus molecular cloud. The vertical axes denote observed fluxes in $\mathrm{erg\,cm^{-2}\,s^{-1}}$. The dotted curve denotes the SED for the WTTS LkCa7, which is a K7-M0 pre-main-sequence star that shows no evidence for accretion. The excess long-wavelength emission is immediately evident, but the short-wavelength excess emission is much more difficult to detect except in the strong-emission star DR Tau. Data from Kenyon & Hartmann (1995).

show changes or breaks in spectral indices at wavelengths near $\sim 100\,\mu\mathrm{m}$ (Figure 6.2). This behavior suggests a transition between optically-thick and optically-thin regimes; from Figure 4.1, an optical depth of order unity at $100\,\mu\mathrm{m}$ corresponds to a visual extinction $A_V \sim 10^3$ (Beckwith *et al.* 1990). This is vastly larger than the $A_V \sim 1$ typically inferred from optical studies. The dust must be distributed so that there are direct lines of sight to the star which do not pass through most of the dusty material; one plausible distribution which accomplishes this is a disk.

While this argument for a dusty disk is very suggestive, it is not absolutely conclusive; any toroidal distribution of dust, such as a flattened collapsing envelope (§§3.3, 4.5) would accomplish the same thing. The essential test depends upon the *spatial* scale of the dust; as discussed in §§4.6 and 4.7, it is difficult for infalling envelopes to contain large amounts of mass within a few hundred AU of the central star, so any mass concentration on such scales is much more likely to represent a disk (Terebey *et al.* 1993).

Similar comments apply to other methods of inferring disk structure relying on extinction. For example, Appenzeller, Jankovics, & Östreicher (1984) and Edwards *et al.* (1987) suggested that the preferential blueshifts seen in optical forbidden emission lines of T Tauri stars indicate disk structure; the forbidden-line emission is flowing outward in winds or jets (Chapter 8), and the receding portion of the flow is hidden by obscuring disks, resulting in a net apparent blueshift. While this picture is consistent with disk structure, it is also consistent with any toroidal distribution of dust, so that other arguments must be used to distinguish between envelope and disk structure. As another example, polarization patterns of optical scattered light have been used to infer disk structure (Bastien & Ménard 1988, 1990), but care must be taken to distinguish between disk and envelope cavity structure (Whitney & Hartmann 1992, 1993; §4.5).

High spatial resolution imaging at optical and near-infrared wavelengths has provided additional evidence for disk structure around T Tauri stars (cf. Malbet *et al.* 1993; Roddier *et al.* 1996). Some of the most compelling results have been obtained at optical wavelengths. As mentioned in §1.5, *HST* images of young stars in the Orion Nebula (O'Dell *et al.* 1993; O'Dell & Wen 1994; McCaughrean & O'Dell 1996) show elongated, oval dark structures around low-mass, pre-main-sequence stars, seen as shadows against the bright nebular background. These dark structures are almost certainly produced by dust on scales of 10^2 AU. The dark dust structures surround central stars that are not highly extincted, showing that the dust is not spherically distributed. The image of the one object in which the central star is highly extincted gives the appearance of a disk viewed edge-on. Similarly, the *HST* images of the nearby Taurus object HH 30 (Burrows *et al.* 1996; Figure 1.4) clearly indicate a flattened absorbing structure viewed nearly edge-on. The simplest intepretation of these observations is that HH 30 is surrounded by a 'flared disk' (§5.7); the curved disk surfaces are illuminated by the central star, which is otherwise hidden from view by the absorbing material in the disk plane. The lack of absorption on large spatial scales for all of these objects demonstrates that the extincting structures are not being fed by a dusty envelope falling in from larger radii.

There is also evidence for disk rotation from optical and near-infrared spectroscopy, first observed in the FU Ori objects discussed in Chapter 7. Both optical and infrared line profiles of these objects provide very strong evidence for both rotation of a flattened structure as well as Keplerian differential rotation (Chapter 7). Infrared spectroscopy of heavily-extincted objects has yielded additional evidence for line emission produced in rotating disks (Carr *et al.* 1993; Najita *et al.* 1996).

In summary, the disk hypothesis provides a unified picture providing an explanation of many features of pre-main-sequence objects which are difficult, if not impossible, to understand with any other model.

6.2   Infrared emission of T Tauri disks

The most generally-applied method for inferring the presence of dust around T Tauri stars has been to measure infrared excess emission. Near-infrared emission is relatively easy to observe with ground-based techniques and can be produced by very small amounts of dust. To see this, consider the case in which only the irradiation of the star heats the (optically-thick) disk. Equation (5.93) indicates that the temperature of the disk will be roughly

$$T_d^4(R) \sim \frac{T_{eff}^4}{\pi}\left(\frac{R_*}{R}\right)^3. \tag{6.1}$$

To obtain strong near-infrared emission at $\lambda \sim 3\,\mu$m, the dust temperature must be $T_d \sim 1000$ K; this temperature will be achieved at disk radii $R_o \sim 4R_*$ for a typical central T Tauri star with $R = 2\,R_\odot$ and effective temperature of $T_{eff} \sim 4000$ K. A large near-infrared excess also requires that the disk be optically thick at $3\,\mu$m. Assuming dust opacities and dust-to-gas ratios characteristic of the interstellar medium for the moment, optical depths of unity at this wavelength will result for a gas column density $\Sigma \gtrsim 0.2\,\mathrm{g\,cm}^{-2}$ (Figure 4.1). Therefore, the (gas) mass required to produce observable near-infrared excess emission under these assumptions is only $\sim \pi\Sigma R_o^2 \approx 10^{-10}\,\mathrm{M}_\odot$. Although there is no reason to suppose that the dust in disks is the same as that in the interstellar medium – indeed, there are reasons to expect grain evolution in disks (§6.3) - this simple estimate demonstrates that extremely small masses of dust can produce sizable near-infrared excesses. In contrast, far-infrared excess emission requires much more dust, as described below, suggesting that the near-infrared emission traces only a very small portion of typical T Tauri disks.

Figure 6.2 shows SEDs of some typical Classical T Tauri stars in the Taurus-Auriga molecular cloud. These SEDs typically peak near $\sim 1\,\mu$m, reflecting the dominance of the stellar photospheric emission with typical stellar effective temperatures ~ 4000 K (the peak flux of a blackbody in λF_λ is $\lambda_{max} = 3670\,\mu\mathrm{m}/T$). For comparison, a K7 stellar photosphere is plotted as a dotted line. The excess infrared emission is apparent, qualitatively as suggested for disk radiation by Lynden-Bell & Pringle (1974).

At very long wavelengths, the excess emission declines steeply. An optically-thick disk with a sharp, fixed, outer edge will have a spectrum with the Rayleigh–Jeans form $\lambda F_\lambda \propto \lambda^{-3}$ (§5.4). However, the spectral slope observed in many objects is much steeper than this. This steep fall-off of emission with increasing wavelength is probably due to the disk becoming *optically-thin*; the spectrum approaches $\lambda F_\lambda \to \lambda^{-3+\beta}$, where β is the slope of the decreasing dust opacity as a function of wavelength (see §6.3).

Let us make an estimate of the disk mass required to account for the observed disk emission at $\lambda_c = 100\,\mu$m. One can analyze the SED more precisely using a parameterization of the temperature distribution (§6.3), but for present purposes it suffices to make an order of magnitude estimate. In optically-thick disk models, most of the emission at λ_c arises from disk regions at temperatures $T_c \sim 3670\,\mathrm{K}\,\mu\mathrm{m}/\lambda_c$. To order of magnitude we may then write

$$\lambda_c L_{\lambda_c} \sim \lambda_c 2\pi[R(T_c)]^2 \sigma T_c^4, \tag{6.2}$$

where $R(T_c)$ is a characteristic size of the region of the disk at T_c. If we adopt a typical flux at the Earth of $\sim 3 \times 10^{-11}\,\mathrm{erg\,cm}^{-2}\,\mathrm{s}^{-1}$ at $\lambda_c = 100\,\mu$m from Figure 6.2, and a distance of 140 pc (for Taurus), $R(37\,\mathrm{K}) \sim 20$ AU. If we further require the disk to be just optically thick, then

$$\tau_{100\mu m} = \kappa(100\ \mu m)\Sigma = 1, \tag{6.3}$$

where the optical depth is written in terms of the opacity κ per unit mass, and it is assumed that the disk is observed face-on. Adopting the opacity curve of Figure 4.1, $\kappa \sim 0.2\,\mathrm{cm^2\,g^{-1}}$, $\Sigma \sim 5\,\mathrm{g\,cm^{-2}}$, so the minimum disk mass which can be optically thick at 100 μm is $M_d \gtrsim 10^{-3}\,M_\odot$. While this estimate is very crude, it demonstrates the need for much larger dust masses to explain far-infrared emission than near-infrared excesses, and suggests that some T Tauri stars may have disks with potential for forming Solar-System-like planets.

As discussed in §5.6 and in the introduction to this chapter, Lynden-Bell & Pringle (1974) predicted that there should be hot boundary layer radiation from T Tauri systems which would be observable at optical and ultraviolet wavelengths. Because T Tauri stars are slow rotators, the boundary layer theory predicts that the luminosity of the hot component should be comparable to that of the disk (§5.6). However, the observations shown in Figure 6.2 suggest that the hot continuum is generally much weaker than infrared excess emission (except possibly in the 'continuum' star DR Tau). Although there may be additional hot continuum emission at unobserved (far-ultraviolet) wavelengths, the most likely explanation of this observation is that much of the infrared disk emission is powered by the irradiation by the central star (§5.7).

In general, the excess infrared disk emission of CTTS is usually a modest fraction of the stellar luminosity. This is demonstrated in Figure 6.3, which compares the total bolometric system luminosity L_b with a luminosity centered at the J band, L_J; the latter should be a good measure of the stellar luminosity L_* in most cases because the stellar photospheric emission peaks near the J-band wavelength $\lambda \sim 1.25\,\mu m$ (see Figure 6.2; Kenyon & Hartmann 1995). The open circles are the WTTS, which scatter around the $L_b = L_J$ relation (solid line). The CTTS, shown as filled circles, generally scatter above the line of equality, showing modest excesses; however, they do not generally exhibit very large bolometric luminosities. For the CTTS sample shown in Figure 6.3, the median value of $L_b/L_J \approx 1.25$, comparable to the disk emission expected for a star with a geometrically-flat, optically-thick, irradiated, and non-accreting disk (§5.7).

The data in Figure 6.3 are really 'apparent' luminosities $\hat{L}$, which for disks depend upon the orientation of the disk with respect to the Earth. The apparent luminosity of a geometrically-flat, optically-thick disk with true luminosity L_d is

$$\hat{L}_d \equiv 4\pi d^2 F = 2L_d \cos i, \tag{6.4}$$

where d is the distance, F is the total disk flux observed at Earth (corrected for extinction), and i is the angle between the disk normal and the line of sight. (This result can be derived from basic definitions; it can also be checked by showing that the average of $\hat{L}_d$ over all solid angles is equal to L_d.) Because system inclinations are poorly known, estimating true disk luminosities in individual cases is difficult.

Although the observed infrared excesses qualitatively resemble that predicted by the optically-thick accretion disk model, and indeed are difficult to explain in any other way, there are quantitative differences. The typical spectral index $s \sim -2/3$ of

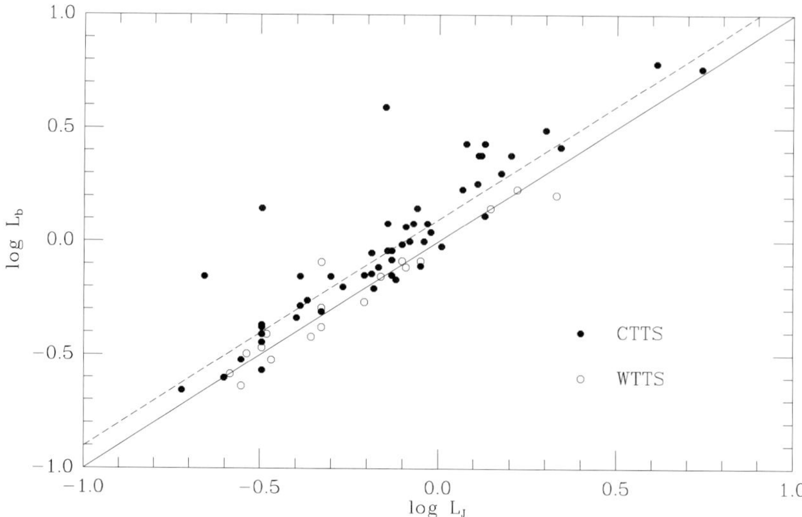

Fig. 6.3. Bolometric (total) luminosity L_b vs estimated stellar luminosity L_J (from the J-band measurements, see text) for pre-main-sequence stars in the Taurus molecular cloud, in units of solar luminosities. Most of the WTTS scatter around $L_b = L_J$ because they have little or no infrared disk emission. The CTTS systematically exhibit higher total or bolometric luminosities, due to the presence of excess emission, mostly at infrared wavelengths. The excess emission of all but a few CTTS is a small fraction of the stellar luminosity. The solid line denotes $L_b = L_J$, i.e. no excess above the stellar luminosity; the dashed line denotes $L_b = 1.25L_J$, the expected average total luminosity for a flat, opaque, non-accreting, irradiated disk surrounding the star (§5.7). It should be pointed out that, for those stars in which $L_b \gg L_J$, the accretion luminosity may well be so large that the emission due to accretion is also important at the J-band, invalidating the use of L_J as an estimate of the stellar luminosity. Data taken from Kenyon & Hartmann (1995).

T Tauri infrared emission ($\lambda L_\lambda \propto \lambda^s$; see Chapter 4) is substantially larger than the $s = -4/3$ predicted by the steady disk (§5.4). This is shown graphically in Figure 6.4, where the mean value of s as a function of wavelength is shown for the Taurus sample of young stars. Over the spectral range observable from the ground, $\lambda \lesssim 20\,\mu$m, the average spectral index is more like $s \sim -0.7$ rather than $s \sim -1.3$; moreover, the spectral index appears to vary systematically with wavelength. At short wavelengths, this variation is produced by the rising contribution of the stellar photosphere, but at long wavelengths stellar emission is unimportant. There are some observational difficulties which make this result somewhat less robust in individual cases than one would like; the large beam sizes used at the longest wavelengths raise the possibility of confusion or inclusion of diffuse (i.e., non-disk) emission; in addition, many systems are binaries, and the individual components can have quite different SEDs. Nevertheless, the generality of the result in Figure 6.4 makes it unlikely that the flattening of the SEDs at long wavelengths can be completely written off as observational error.

For disk emission to reproduce the observed SEDs, it is necessary that the disk temperature fall off more slowly with increasing radius than $R^{-3/4}$. If the temperature distribution in an optically-thick disk is a power law, $T \propto R^{-q}$, then it is straight-

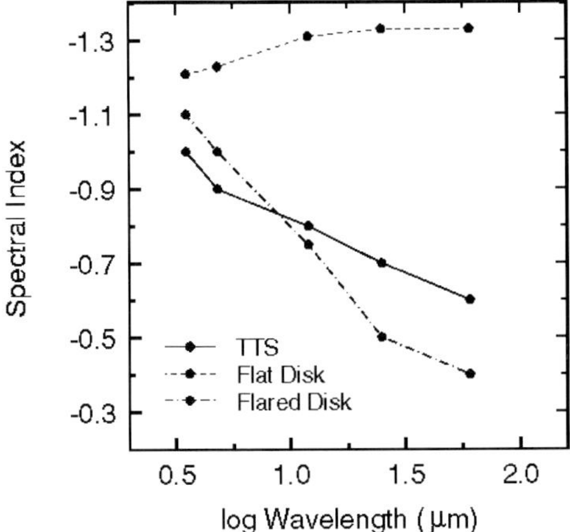

Fig. 6.4. Observed average spectral indices s (where $\lambda L_\lambda \propto \lambda^s$) as a function of wavelength for T Tauri stars in the Taurus cloud complex, compared with values expected for various models. The flat disk model exhibits $s = -4/3$ except at the shortest wavelengths, where the contribution of the stellar photosphere changes the spectral slope. The observed averages are comparable to that predicted for a flared disk model (see text). From Kenyon & Hartmann (1995).

forward to show (following an analysis similar to that of §5.4) that $s = (2/q) - 4$. Thus, the typical observed value $s \sim -2/3$ near $\lambda \sim 10\,\mu$m implies an intermediate temperature distribution $q \sim 3/5$, while the typical value $s \to 0$ at the longest infrared wavelengths suggests $q \to 1/2$.

What is responsible for the relatively flat temperature distributions of the emitting dust around T Tauri stars? One possibility is that we simply do not understand the way accretion heating should affect disk temperatures, or that mass accretion rates are not steady as a function of radial distance (e.g., discussion in Kenyon & Hartmann (1987)). Proposals have been made that disks can be somehow 'active', i.e. self-luminous with relatively flat temperature distributions (Adams, Lada, & Shu 1988). Shu *et al.* (1990) suggested that low-azimuthal order gravitational instabilities in a modest-mass disk could dissipate energy over large distance scales as required to account for flat spectrum sources in particular. As mentioned in §5.3, the disk may need to have a sharp outer edge in order to amplify the instability properly, and it is not clear that this is a likely boundary condition, given the tendency of the disk to diffuse. Heating rates have not yet been calculated for this mechanism, and it is difficult to see how large amounts of heating can occur in the outer disk without resulting in outer disk accretion rates that seem implausibly large (Kenyon & Hartmann 1987).

Because the infrared excesses produced by most disks are relatively large in comparison with hot continuum excesses, yet are a modest fraction of the stellar luminosity, irradiation heating of the disk can be important in establishing the disk temperature distribution and resulting SED (§5.7). In the 'flared disk' hypothesis of Kenyon & Hartmann (1987), disks generally become proportionately thicker with increasing radius, i.e. the ratio of the disk scale height H to the radial distance R increases with R;

this 'tilts' the disk surface more toward the star, so that it can absorb more radiation and be heated to higher temperatures.

The magnitude of the effect of flaring depends upon the shape of the disk 'photosphere', i.e. the region where the disk can effectively absorb stellar photons. Kenyon & Hartmann (1987) showed that for plausible surface densities in T Tauri disks (see next section), the disks can be so optically-thick at $1\,\mu$m (where most of the stellar energy is radiated) that the 'disk photosphere', i.e. the level at which the vertical optical depth from the outside reaches unity, could easily be at 2–3 times the scale height H above the midplane. Figure 6.4 shows the predicted spectral index s as a function of wavelength for typical T Tauri parameters and a disk with a photosphere at $3H$. This simple model does a remarkably good job of explaining the average infrared spectral slope in CTTS.

The flared disk model also accounts for the second observational clue to the mechanism producing the needed disk heating; namely, s generally becomes larger in magnitude (less negative) with increasing wavelength. This occurs simply because the scale height $H \propto T_d^{1/2}R^{3/2}$; thus in all reasonable disk models $H << R$ close to the star, and can only become large at large distances. This means in effect that the disk must be effectively flat in its inner regions, and can become flared only at large scales. In consequence, simple irradiated flared disk models have $T_d \propto R^{-3/4}$ at small R, but can approach $T_d \propto R^{-1/2}$ at large distances.

The difficulty with the flared disk picture is that it requires that dust be suspended vertically along with the gas rather than concentrated in a thin layer at the midplane (dust contains essentially all the opacity of the disk). Models of the solar nebula (e.g., Weidenschilling & Cuzzi 1993, and references therein) predict that dust grains should settle to the midplane of a disk fairly rapidly. This is because grains feel the same gravity per unit mass as they would if they were freely dispersed in a gas, but feel much less gas pressure per unit mass to support them in the vertical direction. Indeed, this settling is desired for models in which coagulation of dusty planetesimals provides the cores onto which planets accrete. However, the *HST* image of HH 30 (Figure 1.4) suggests that dust settling does not proceed too rapidly, at least on large scales.

Although calculations of dust settling suggest typical timescales of 10^3 yr, this generally refers to disk conditions at a radius of 1 AU; i.e., the settling occurs over $\sim 10^3$ orbits. At radial distances 10–100 AU, the same number of orbits corresponds to 3×10^4–10^6 yr, more comparable to T Tauri star ages. Moreover, the physics of the problem is complicated; only a small amount of the total dust mass needs to be suspended for a flaring photosphere which can absorb $\sim 1\,\mu$m starlight. Collisions between particles might result in some fragmentation and ejection of smaller particles even as larger grains coalesce. The rates of settling also depend upon the structure of grains; if dust particles are 'fluffy', they are more easily supported by gas pressure forces against gravity. This subject is complicated and one in which theory must be complemented by observational constraints.

Other models for the infrared SEDs of T Tauri stars have been advanced in which irradiation also dominates, but which invoke another dusty component in addition to the disk. Safier (1993) has suggested that an opaque, dusty disk wind can absorb light from the central regions and reradiate it as required. An interesting variant of this has been suggested by Natta (1993), who pointed out that even a dusty envelope of small optical depth can still be important in the far-infrared radiation process if it can scatter radiation into an optically-thick disk, thereby heating the disk in excess of the direct radiation from the central star. Natta suggested that this envelope could also be disk wind, but much less dense than invoked by Safier (see further discussion in Chapter 8).

None of these complications generally challenge the idea that disks are responsible for the excess infrared emission of most T Tauri stars. Studies suggest (Strom *et al.* 1989; Skrutskie *et al.* 1990) that roughly 60% of T Tauri stars younger than about 3 Myr have excess near-infrared emission which may signify the presence of disks. These observations suggest that particulate disks are quite common in early stellar evolution.

6.3   Long-wavelength emission; disk masses, sizes

At long wavelengths, the dust around T Tauri stars becomes optically thin and thus its emission becomes sensitive to the total mass present. Improvements in mm-wave interferometry have been exploited to provide the first systematic estimates of T Tauri disk masses (Beckwith *et al.* 1990). Here we discuss the basic ideas behind these estimates and their limitations.

Suppose for simplicity that the disk is geometrically thin and isothermal in the vertical direction (perpendicular to the disk plane), and that the emission is purely blackbody at the local temperature (which is a function only of cylindrical radius R). Then the apparent luminosity is

$$v\hat{L}_v = 4\pi d^2 v F_v = 4\pi\mu \int_{R_i}^{R_d} v B_v [1 - \exp(-\tau_v/\mu)]\, 2\pi R dR, \tag{6.5}$$

where F_v is the flux observed at Earth (corrected for extinction), B_v is the Planck function, $\mu = \cos i$, and i is the inclination angle of the disk to the line of sight, and τ_v is the vertical optical depth at cylindrical radius R at frequency v. The outer and inner radii of the disk are R_d and R_i, respectively. If the optical depth $\tau_v \gg 1$, then the emission depends solely on the Planck function and thus on the temperature distribution as a function of radius, and is independent of the disk mass, as discussed in the previous chapter. In contrast, when $\tau_v \ll 1$, the emission does depend upon the disk surface density distribution,

$$v\hat{L}_v = 4\pi \int_{R_i}^{R_d} v B_v k_v \Sigma(R)\, 2\pi R dR, \tag{6.6}$$

where we have written the optical depth in terms of the opacity k_v per unit mass,

$$\tau_v = k_v \Sigma. \tag{6.7}$$

If the observed emission occurs at wavelengths where the disk is optically thin, and if the disk temperature as a function of radius is known, measurements of the long-wavelength flux allow one to estimate a quantity $\propto \Sigma R^2$, which in turn is proportional to the disk mass. Given an estimate of the dependence of the surface density on radius, the total disk mass M_d can be determined. To illustrate this more explicitly, we make the usual simplifying assumptions that the surface density and temperature can be represented by power laws:

$$T = T_\circ (R/R_\circ)^{-q}, \qquad (6.8)$$

$$\Sigma = \Sigma_\circ (R/R_\circ)^{-p}. \qquad (6.9)$$

To simplify the analysis further we take the long-wavelength limit where $hv \ll kT$, so that $B_v = 2v^2 kT/c^2$, and assume that the inner radius $R_i \ll R_d$. With these assumptions the apparent luminosity becomes (e.g., Beckwith *et al.* 1990)

$$v\hat{L}_v = \frac{16\pi^2 k}{c^2} v^3 k_v \Sigma_\circ T_\circ R_\circ^2 \frac{(R_d/R_\circ)^{2-p-q}}{2-p-q}. \qquad (6.10)$$

The disk mass is

$$M_d = \int_{R_i}^{R_d} 2\pi \Sigma R dR = 2\pi \Sigma_\circ R_\circ^2 (R_d/R_\circ)^{2-p}/(2-p), \qquad (6.11)$$

and therefore

$$v\hat{L}_v = \frac{8\pi k v^3 k_v}{c^2} M_d T_\circ \left(\frac{R_d}{R_\circ}\right)^{-q} \frac{2-p}{2-p-q} = \frac{8\pi k v^3 k_v}{c^2} M_d T(R_d) \frac{2-p}{2-p-q}. \qquad (6.12)$$

If the dust opacity k_v and the outer disk temperature $T(R_d)$ (or an equivalent scaling temperature $T_\circ$) are known, and if p and q can be determined in some fashion, then disk dust masses can be determined from observed long-wavelength fluxes.

The most extensive attempts to estimate disk masses so far are those of Beckwith *et al.* (1990 = BSCG) and Osterloh & Beckwith (1995) from continuum observations at 1.3 mm. As discussed in the previous section, the disk temperature distribution can be inferred from the spectral index in the wavelength regions where the disk is optically thick. BSCG and Osterloh & Beckwith used this idea to derive appropriate values of q. Although, as discussed above, the infrared emission is not generally representable by a single power law (e.g., Figure 6.4), these investigators concentrated on fitting the longest-wavelength optically-thick emission to fix the outer disk temperature; this is a reasonable approach since most of the disk mass is likely to reside at large radii.

The determination of p is more problematic. BSCG and Osterloh & Beckwith assume $p = 3/2$ for most objects, attributing this to the general result for α disks. However, as described in Chapter 5, the values for steady $\alpha = $ constant disks range from $p = 3/4$ for $T \propto R^{-3/4}$ to $p = 1$ for $T \propto R^{-1/2}$. Moreover, as shown in the discussion of the similarity solution (§5.2), the outer surface density structure of a viscously evolving disk is *not* generally a power law.

While these problems introduce some error in disk mass estimates, the largest uncertainties probably arise from lack of knowledge of the dust opacity. At long wavelengths or low frequencies, where the disk is optically thin and $hv \ll kT$ for any reasonable T, the disk spectrum should vary as $k_v v^3$. For the Draine & Lee (1984) opacities, $k_v \propto v^\beta \propto v^2$ since one is observing at wavelengths much greater than the particle size. However, the long-wavelength observations of many T Tauri stars indicate values of β closer to 1 or even 0 (Beckwith & Sargent 1991). This is not particularly surprising; indeed, it may be evidence that disk grains are evolving to larger sizes than present in the interstellar medium, as predicted/desired for models of planet formation. However, the possibility of the evolution of dust properties

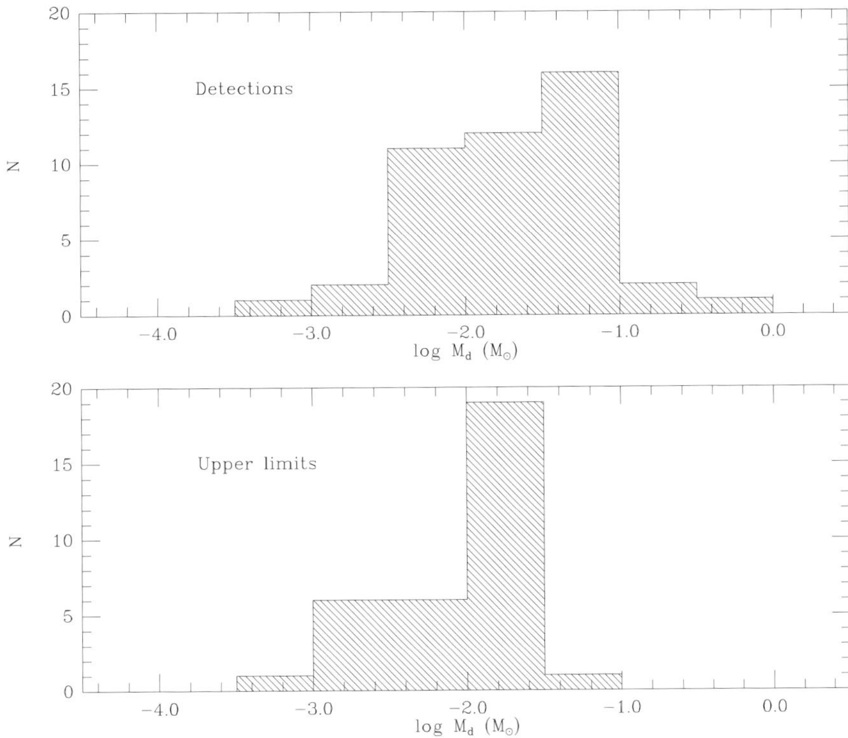

Fig. 6.5. Disk masses estimated from 1.3 mm fluxes and fits to SEDs for stars in Taurus by BSCG and Osterloh & Beckwith (1995) (masses from the latter reference have been modified (Hartmann *et al.* 1998)).

necessarily introduces a significant uncertainty into the mass estimates, which are essentially inversely proportional to the assumed opacity.

In many cases researchers have adopted the parameterization of BSCG,

$$k_v = 0.1(v/10^{12} Hz)^{\beta} \, \text{g cm}^{-2}, \tag{6.13}$$

with $\beta = 1$. This parameterization matches the form estimated by Hildebrand (1983) at wavelengths $\lambda \lesssim 250 \, \mu m$, and is roughly a geometric mean of suggested values at 1 mm (see also André *et al.* (1993); Beckwith & Sargent (1991); Mannings & Emerson (1994)). Pollack *et al.* (1994) argued that the total uncertainty in the 1 mm opacity is roughly a factor of four (see also Henning & Stognienko (1996)); however, this argument necessarily depends upon how much coagulation of grains has taken place, a point which is hardly well understood at present.

BSCG and Osterloh & Beckwith estimated disk masses for a number of T Tauri stars from fluxes measured with the IRAM interferometer with a beam size ~ 11 arcsec in diameter, using equation (6.13) with $\beta = 1$, and the power-law fitting for p and q described above. Figure 6.5 shows their results for young stars in the Taurus molecular clouds. Approximately half of the stars with excess infrared emission were detected at 1.3 mm. As discussed further in §6.5, roughly half of the young stars with ages ~ 1 Myr do not show infrared excesses; of these, only about 10% were detected at mm wavelengths. The results further support the findings from infrared studies (§6.2) that dusty disks are a common feature in young stars. Although the results in Figure 6.5 are subject to large uncertainties due to the problematic dust opacity, they indicate that many T Tauri disk masses are of the order of magnitude of a so-called 'minimum mass solar nebula', $M_d \sim 10^{-2}\,M_\odot$, further suggesting that these disks can potentially build planetary systems like the Sun's.

Disk sizes are more difficult to measure with current techniques. However, some interferometric observations at mm wavelengths (Koerner & Sargent 1995; Dutrey *et al.* 1995), sub-mm wavelengths (Lay *et al.* 1994) and even optical observations (Roddier *et al.* 1996) have begun to provide some resolved disk structure. Current results suggest some disks have sizes $\sim$ 100–300 AU, but these probably represent the upper end of the distribution because of limitations of resolution. The 'shadow disks' observed in the Orion Nebula by McCaughrean & O'Dell (1996) have sizes ranging from around 50 AU to 1000 AU.

6.4   Disk accretion

As outlined in §6.2, the disk emission of many T Tauri stars can be attributed to heating by the central star, implying that the energy dissipation due to accretion is not very large. Another diagnostic of accretion energy release must be used to measure T Tauri disk mass fluxes.

Lynden-Bell & Pringle (1974) originally suggested that dissipation of energy in the narrow boundary layer between the rapidly-rotating disk and the slowly-rotating star was responsible for the hot excess continuum emission observed in T Tauri stars. As outlined in §5.6, in this model the boundary layer radiates nearly as much energy as the entire disk, but over a much smaller area $2 \times 2\pi R_* \Delta R << \pi R_d^2$, and therefore the boundary layer emission is characterized by a much hotter spectrum than is produced by the disk.

Figure 6.6 displays short-wavelength emission of some typical T Tauri stars. The excess hot emission is evident. Typically the Balmer continuum at $\leq 0.365\,\mu$m is observed in emission. It is clear that this emission comes from gas at much higher temperatures than that of the stellar photosphere, as predicted by the boundary layer model.

However, it now seems probable that T Tauri stars do not have boundary layers at the inner edges of their disks. Instead, it is believed that T Tauri disks are disrupted by stellar magnetic fields, which hold off the disks at several stellar radii away from the stellar surface (Königl 1991; Camenzind 1990). In this magnetospheric model, the accreting material is channelled along the magnetic fields lines out of the disk and onto the star (Figure 6.1). The accreting gas falls in under gravity at nearly free-fall velocities; it shocks when it crashes into the stellar surface, producing the hot continuum radiation.

The lines of evidence which support this picture may be summarized as follows:

Spots and magnetic activity. T Tauri stars which apparently do not have accretion disks (§6.5) show chromospheric and coronal emission like that produced on the Sun by magnetic field activity, but at much higher levels than the Sun, suggesting enhanced surface magnetic fields (e.g., Walter *et al.* 1988, and references therein). In addition, some of these stars provide indications of dark cool spots on their photospheres, suggestive of sunspots, but covering much larger areas (Bouvier *et al.*

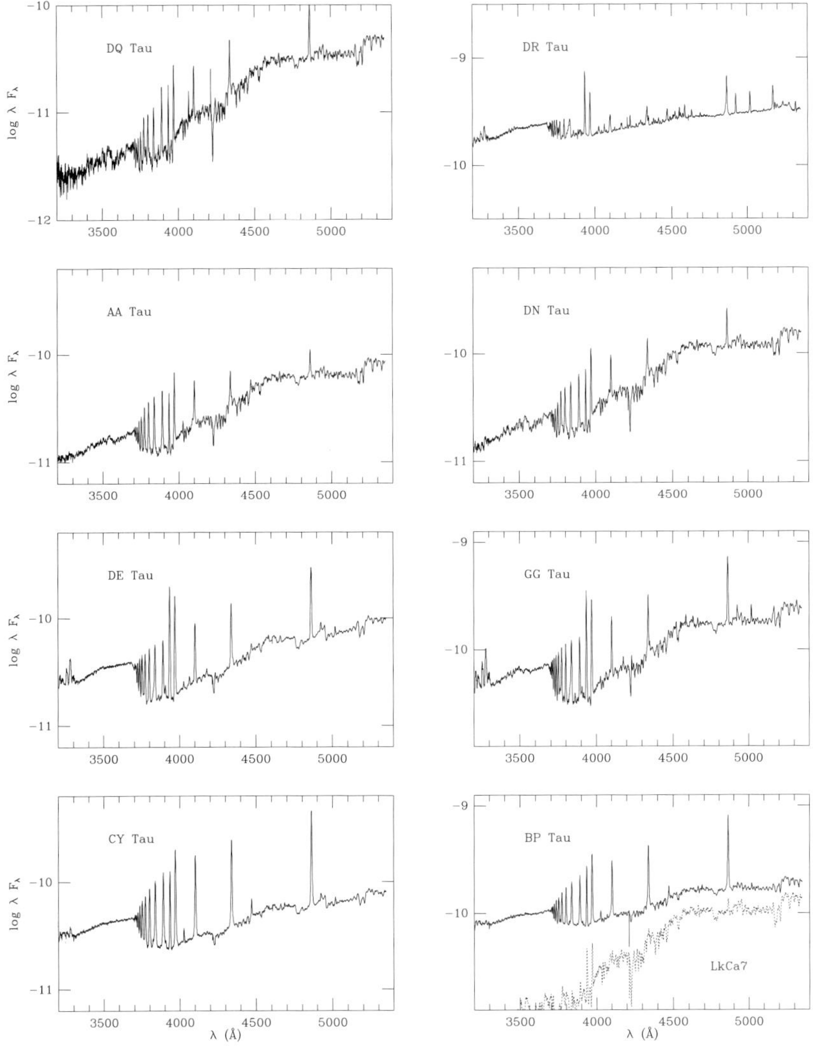

Fig. 6.6. Optical spectra of selected Taurus pre-main-sequence stars (cf. Figure 6.2), showing the emission lines and hot continuum emission. The excess hot emission is especially apparent in the Balmer continuum shortward of 3650 Å; however, many objects exhibit strong emission at longer wavelengths (DR Tau is an extreme example of this). Comparison with the spectrum of the WTTS LkCa7 (dotted line, bottom right panel) shows how the hot continuum emission makes the overall spectrum bluer, and that it reduces the apparent equivalent widths of photospheric absorption features. From Gullbring *et al.* (1997).

1993 and references therein). The magnetic field strengths of pre-main-sequence stars are not well known (see Basri *et al.* (1992)), but the limited measurements existing so far plus plausible extrapolations from the Sun suggest that they could be strong enough to perturb accretion disks (see Chapter 8 for additional discussion). There is also some evidence for stellar magnetic fields from polarized radio emission (André *et al.* 1991, 1992; Skinner 1993).

Periodic variations of hot continuum emission. In some accreting T Tauri stars, hot spots are inferred from periodic or quasi-periodic variation of the hot continuum emission. The inferred timescales are consistent with the rotational periods of the central stars predicted from observations of rotational line broadening coupled with estimates of the stellar radii. The most likely explanation of this behavior is that the stellar magnetic field is channelling the accretion onto specific locations on the stellar surface which rotate into and out of view at the stellar rotational period (Bertout *et al.* 1988).

Disk holes inferred from SEDs. In some rapidly-accreting T Tauri stars, such as DR Tau (Bertout *et al.* 1988; Kenyon *et al.* 1994), the accretion energy is much larger than the stellar luminosity, and so one can trace the high-temperature disk emission more clearly. This disk emission tends to peak at longer wavelengths than would be expected if the inner disk edge lies close to the stellar surface; the maximum disk temperatures appear to be smaller than would be expected from a disk with inner radius $R_i = R_*$. The inference is that there is a hole in the inner disk, which eliminates the high-temperature disk radiation.

Infall line profiles. In the boundary layer model, the accreting material slows down from rapid rotation to join the star; radial velocities are small in this picture. In contrast, the magnetospheric model predicts that the accreting material falls onto the star with a radial velocity

$$V_{in}^2 \sim \frac{2GM}{R_*} \left(1 - \frac{R_*}{R_m} \right) , \qquad (6.14)$$

where R_m is the 'magnetospheric radius' at which the disk is disrupted by the stellar magnetic field (Chapter 8). The infall velocity thus approaches the surface escape velocity when $R_m >> R_*$.

Figure 6.7 shows a sample of Hα line profiles for T Tauri stars. One notes that the emission lines are quite wide, with velocities typically $\sim \pm 200 \, \mathrm{km \, s^{-1}}$ from line center. Inserting typical T Tauri parameters $M_* \sim 0.5 \, \mathrm{M_\odot}$ and $R_* \sim 2 \, \mathrm{R_\odot}$ (cf. Figure 1.2), $V_{in} \lesssim 300 \, \mathrm{km \, s^{-1}}$, so the infall velocities can in principle explain much or all of the line profile widths (though for some strong lines opacity broadening must also be taken into account).

The magnetospheric model thus can potentially explain the large and puzzling velocity widths of emission lines of T Tauri stars. The large velocity widths were initially attributed to wind expansion, especially for lines like Hα which show blue-shifted wind absorption (Figure 6.7). However, as discussed in more detail in Chapter 8, it is difficult to explain many details of the permitted emission line profiles of most CTTS, particularly those that show no blueshifted absorption at all (Hartmann *et al.* 1990). On the other hand, infall models are able to reproduce many observed profile features with relative ease; for example, 'infall' asymmetries can be detected in many optical emission lines (§8.12), qualitatively similar to the asymmetries used to infer protostellar infall (§4.7). Careful studies have now turned up common evidence for *redshifted* absorption as direct evidence for infall (Figure 6.7; Edwards *et al.* 1994). It now appears that the 'YY Ori stars', which show strong evidence for redshifted absorption, are not a separate class of objects distinct from CTTS. The YY Ori objects are experiencing magnetospheric infall like many other T Tauri stars, but physical

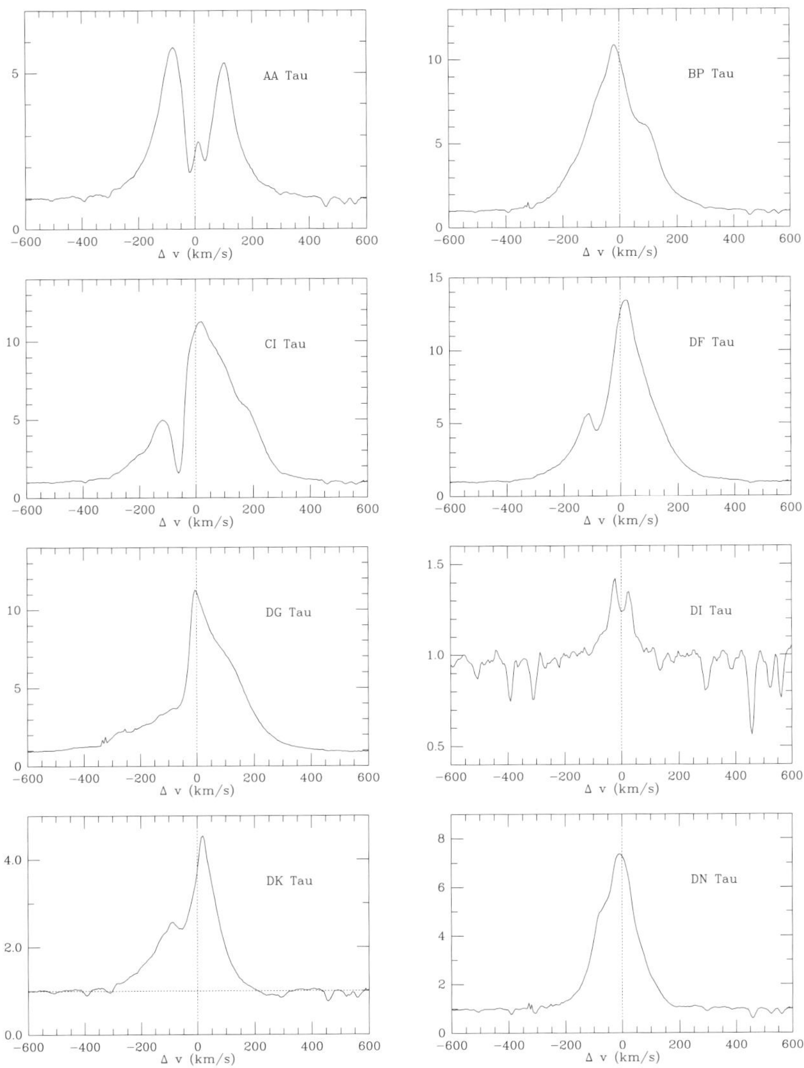

Fig. 6.7. Hα line profiles for T Tauri stars. Observed fluxes are normalized to the stellar continuum, and are plotted as a function of the velocity shift from line center (in the star's rest frame). Most objects shown broad emission over ranges $\sim \pm 200\,\mathrm{km\,s^{-1}}$ from line center. Many stars (CI, DF, DG, DK Tau) show blueshifted absorption probably characterizing mass ejection (Chapter 8), but others (BP, DN Tau) show no evidence for blueshifted wind absorption, and one object (DK Tau) shows evidence for faint redshifted absorption at $\sim +200\,\mathrm{km\,s^{-1}}$. DI Tau is a WTTS and therefore does not show the broad Hα profile produced in the infalling magnetospheric gas observed in accreting CTTS (see also Chapter 8).

conditions are such that this accretion is more easily noticable in particular spectral lines (§§8.11, 8.12).

Rotational braking. Among T Tauri stars of roughly the same age and stellar properties, surveys indicate that the objects without close circumstellar disks rotate more slowly than the stars with close circumstellar disks (Bouvier *et al.* 1993; Edwards *et al.* 1993). One explanation of this result is that the interaction of the stellar magnetic field with the disk helps slow down the star (König 1991; Shu *et al.* 1994). While this proposal is plausible, the argument is subtle and complicated; it is not a simple matter to spin down a star while it is accreting (§8.12).

Given all these arguments in favor of magnetospheric accretion, it is clear that this model is to be preferred over the old boundary layer picture for the accreting CTTS. The magnetospheric model may not be applicable at high disk accretion rates, which may overwhelm the stellar magnetic field; but, as shown in Chapter 7, other effects come in to play at high mass accretion rates which can invalidate the simple boundary layer model.

Assuming that the hot continuum emission of CTTS is produced by the accreting gas as it shocks at the stellar surface, the steady-state, hot continuum luminosity is

$$
L_{hot} \approx \frac{GM\dot{M}}{R_*} \left(1 - \frac{R_*}{R_m} \right),
\tag{6.15}
$$

while the disk luminosity is*

$$
L_d \approx \frac{GM\dot{M}}{2R_m} + L_{diss} + L_{irrad}.
\tag{6.16}
$$

Here R_m is once again the magnetospheric radius, where the stellar magnetic field truncates the disk, L_{diss} is the energy dissipated by the stellar magnetic fields passing through the disk if the angular velocity of the magnetic field lines differs from the angular velocity of disk material, and L_{irrad} is the luminosity due to heating by the central star's radiation. The magnitude of L_{diss} is subject to substantial theoretical uncertainty. At one extreme, if the star is not rotating, and disk material is spun down to match the star before beginning to fall in along the magnetospheric field lines, then $L_{diss} = GM\dot{M}/2R_m$, i.e. the equivalent result for the boundary layer (§5.6) but with the magnetospheric radius now playing the role of the inner disk radius. If the inner disk radius is not too far from co-rotating with the stellar magnetic fields, the amount of energy dissipated in the disk is likely to be modest (Kenyon, Yi, & Hartmann 1996). For typical T Tauri stars, the corotation radius is probably $R_i \approx 3$–$5\ R_*$, in which case, to first order most of the energy of accretion is radiated in hot continuum emission, not the disk.

Thus, it would seem that all that is needed to measure mass accretion rates is to measure the hot continuum luminosity, knowing the stellar mass and radius. In practice this is not very easily done, because much of the energy release is thought to emerge at ultraviolet wavelengths, necessitating (a) large and uncertain extinction corrections and (b) extrapolations to wavelength regions not generally observed (i.e. below the atmospheric cutoff).

To consider the first of these problems, even in lightly-reddened, low-density star-forming regions like Taurus, most T Tauri stars have estimated extinctions $A_V \sim 1$

---

* The exact amount of energy dissipated in disk and accretion shock depends upon the amount of energy which goes into spinning up (or down) the star, the fraction which goes into a wind, and the potential dissipation of energy as the magnetic field interacts with the disk. The boundary conditions at the inner edge of the disk are uncertain and affect the total disk luminosity (cf. §5.4).

mag. Using a typical interstellar extinction law (Mathis 1990; cf. also Figure 4.1), the extinction at $0.33\,\mu m$ is ~ 1.65 mag, i.e. a correction of almost a factor of five is needed to convert from the observed fluxes to the true values. If the *intrinsic* colors (relative fluxes as a function of wavelength) of the star are known, for example, by using photospheric line ratios to determine the spectral type or effective temperature, and then adopting the colors or fluxes of a standard star of the same spectral type, then the difference between the observed colors and the standard star's colors yields an estimate of the interstellar extinction (which reddens the star as it makes it fainter; Figure 4.1 (Mathis 1990)). Unfortunately, the optical colors of the star are not known *a priori* because the emission is not simply that of a normal stellar photosphere; the hot continuum makes the system fluxes bluer. If this extra hot emission is not accounted for, the extinction will be underestimated, resulting in an underestimate of the ultraviolet continuum luminosity. Unfortunately, it is not easy to account for the hot continuum; the range of emitting temperatures is not known, and the emission cannot be approximated by a simple single-temperature blackbody.

The most successful way of estimating the shape of the hot continuum spectrum uses simultaneous or nearly-simultaneous high-resolution optical spectroscopy plus intermediate resolution spectrophotometry. If the effective temperature or spectral type of the star is known, and if the excess emission is a pure continuum, then the observed depth of the photospheric lines from high-resolution spectra can be used to determine the ratio of photospheric to excess emission (Hartigan *et al.* 1991). The stronger the excess continuum relative to the stellar photosphere, the less deep the absorption lines appear to be. From measurements of the photospheric line 'veiling', one has an estimate of the hot continuum as a function of wavelength relative to the stellar photospheric emission. If then one further assumes that the stellar photosphere is that of a normal star for the given spectral type or effective temperature, so that the optical photospheric spectrum is known, the ratios derived from high resolution spectroscopy can be used in concert with the spectrophotometric fluxes to produce an observed emission spectrum for the hot continuum. The remainder is the stellar photospheric spectrum; this latter component can then be dereddened to produce a better estimate of the dust extinction. Finally, this extinction correction can be applied to the observed emission of the hot component.

One then must make corrections for the emission at unobserved wavelengths. Crude estimates have been made by trying to model the derived optical excess continuum in terms of hot gaseous slabs emitting at a constant temperature and density. Typical model results suggest temperatures $\sim 10^4$ K, electron densities $\sim 10^{14}\,cm^{-3}$, implying that about half of the hot continuum luminosity emerges at wavelengths below the atmospheric cutoff at ~ 3100 Å (Hartigan *et al.* 1991). Unfortunately, although the extrapolation to short wavelengths appears reasonable, the method of fitting is physically suspect. The region where the 'hot continuum' radiation is produced must have substantial temperature and density gradients. Gas falling in at ~ 200–300 km s^{-1} will initially shock and be heated to temperatures $\sim 1 - 3 \times 10^6$ K. This hot gas is very optically thin, and it will radiate energy away into the column of accreting gas on the one side and into the star on the other. Because the regions on either side have appreciable optical depth, the original radiation field will be modified substantially. To the extent that the hot continuum emission is produced in regions of high optical depth, it may be approximated as blackbody radiation with a

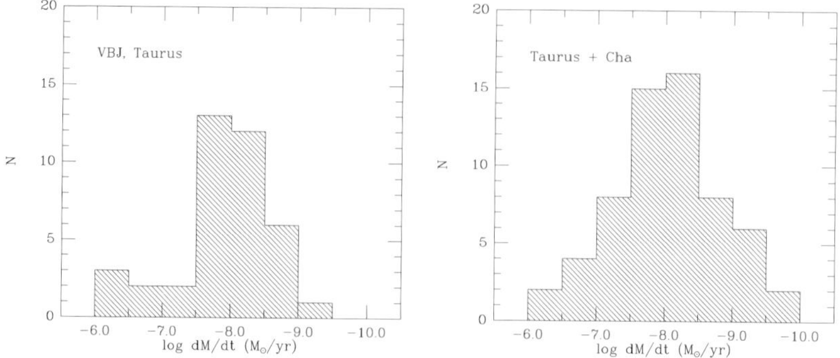

Fig. 6.8. Estimated accretion rates for T Tauri stars. The left-hand panel shows results from Valenti *et al.* (1993) for CTTS mostly in the Taurus cloud, while the right-hand panel illustrates the results from Gullbring *et al.* (1997) and Hartmann *et al.* (1998) for stars in the Taurus and Chamaeleon I regions. Upper limits to mass accretion rates for WTTS are generally $\sim 10^{-9}\,M_\odot\,\mathrm{yr}^{-1}$. See text.

characteristic temperature of

$$T_{hot} \sim \left(\frac{GM\dot{M}}{R_* A \sigma} \right)^{1/4}, \qquad (6.17)$$

where A is the area of the accreting material. There is presently no theory which explains how much of the star should be covered by accreting material, although clearly it is some modest fraction of the star. It is plausible that this material covers a few per cent of the stellar surface, comparable to the area predicted by the boundary layer model, and therefore the temperatures of the hot continuum are similar in either the magnetospheric or boundary layer pictures. Accretion shocks covering 5% of the stellar surface at $\dot{M} = 10^{-7}\,M_\odot\,\mathrm{yr}^{-1}$ would result in a characteristic temperature of emission of ~ 8000 K, consistent with estimates of hot continuum properties in CTTS. This temperature is vastly lower than that of the initial temperature of the shocked gas, demonstrating that large temperature gradients must be present and may need to be considered in understanding the emergent radiation field.

It is clear that present estimates of hot continuum luminosities are subject to substantial uncertainties, and have only been applied to a small number of stars where detailed information is available. The largest samples which include Balmer continuum emission are those of Valenti *et al.* (1993) for about 40 stars in the Taurus-Auriga molecular cloud, and Gullbring *et al.* (1997) and Hartmann *et al.* (1998) for approximately 60 stars in the Taurus and Chamaeleon I regions. The rates, as shown in Figure 6.8, scatter widely around a median accretion rate $\approx 10^{-8}\,M_\odot\,\mathrm{yr}^{-1}$ (although Hartigan *et al.* (1995) find much larger values from observations weighted toward Paschen continuum measurements).

In general, one would expect that the disk masses should be of the order $\sim \dot{M}t_*$, where t_* is the age of the star. For example, in the special case of the similarity solution discussed in §5.2, the mass accretion rate in the inner disk varies as $\dot{M} \propto t^{-3/2}$, and so the mass remaining in the disk at time t_* is related to the mass accretion rate at

t_* by

$$M_d = \int_{t_*}^{\infty} dt\, \dot{M} = 2\dot{M}(t_*)t_*. \qquad (6.18)$$

For the sample of objects shown in Figure 6.8, the median mass accretion rate is $\dot{M} \sim 10^{-8}\,M_\odot\,\mathrm{yr}^{-1}$ and the median age is $t_* \sim 5 \times 10^5$ yr (Kenyon & Hartmann 1995); thus, application of the similarity solution would suggest that the median disk mass in this sample is approximately $M_d \approx 0.01\,M_\odot$, comparable to that estimated from the mm-wave dust continuum measurements (§6.3).

It is also useful to consider the constraints implied by Figure 6.3. For typical stellar parameters $M_* = 0.5\,M_\odot$, $R_* = 2\,R_\odot$, the predicted disk accretion luminosity for a median mass accretion rate of $10^{-8}\,M_\odot\,\mathrm{yr}^{-1}$ is $L_d(acc) = GM_*\dot{M}/2R_* \sim 0.08\,L_\odot$. This value is consistent with the modest infrared excesses generally found in CTTS (Figure 6.3), and suggests that the disk emission in most CTTS is powered by irradiation, especially if inner disks are truncated by magnetospheres.

The situation is different when the accretion luminosity greatly exceeds that of the central star, as occurs with the most-rapidly accreting T Tauri stars, or the FU Ori objects (Chapter 7). In such cases irradiation from the central star can be ignored, and accretion rates can be inferred from disk luminosities directly, as originally envisaged by Lynden-Bell & Pringle (1974).

Intrinsic time variability of CTTS accretion, which may be responsible for much of the spread shown in Figure 6.8, is poorly understood (Herbst *et al.* 1994; Gahm 1994, and references therein; Safier 1995). Timescales of potential accretion events range from less than one hour to years with amplitudes at the visual band ($\lambda \sim 0.55\,\mu$m) V < 0.05, for the fast events, to several magnitudes in V for longer-term variations (Vrba *et al.* 1993; Gahm *et al.* 1995; Gullbring 1994; Gullbring *et al.* 1996). The variable brightness of the hot continuum may have an analogue in the accreting intermediate polars, which are close binary systems also accreting through magnetospheres (cf. Frank *et al.* 1992).

Herbig (1977b, 1989) has called attention to relatively long-term, substantial variations in the optical emission of CTTS. Typically these 'EXor' outbursts involve increases in optical brightness of a few magnitudes and may last for fractions of a year to decades. For example, DR Tau became brighter by about two magnitudes in the B photometric band ($\lambda \sim 0.45\,\mu$m) in the 1970s and has remained relatively bright since. Because DR Tau is currently a very-strong-emission CTTS (Figures 6.2, 6.6), it is likely that this brightening is due to an increase in the accretion rate, which made the hot continuum emission much brighter. However, at present very little is known about the statistics of such outbursts (see also Chapter 7).

6.5 The WTTS

T Tauri stars were originally recognized as young objects found in regions of star formation with strong emission lines as seen in objective-prism spectra (e.g., Haro, Iriarte, & Chavira 1953; Herbig 1954, 1957; Herbig & Kuhi 1963). Using such low-resolution spectra it was difficult to detect emission at Hα with equivalent widths less than about 5-10 Å. Later, other techniques such as X-ray surveys were used which could find young stars, many with much lower optical emission line strengths than those of the CTTS (Montmerle *et al.* 1983; Walter 1986; André, Montmerle,

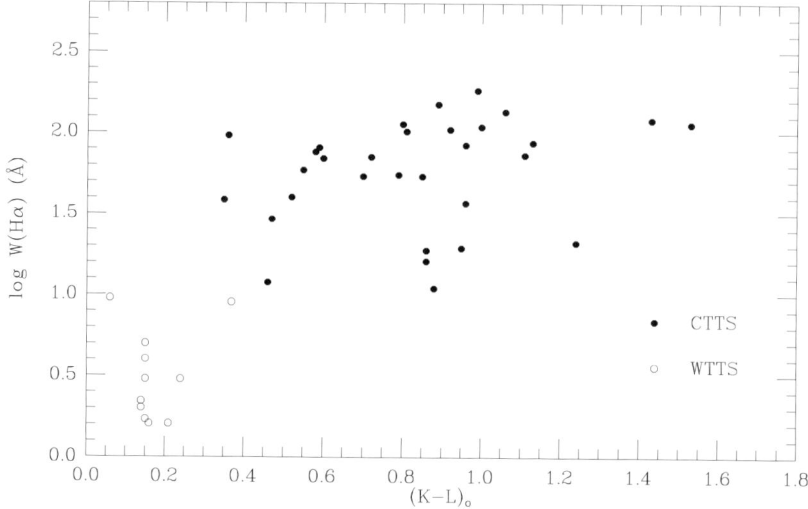

Fig. 6.9. Plot of Hα equivalent widths of T Tauri stars. The photospheric colors of normal stellar photospheres are not expected to exceed $K - L \sim 0.3$ (K $\sim 2.25\,\mu$m; L $\sim 3.4\,\mu$m); thus, only those T Tauri stars with near-infrared excess emission exhibit Hα line emission larger than about $W \sim 10$ Å(see text). Equivalent width data from Herbig & Bell (1988); reddening-corrected $(K-L)_o$ colors from Hartigan *et al.* (1995).

& Feigelson 1987; Feigelson *et al.* 1987; Walter *et al.* 1988; Feigelson *et al.* 1993; Huenemoerder, Lawson, & Feigelson 1994; Walter *et al.* 1994). Extended discussions of this topic are given in Montmerle *et al.* (1993) and Stahler & Walter (1993). Although a few WTTS had been detected in previous optical studies, the X-ray surveys added greatly to the known population.

The nominal definition of a WTTS is a young star with $W_\lambda(H\alpha) < 10$ Å (cf. Herbig & Bell 1988; §1.2), although we shall argue below that this definition should be refined in certain applications. Many of the WTTS exhibit similar stellar luminosities and spectral types as those of the strong-emission CTTS, which have $W_\lambda(H\alpha) \geq 10$ Å. This implies that there are both WTTS and CTTS with the same age (Chapter 9). The young WTTS with ages similar to those of CTTS seem also to be spatially concentrated near molecular material.

The current interpretation of WTTS is that they are not accreting from circumstellar disks; their excess emission is powered by solar-type magnetic energy dissipation, not accretion. The reasons for arriving at this conclusion may be summarized as follows:

First, WTTS do not show near-infrared excess emission characteristic of disks (see also Wolk & Walter (1996)). As shown in Figure 6.9, stars with red $K - L$ colors (large flux ratios of 3.5 μm emission to 2.25 μm emission) expected from disks generally exhibit strong Hα emission. Conversely, stars with $K - L < 0.3$, which corresponds basically to photospheric emission, i.e. no near-infrared excess at 3 μm, are WTTS, with $W_\lambda(H\alpha) < 10$ Å.

Second, WTTS do not show optical hot continuum emission expected for disk accretion. This is shown in Figure 6.10, where objects with $K - L < 0.3$ show no

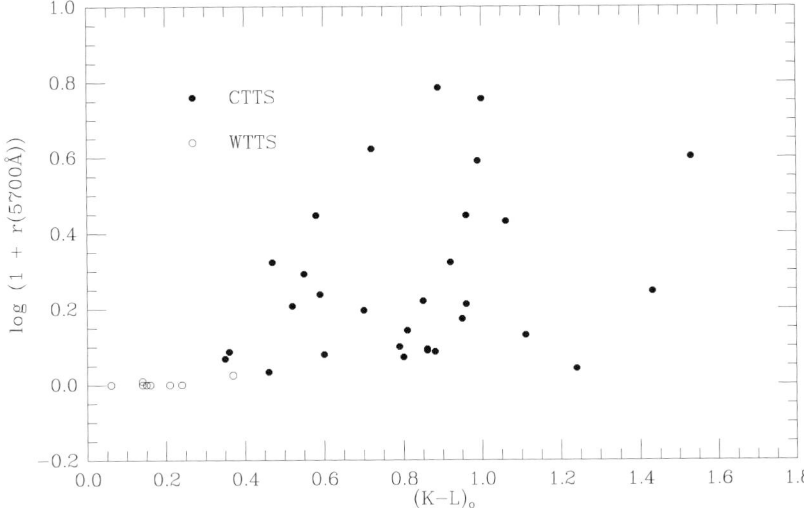

Fig. 6.10. Plot of hot continuum optical emission vs $K - L$ (see Figure 6.9) for T Tauri stars. The hot continuum is presented in terms of r, the ratio of hot continuum to stellar continuum emission at 5700 Å. Data are taken from Hartigan *et al.* (1995).

evidence for hot continuum emission which would 'veil' the normal photospheric spectrum at $0.57\,\mu$m.

Taken together, these results strongly suggest that the difference between WTTS and CTTS is that the latter have disks reaching close to the central star; such disks can be hot enough to emit significantly in the near-infrared, and can accrete onto the star, accounting for the hot continuum.

A further consideration which is not generally applied, but which probably should be taken into account, is the velocity width of the Hα emission line. WTTS generally have narrow Hα lines with FWHM $\lesssim 100\,\mathrm{km\,s^{-1}}$, unless the star is rapidly rotating (cf. DI Tau in Figure 6.7). In contrast, CTTS generally have very large Hα velocity widths, FWHM $> 100\,\mathrm{km\,s^{-1}}$ (Johns & Basri 1995a; Figure 6.7). This observation, derived from the author's own spectra, is completely consistent with the magnetospheric accretion picture; the CTTS have large emission line widths because the emitting gas is falling in at large velocities (from a disk disrupted at several stellar radii). In contrast, the WTTS have *chromospheric* Hα emission, which is formed in a dense layer near the stellar surface and therefore does not have large velocity widths.

In a few 'transition' cases it is difficult to tell whether a star is accreting or not from the Hα equivalent width alone. In the author's experience, no confusion results if one uses the additional criterion of the Hα velocity width for slowly-rotating stars. There appear to be cases where the infall rate in the magnetosphere is low, so the Hα equivalent width is also low, such as IP Tau; this object has an equivalent width ~ 11 Å, even though it has a large near-infrared excess, $K - L \sim 0.9$, and therefore presumably has an inner disk. However, IP Tau is observed to have a broad Hα emission line; in fact, it exhibits an inverse P Cygni profile, with redshifted absorption (Hartmann, Soderblom, & Stauffer 1988), indicating infall at high velocities (Chapter 8).

This point is important for understanding the T Tauri stars of earlier spectral types (F–G), with higher photospheric effective temperatures. Theoretical studies show that larger emission measures of hot gas are required to produce Hα emission as the underlying photospheric emission becomes hotter. This means that the amount of infalling magnetospheric gas which can produce strong Hα emission in a cool M star might end up producing much weaker Hα emission, or even *absorption*, in a hotter G or F star (Cram & Mullan 1985). The criterion of $W_\lambda(\mathrm{H}\alpha) < 10$ Å was originally developed for M stars because these objects constitute the majority of well-studied T Tauri stars, and is useful in this context, but it is not necessarily a good indicator of accretion in stars of earlier spectral types. For example, the G2 star SU Aur clearly has an accretion disk, and yet its Hα equivalent width is only 4 Å; the gas responsible for this emission would be able to produce very strong Hα emission around an M star. On the other hand, SU Aur exhibits very broad Hα emission, spreading over much larger velocities than would be expected if stellar rotation were the only broadening mechanism (Giampapa *et al.* 1993; Johns & Basri 1995b). Stars may be WTTS in terms of a strict spectroscopic definition, but the utility of this definition to distinguish physically-different situations is limited.

The chromospheric and coronal emission of WTTS is strong, but in the same range as observed in other stars of relative youth, such as the young main sequence stars in the Pleiades or α Persei clusters, which have ages $\sim$ 50–100 Myr (Randich *et al.* 1996; Stauffer *et al.* 1994). WTTS may have even stronger coronal X-ray emission than CTTS (Neuhauser *et al.* 1995). However, these emission levels are small fractions of the stellar luminosity, and are easily distinguished from the much larger energy release of disk accretion at ages of $\sim$ 1 Myr.

Roughly 40% or so of the known T Tauri stars with ages less than about 3 Myr in Taurus are WTTS, i.e. nearly half of the pre-main-sequence population show no evidence for accretion disks or near-infrared excess emission from disks. Why do some million-year-old stars have large circumstellar (accretion) disks while others do not? Although the situation is not entirely clear, it is likely that many, if not most, of the WTTS have binary stellar companions. Most of the spectroscopic binaries known among the T Tauri stars, i.e. most of the objects with stellar companions close enough to cause substantial velocity shifts, are WTTS (e.g., Mathieu 1994). Moreover, there is a tendency for binaries with separations less than about 100 AU to have lower mm-wavelength dust emission (Jensen *et al.* 1994, 1996; Osterloh & Beckwith 1995).

Some binaries do exhibit disk emission at either infrared or mm wavelengths, or even disk accretion (see Mathieu (1995) for a fuller discussion). However, the effects of binary companions on circumstellar disks are complicated. The gravitational attraction of a companion star will prevent stable orbits of other particles on size scales comparable to that of the binary orbit. Typically it is estimated that orbits within a factor of three or so smaller or larger than the binary semi-major axis will be disrupted (Artymowicz & Lubow 1996a). Thus, the action of a companion star will be to cause any disk material to fall in rapidly or be ejected outward. Wide companions with orbits $\gg$ 100 AU are unlikely to disrupt disks on scales such that accretion is unimportant or mm-wave fluxes are unobservably low; in essence, the situation is that of a single star with a small external gravitational perturbation. A close companion at a distance $\ll$ 10 AU would not disrupt an external disk on scales $\gtrsim$ 100 AU; thus, one might have a circumbinary disk. The close companion will tend to prevent the circumbinary disk from accreting onto one or both of the binary system (although under certain conditions accretion may proceed (Artymowicz & Lubow 1966b)). In the absence of material from the circumbinary disk penetrating across the binary orbit, any inner disk much smaller than the binary orbit may accrete on a relatively rapid timescale (because material from disk regions outside of the orbit, where much of the disk mass may reside, may not be able to accrete past the binary companion).

From the above discussion it seems quite possible that binary companions with orbits comparable to typical disk sizes ~ 100 AU will tend to disrupt typical disks. Close binaries, with orbits much smaller than this, may have circumbinary disks but may not be able to accrete from these disks. These trends may be confused in individual objects by accretion overpowering binary companions, or by the finite timescales for disk clearing, but should serve to outline general expectations.

The observed distribution of binary companion stars among solar-type stars peaks at periods of $\sim 6 \times 10^4$ days, or semi-major axes ~ 30 AU (Duquennoy & Mayor 1991). Approximately 30% of all stars have binary companions at least as close as this. There is some evidence that the binary frequency may be even higher among young stars in Taurus than in the field population (Simon *et al.* 1995; Ghez *et al.* 1993; Leinert *et al.* 1993; Reipurth & Zinnecker 1993; Mathieu 1994). Thus, it seems quite likely that the perturbing effects of binary companions can account for the lack of disk accretion, and lower-mass external disks, in many, if not most, WTTS.

6.6 Infalling envelopes and flat spectrum sources

Some T Tauri stars exhibit very much flatter infrared spectra than are easily explained by disk emission. There is a small class of optically-visible T Tauri stars called 'flat spectrum' sources, which have spectral indices $s \sim 0$ between ~ 3 and $100\,\mu$m. The classic example of this group is HL Tau (Figure 6.11). As discussed in §6.2, a spectral index near zero, which is much different than that exhibited by most T Tauri stars (compare GG Tau in Figure 6.11), requires a temperature power law index $q \sim 1/2$ *throughout* the disk, not just in its outer regions (§6.4). Such temperature distributions cannot be obtained through irradiation unless the disk flares very strongly over a very large range of distance scales. This is very difficult to achieve for reasonable parameters (Adams *et al.* 1988). Alternative proposals for obtaining such flat temperature distributions have included disk density waves, dusty winds, or some kind of optically-thin dusty envelope (see §6.2).

However, if one simply compares the SED of HL Tau to a Taurus Class I object of comparable luminosity, IRAS 04016+2610, the similarity is immediately evident (Figure 6.11). The comparable properties of these objects extend beyond the infrared SEDs. Both 04016+2610 and HL Tau are completely hidden by dusty nebulae at optical wavelengths, and can only be observed in scattered light in the optical spectral region (Figure 6.12). This comparison suggests that the far-infrared emission of HL Tau might also arise from an infalling dusty envelope rather than a disk. Indeed, infall was independently suggested in HL Tau on the basis of scattered light analyses and high-resolution spectroscopy (Beckwith *et al.* 1989; Grasdalen *et al.* 1989), and from interferometry at radio wavelengths (Hayashi *et al.* 1993). The spatially-resolved mapping in ^{13}CO by Hayashi *et al.* suggested that an envelope of radial extent ~ 2000 AU was mostly falling in, with a small amount of rotation. The mass mass infall rate found for this CO cloud was very similar to that found for typical Taurus Class I

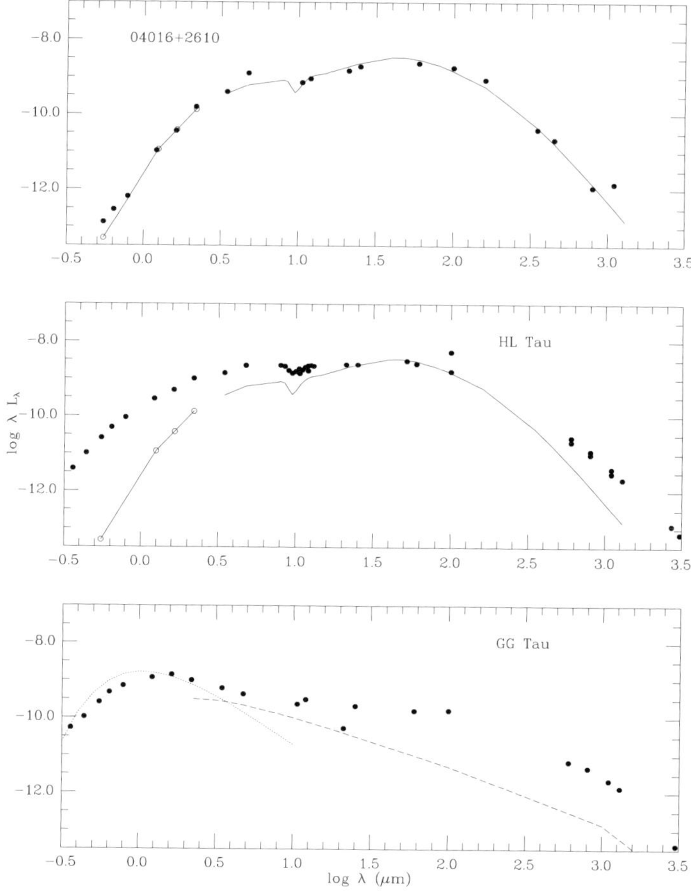

Fig. 6.11. SEDs of three representative young objects in Taurus: the Class I or protostellar source IRAS 04016+2610 (Figures 1.3, 4.9), and the 'flat-spectrum' source HL Tau, compared with the CTTS GG Tau. The SED of HL Tau clearly resembles that of the Class I source, but is qualitatively different from the 'disk' excess of the T Tauri star. The solid line connected by open circles is a calculation which includes scattering of short-wavelength light out of an envelope hole (§4.5). The model calculated for 04106+2610 (Figure 4.7) is repeated again in HL Tau to suggest that the basic envelope model can explain the latter source as long as additional escape of scattered light occurs at short wavelengths; at long wavelengths, there is an excess due to a small dense structure on scales $\sim$ 100 AU which probably is the disk (Lay *et al.* 1994; Mundy *et al.* 1996). The two curves in GG Tau are schematic representations of star and disk spectra. Modified from Calvet *et al.* (1994).

sources from modelling SEDs (§4.3). Detailed fitting of the HL Tau SED (Calvet *et al.* 1994; Hartmann *et al.* 1996) showed that the overall infrared emission plus the scattered light nebula of HL Tau could be matched quite well with the median infall rate of $\dot{M} \sim 4 \times 10^{-6}\,\mathrm{M_\odot\,yr^{-1}}$ typical of Taurus protostellar sources.

These results suggest that the flat-spectrum sources are generally protostellar-like sources, with dense infalling envelopes. Although HL Tau is optically visible, it is seen only in scattered light, as demonstrated by its large polarization and its extended

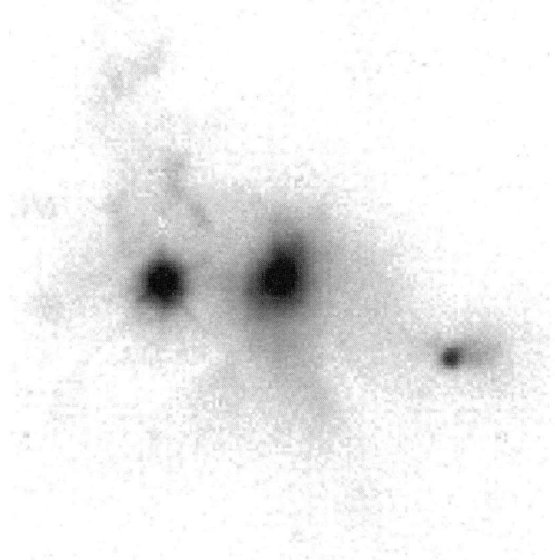

Fig. 6.12. Optical image of the HL Tau region. The distance between HL Tau (central source) and XZ Tau (left center source) is about 23 arcsec $\sim$ 3200 AU. HL Tau is not a point source, but is seen only in light reflected from its dusty envelope. Compare the triangular reflection nebula of HL Tau with the image of 04016+2610 in Figure 4.8. CCD image courtesy of C. Briceño.

optical scattering nebula, where no central point source can be discerned (Stapelfeldt *et al.* 1995). This suggests that if the HL Tau dusty envelope did not have hole(s) in it, or if we were positioned at a different viewing angle, it might have been optically invisible and thus identified as a Class I source through infrared surveys rather than a T Tauri star.

At sub-mm and mm wavelengths, HL Tau exhibits very strong compact emission on a size scale $\sim$ 100 AU which is almost certainly due to the Keplerian disk (Lay *et al.* 1994; Mundy *et al.* 1996). The disk may well have a nearly flat-spectrum temperature distribution, due to the backwarming effect of the infalling envelope (e.g., Keene & Masson 1990; Natta 1993; Butner *et al.* 1994; D'Alessio, Calvet, & Hartmann 1997). The opaque envelope acts like a blanket, with a tendency to equalize the disk temperature with the local envelope temperature. The resulting extra disk heating produces higher temperatures which accounts for the strong excess emission at wavelengths $\gtrsim$ 100 μm, while the envelope probably dominates the system emission between $\sim$ 3 and 100 μm (D'Alessio *et al.* 1997).

The scenario of flattened collapse discussed in §4.5 suggests that, as collapse proceeds, the infalling envelopes of protostars should become even more flattened, with larger cavities through which short-wavelength radiation can escape. In this picture, the flat-spectrum sources, which are optically visible, may generally be somewhat 'older' protostars, where much of the collapse has already occurred, revealing the central accreting T Tauri star and disk over a wider range of viewing angles than possible earlier (Calvet *et al.* 1994).

In some other cases, the flat spectrum appears to be the result of combining the

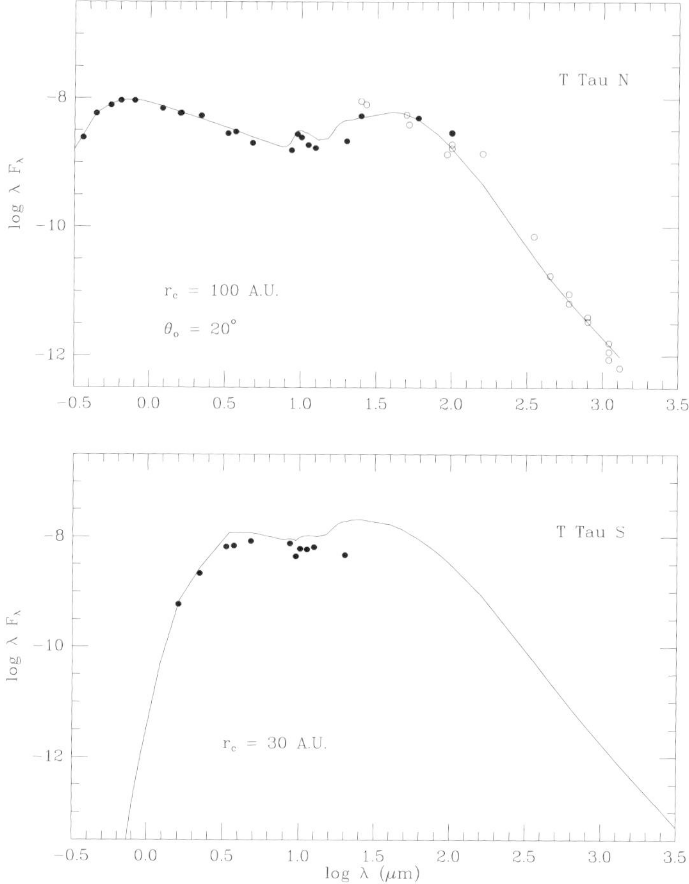

Fig. 6.13. The T Tau binary system. The measurements from $\sim$ 1–20 μm are obtained by speckle interferometry (Ghez *et al.* 1991). It is assumed that the optical fluxes are dominated by the northern component. It is not possible to distinguish between sources at wavelengths $> 20\,\mu$m. The curves represent models including infalling envelopes. In both cases, the models have mass infall rates $\dot{M} = 2 \times 10^{-6}(M/2\ \mathrm{M_\odot})^{1/2}\,\mathrm{M_\odot}\,\mathrm{yr}^{-1}$ (the radiative transfer models depend directly only upon the density; the mass of the companion is not known, but the optical star has a mass $\sim 2\ \mathrm{M_\odot}$). For T Tau N, the model assumes that we are viewing the object along a *hole* in the envelope; the excess emission between 2 and 10 μm is due to a disk, while the emission at longer wavelengths, including the silicate emission 'bump' at 10 μm, is due to the infalling envelope. The model for the southern component assumes no envelope hole. Fitting the two components requires different centrifugal radii in the models. which may reflect differing distances of the (orbiting) stars from the edges of what is presumably a common infalling envelope. Modified from Calvet *et al.* (1994).

emission of an optically-visible T Tauri star with that of a heavily-extincted, possibly protostellar, close companion. The classic case of this is T Tauri itself, which is a binary system of projected separation = 0.6 arcsec $\sim$ 100 AU. The total system spectrum appears to be of flat-spectrum type. However, as shown in Figure 6.13, the actual spectrum of the optical (northern) object falls off much more rapidly with increasing wavelength. The extra infrared emission comes from the companion, which is optically-invisible and but is very much brighter at wavelengths $> 1\,\mu$m (Ghez *et al.* 1991).

The spectrum of the southern infrared component is reminiscent of the embedded protostellar sources in Taurus (Chapter 4; Calvet *et al.* 1994), although it is not possible to apportion the far-infrared fluxes measured in large beams between the two sources. One might suggest that the southern source is still surrounded by an infalling envelope like the other embedded Taurus sources. However, as discussed in Chapter 4, the envelopes of such sources are very likely to be much larger than 100 AU in extent, so it is difficult to imagine that such an envelope could be around only one of the sources. Instead, the suggested infalling envelope would have to surround both stars, with a clearer line-of-sight path to the optical component. This requires a complicated envelope geometry near the stars, but this may not be too surprising, given that there are two centers of gravitational attraction in the system (e.g., Bonnell & Bastien 1992).

7

The FU Orionis objects

In the 1970s special attention was drawn to the FU Orionis systems, a small but remarkable class of variable YSOs. As summarized by Herbig (1977b) in his Russell Lecture, FU Ori variables undergo outbursts in optical light of 5 mag or more. With the discovery of three additional objects in that decade, many observational studies were made in an effort to understand the nature of these highly variable young objects. Several theoretical explanations were advanced for the light outbursts, including the possibility that some circumstellar material had been accreted onto the young star (Herbig 1977b; Larson 1980, 1983).

It soon became apparent that FU Ori objects exhibit substantial near-infrared excess emission (e.g., Cohen & Woolf 1971; Rieke, Lee, & Coyne 1972; Simon *et al.* 1972; Grasdalen 1973; Simon 1975). With our current understanding, this immediately suggests the presence of a disk; but the picture was confused by the optical and infrared spectra, which looked much like stellar spectra, with atomic and molecular absorption features. It was also puzzling that while the optical F-G supergiant spectral types of the FU Ori objects correspond to effective temperatures $\sim 7000 - 6000$ K, the near-infrared molecular absorption bands were typical of M supergiants with much lower effective temperatures ~ 3000 K (Mould *et al.* 1978). Mould *et al.* suggested that the FU Ori objects might be extremely rapidly-rotating stars; in this model the near-infrared spectra would be produced in the cooler, rapidly-rotating equatorial regions of the highly-flattened star, while the hotter polar caps would be responsible for the optical emission. However, the amount of flattening required to make this model work is implausibly large; furthermore, we now know that the cool infrared-emitting regions are rotating *slower*, not faster, than the optical photospheres (Hartmann & Kenyon 1987a,b).

Outbursting behavior had been recognized in close binary systems, and interpreted as the result of some kind of instability in accretion disks. Researchers who studied such accreting binary systems were the first to suggest that FU Ori outbursts were similarly the result of some temporary rapid increase in the rate of mass accretion in a circumstellar disk (Paczynski 1976; Lin & Papaloizou 1985; Hartmann & Kenyon 1985). The key to making sense of the observations was the realization that, for a high enough disk accretion rate, the disk itself can shine strongly in the optical spectral region, completely overwhelming the stellar photospheric light during outburst. The peculiar FU Ori spectrum is naturally explained by an accretion disk model; the outer cooler regions of the disk contribute more strongly at longer wavelengths, producing precisely the variation in spectral type with wavelength observed.

Although the disk interpretation seems evident in retrospect, it is difficult to overestimate the general reluctance to believe that a disk spectrum could look so much like that of a normal star with strong absorption lines, at least when observed at low spectral resolution. This similarity arises because both the stellar atmosphere and the FU Ori disk atmosphere have accidentally similar effective temperatures and surface gravities. In addition, both regions passively transmit energy generated in the deep interior; this produces the outwardly decreasing temperature gradient needed to produce an absorption line spectrum.

High-resolution spectra confirm the difference between FU Ori objects and normal stars. Double-peaked absorption line profiles characteristic of a rapidly-rotating disk are observed at both optical and infrared wavelengths (Hartmann & Kenyon 1985, 1987a,b). Moreover, the slower rotation observed at infrared wavelengths is precisely what is predicted by a differentially-rotating Keplerian disk model, and opposite of what would be observed if the cool long-wavelength spectrum originated in the equatorial regions of a rapidly-rotating star.

Because the rapid accretion rates of FU Ori disks produced emission at easily-observed optical wavelengths, the first detailed application of accretion disk models for pre-main-sequence objects was made for FU Ori objects (Hartmann & Kenyon 1985). The FU Ori systems, though rare, are especially important for our current understanding of pre-main-sequence disks; the time-variability provides constraints on disk viscosities, while the large accretion energy release has important implications for mass ejection (Chapter 8) and for the evolution of the central star (Chapter 9).

Using equation (5.74) for the maximum disk temperature, and adopting fiducial parameters typical of T Tauri stars,

$$T_{max} \sim 6500 \, M_{0.5}^{1/4} \, \dot{M}_{-5}^{1/4} R_2^{-3/4} \, \text{K}, \tag{7.1}$$

where $M_{0.5}$ and R_2 are the central star mass and radius in units of $0.5 \, \text{M}_\odot$ and $2 \, \text{R}_\odot$, respectively, and $\dot{M}_{-5}$ is the mass accretion rate in units of $10^{-5} \, \text{M}_\odot \, \text{yr}^{-1}$. Thus, the optical emission from FU Ori disks requires a much higher accretion rate than typical of T Tauri stars. Observations suggest that the inner disk radii of FU Ori objects are a factor ~ 2–3 larger than the value adopted above, requiring mass accretion rates closer to $\sim 10^{-4} \, \text{M}_\odot \, \text{yr}^{-1}$ to explain the observed spectra and accretion luminosities.

FU Ori objects are rare because outbursts at the peak accretion rates $\sim 10^{-4} \, \text{M}_\odot \, \text{yr}^{-1}$ cannot last for long. On the other hand, the frequency of FU Ori outbursts appears to exceed the expected frequency of star formation in the solar neighborhood by a factor of five or more; thus, at least some stars must have repetitive outbursts, though it is not known whether most low-mass stars undergo FU Ori outbursts. The outbursts are thought to be short-lived phenomena, lasting $\sim 10^2$ yr; however, this is quite uncertain, since no known FU Ori has actually yet returned to its pre-outburst state.

The total current mass accretion rate for the known FU Ori objects in the solar neighborhood is estimated to be ~ 5–10% of the rate at which interstellar matter is being converted into stars. This suggests that FU Ori events are responsible for accreting a modest fraction of the total mass of low-mass stars. However, it is quite likely that the present census of FU Ori objects is incomplete, due to difficulties in detecting heavily-extincted sources. Large-scale infrared surveys are needed to understand the true place of FU Ori objects in early stellar evolution.

It is not clear why disk accretion rates are so variable, although there are some promising hypotheses, involving intrinsic disk thermal instability, perturbations by companion stars or planets, or possibly a combination of both. FU Ori outbursts appear to be a phenomenon of the earliest stages of stellar evolution, consistent with the original suggestion by Lynden-Bell & Pringle (1974) that protostellar disks may outshine their central stars at very early ages. The large extinctions and far-infrared emission of many FU Ori objects suggests that they are still experiencing infall from protostellar envelopes. Mass infall to the disk may pile up material until the disk can adjust itself in a violent manner to rid itself of excess material by accreting onto the central star.

The material in this chapter draws on the review by Hartmann & Kenyon (1996) and other references listed therein, including reviews by Herbig (1966, 1977b, 1989) and Reipurth (1990).

7.1 Basic observational properties

FU Ori objects were originally identified by their large outbursts in optical light (Herbig 1966, 1977b; Kolotilov & Petrov 1983, 1985). Figure 7.1 illustrates optical light curves of the three best-studied objects, which all exhibit large increases in optical brightness of ~ 4 mag or more and remain luminous for decades. Though similar in many respects, differences in the light curves suggest that these objects are not indentical. The two best-studied objects, FU Ori and V1057 Cyg, exhibited very short rise times to maximum light (~ 1 yr), while the rise of V1515 Cyg to maximum has taken more than a decade. The decay timescales can differ dramatically; FU Ori shows a slow decline with an e-folding time ~ 50–100 yr; V1057 Cyg initially faded about ten times faster than FU Ori, but has slowed its rate of decay in recent years; and V1515 Cyg shows no real evidence of becoming fainter yet.

The FU Ori objects are clearly young systems. They are all spatially and kinematically associated with star-forming regions, and all have reflection nebulae (cf. Goodrich 1987; Figures 7.2, 7.3). FU Ori systems are often heavily extincted, and all have large infrared emission excesses (Weintraub, Sandell, & Duncan 1991). Furthermore, all FU Ori objects exhibit strong Li I $0.6707\,\mu$m absorption characteristic of young stars.

FU Ori systems exhibit distinctive spectroscopic properties. Optical spectral types are late F to G (effective temperatures ~ 7000–6000 K), and line ratios indicate surface gravities much lower than those typical of T Tauri stars. Broad blueshifted absorption is observed in the Balmer and Na I resonance lines (Chapter 8). Infrared spectra of FU Oris show strong CO absorption at $2.2\,\mu$m and water vapor bands in the near-infrared (~ 1–$2\,\mu$m) region, characteristic of K-M giant-supergiant atmospheres (effective temperatures ~ 3000 K). When observed at high spectral resolution, FU Ori objects are generally seen to be rapidly rotating, and exhibit double-peaked absorption line profiles (§7.2).

A pre-outburst spectrum is available for one FU Ori object (V1057 Cyg), showing the Balmer, Ca II, Fe I, and Fe II emission lines characteristic of an accreting T Tauri star (Herbig 1977b; cf. Figure 6.6). During the outburst of V1057 Cyg, most of the emission lines disappeared. Immediately after the outburst, the initial spectral type was early A ($T_{eff} \sim 9000$ K); the spectrum became later (cooler) as V1057 Cyg faded

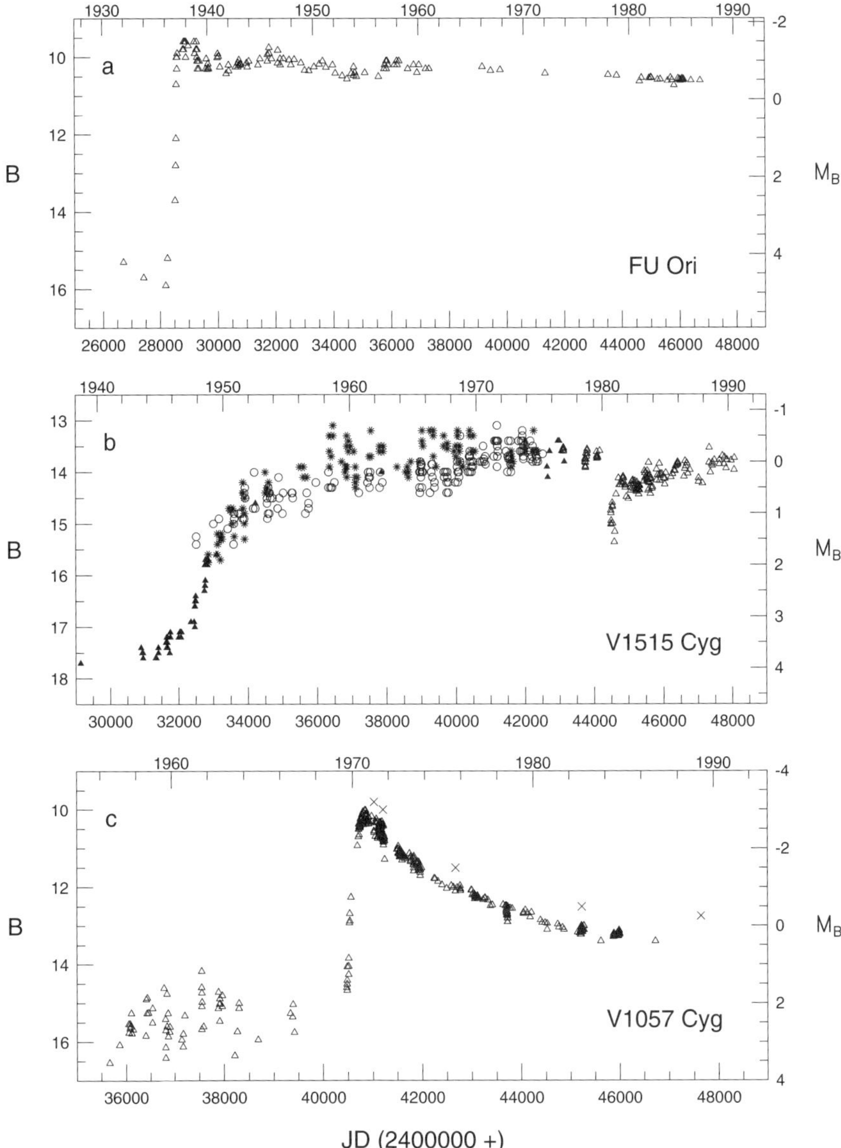

Fig. 7.1. Optical (B) photometry of outbursts in three FU Ori objects. From Bell *et al.* (1995), where references are given.

from maximum light. These spectral changes demonstrate that the outburst was not caused by the removal or dispersal of obscuring dust.

Some objects are known which are spectroscopically similar to the first three identified FU Ori objects (Figure 7.1), but for which outbursts have not been observed. Many of these objects are heavily extincted, making it difficult or impossible to detect historical outbursts even if they had occurred. For example, the optical spectrum of the prototypical Class I source in Taurus, L1551 IRS5 (Chapter 4; Figure 4.7), which

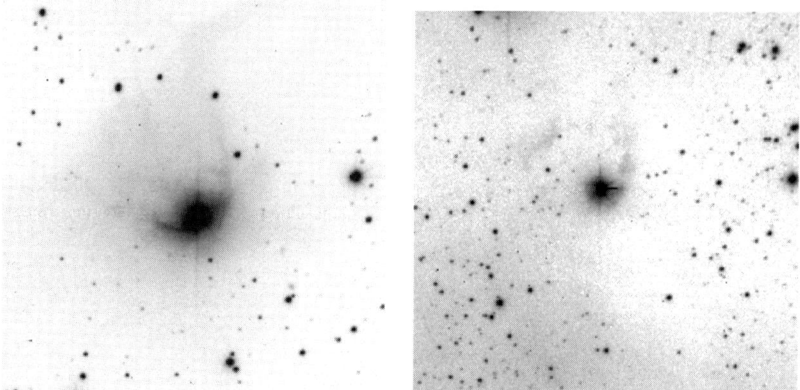

Fig. 7.2. FU Ori (left) and V1057 Cyg (right) at optical wavelengths. Courtesy C. Briceño.

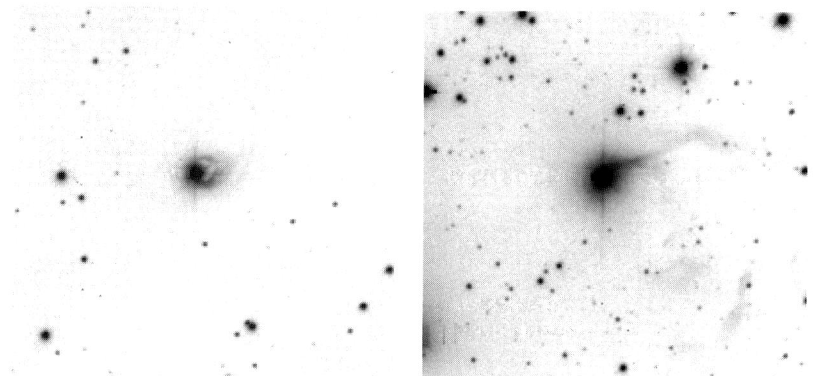

Fig. 7.3. V1515 Cyg (left) and Z CMa (right) at optical wavelengths. Courtesy C. Briceño.

can only be observed in scattered light, is typical of FU Ori objects (Stocke *et al.* 1988). In addition, the 2 μm CO absorption of this source is also typical of FU Ori systems (Carr, Harvey, & Lester 1987). As discussed in the next section, these spectroscopic properties are consistent with a luminous disk; when considering issues beyond the narrower question of optical outbursts, it is important to include such heavily-extincted objects when considering rapid pre-main-sequence disk accretion.

7.2   The steady accretion disk model

FU Ori objects have been identified as accretion disks because many properties can be explained with the predicted emission of steady optically-thick disks (§5.4). Although it may seem surprising that outbursting objects can be treated as steady disks, most investigations of the best-studied objects FU Ori and V1057 Cyg have been conducted during epochs considerably after outburst, when the objects were much less variable and therefore the changes in the mass accretion rates were slower. Furthermore, most of the detailed observations used to test the disk model have been made in the optical and near-infrared spectral regions, which probe only a limited region of the inner

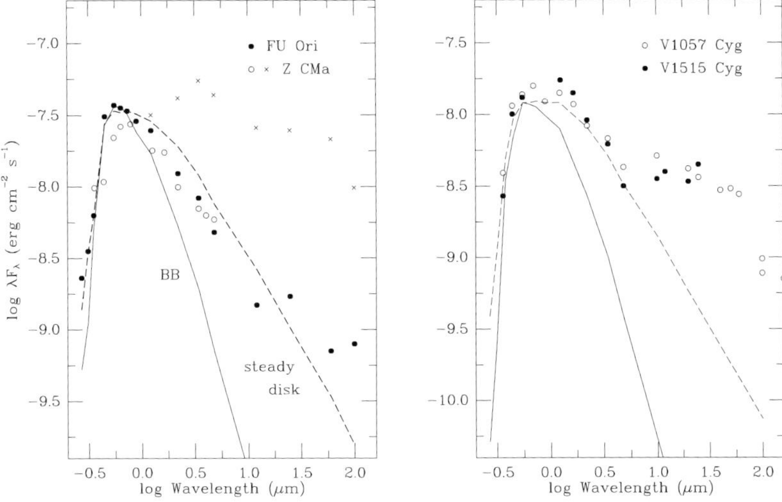

Fig. 7.4. Dereddened SEDs of four FU Ori objects, compared with single temperature blackbody (stellar photosphere) distributions (solid line) and steady disk models (dashed lines). The open circles denote the SED of the optical primary of the Z CMa binary (Koresko *et al.* 1991). From Hartmann & Kenyon (1996).

disk. Thus, the available data do not span a large enough region of time and physical extent to be very sensitive to time-dependent effects.

The principal tests of the disk hypothesis employ the standard steady, optically-thick accretion disk temperature distribution (equation (5.73)),

$$T_d^4 = \frac{3GM_* \dot{M}}{8\pi\sigma R^3}[1 - (R_i/R)^{1/2}], \qquad (7.2)$$

where R_i is the inner disk radius. As discussed in Chapter 5, the term in square brackets depends upon the choice of inner boundary condition; this particular form may well not be appropriate for FU Ori objects, as discussed in §7.6, but we ignore this complication for the present.

The observed SEDs of four FU Ori objects are compared with steady optically-thick disk spectra in Figure 7.4. The disk models agree reasonably well with the observations at wavelengths $\lambda \lesssim 10\,\mu$m); the observed emission cannot be explained by a single-temperature blackbody or normal star. At wavelengths longer than 10 μm, the SEDs of V1057 Cyg and V1515 Cyg lie far above the disk model. This long-wavelength excess emission probably arises from a circumstellar dust shell powered by light absorbed from the central regions (§7.4). The SED of Z CMa seemed quite peculiar until it was recognized that this object is *binary*; the optically-visible FU Ori object has a disk-like infrared excess, while the optically-invisible, infrared bright companion has a SED suggestive of protostellar envelope emission (Koresko *et al.* 1991; Haas *et al.* 1993).

The apparent temperature of the emitting gas in a steady accretion disk varies with the wavelength of observation. At longer wavelengths, outer cool regions dominate the emission; this is why the disk has a shallower spectral slope at long wavelengths than a single-temperature blackbody. This property enables disk models to explain the

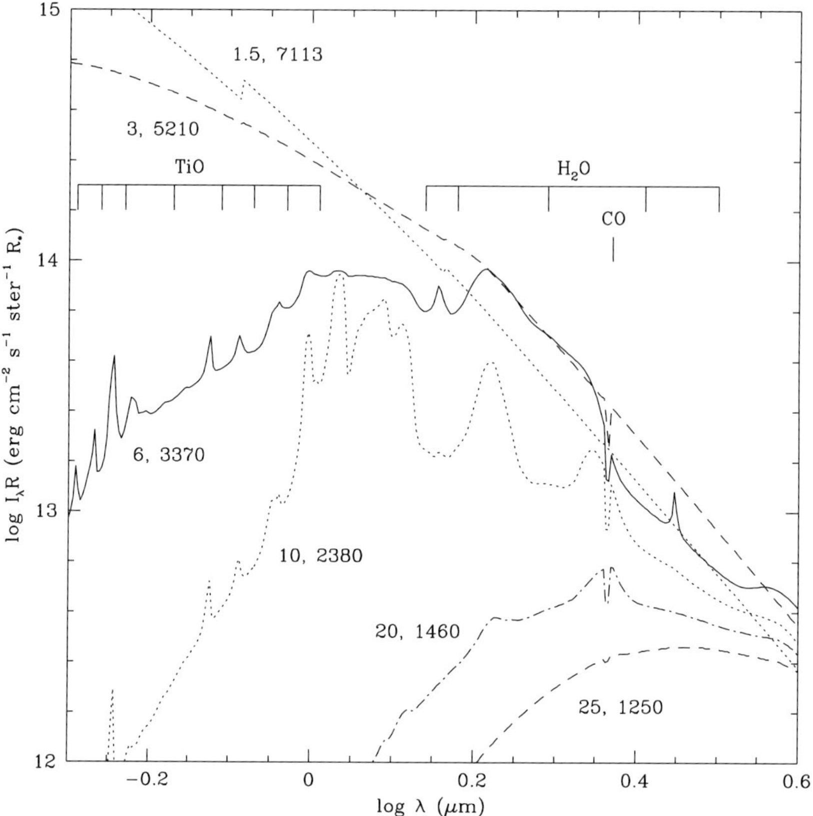

Fig. 7.5. Emission of individual disk annuli for a steady disk model which approximately reproduces the SED of FU Ori (Figure 7.4). The intensity at each annulus is weighted by its (cylindrical) radius R to provide an indication of the amount of contribution of each region to the final spectrum (Figure 7.6). Each curve is labeled by the radius of the annulus in units of the inner disk radius, and the effective temperature (in K) of the annulus. The hot inner disk annuli dominate the optical spectrum but do not contribute molecular absorption features in the near infrared. Conversely, disk regions at $R \sim 5$–$10\,R_i$ contribute strong molecular absorption in the 1–2 μm region. The lowest-temperature annulus shows no molecular absorption because continuous dust opacity is assumed to dominate. Calculation courtesy N. Calvet.

variation of spectral type with wavelength observed in FU Ori objects. The difference in temperature of the emitting gas can be recognized from the change in the spectral features.

To demonstrate this more clearly, consider the emission from differing disk annuli. Assume that the emergent spectrum of each annulus is that of an independent stellar atmosphere, with local effective temperature $T_d(R)$ and vertical surface gravity $g(R)$.[*]

---

[*] This assumes that all of the energy flowing through the atmosphere is generated at larger optical depths in the disk interior. As discussed in §7.3, this approximation is reasonable for FU Ori disks, which should be extremely dense and optically thick; the viscous energy generation should occur mostly at the high-density internal disk layers, not in the low-density, outer disk atmosphere. External heating by the central star (§5.7), which acts in the direction of making the disk more vertically isothermal, can be neglected because the star is much less luminous than the disk in outburst.

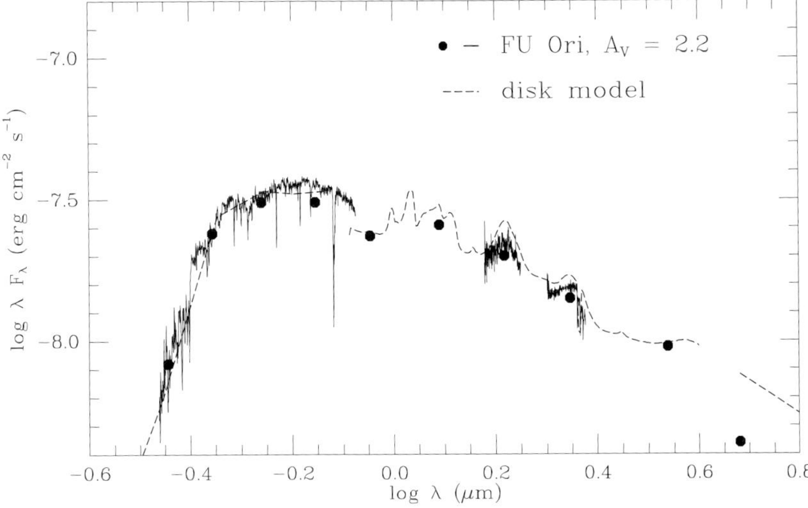

Fig. 7.6. Optical and infrared spectrophotometry of FU Ori (solid lines: Kenyon, Hartmann, & Kolotilov 1991; Mould *et al.* 1978) and broad-band photometery (filled circles), compared with a steady accretion disk model incorporating infrared water vapor and other absorption bands (but which does not include optical absorption features (Calvet, Hartmann, & Kenyon 1991)). The disk model naturally explains both the optical spectrum as well as the deep water vapor absorption (irregular continuum in the 1–2 μm spectrum) and the CO first overtone absorption bands (at 2.2 μm; log $\lambda \simeq 0.34$) characteristic of much cooler stars (compare Figure 7.5). From Hartmann & Kenyon (1996).

The spectra are much more sensitive to the temperature than to the precise gravity, so it is adequate to make a rough estimate $g(R) \sim (GM/R^2)(H/R)$ with $H \sim 0.1R$.

Figure 7.5 shows the results of a typical calculation for a steady disk temperature distribution which approximately reproduces the observed SED of FU Ori. One observes that the hot inner disk regions dominate the optical spectrum, as expected, but do not contribute any molecular features in the near-infrared. In contrast, disk annuli at $R \gtrsim 6\,R_i$ contribute strong water vapor features in the 1–2 μm region, and substantial CO first-overtone absorption at 2.2 μm.

By adding up the intensities of the disk annuli, and weighting them by the appropriate area $2\pi R dR$ (cf. equation (5.75)), the spectrum of the disk model can be synthesized. The resulting spectrum (dashed line in Figure 7.6) can account for the optical G-type spectra of FU Ori objects at the same time that it reproduces the near-infrared water vapor and first overtone CO absorption features.

Kinematic studies generally provide the clearest tests for the presence of disks. One such test involves line profiles, which are qualitatively different in a disk than in stars. The absorption line shapes of rotating stars are typically parabolic, because there is a large contribution to the total flux from regions near disk center, where projected rotational velocities are small. In contrast, the only slowly-rotating regions in a disk lie at large radii. If the outer regions of the disk are too cold to emit effectively at the wavelength of observation, then the resulting profile lacks contributions at low

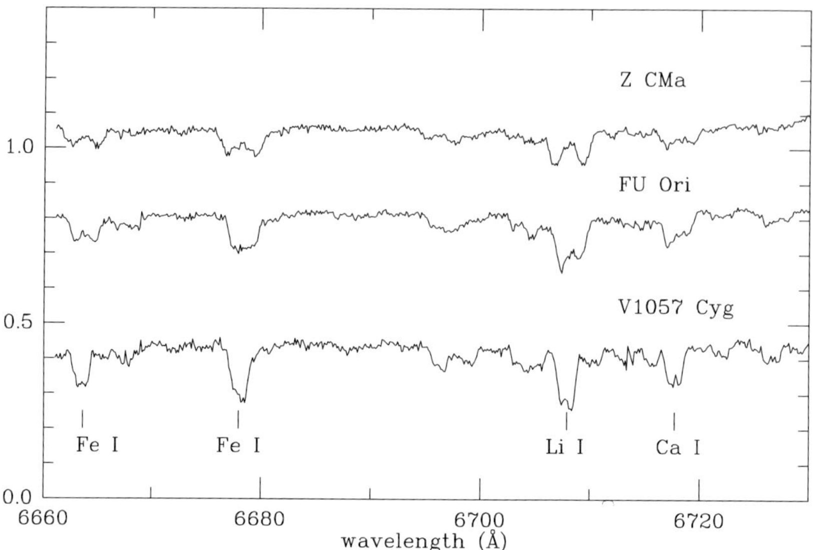

Fig. 7.7. High-resolution red (optical) spectra of three FU Ori objects. The absorption line profiles show evidence for doubling. The sequence from V1057 Cyg → FU Ori → Z CMa illustrates the effect of increasing rotational velocity. The doubling is difficult to detect at the low rotational velocity of V1057 Cyg in the $\lambda 6678$ feature, because it is a blend of several individual lines. Modified from Hartmann & Kenyon (1996).

velocities. The result is a profile in which the absorption is stronger at some velocity than at line center, i.e. a double-peaked profile.

The line profile produced by a rotating, flat, narrow annulus as a function of velocity shift from line center Δv is of the form

$$\phi(\Delta v) = [1 - (\Delta v/v_{max})^2]^{-1/2} , \quad -v_{max} < \Delta v < v_{max}, \tag{7.3}$$

where $v_{max} = v_K(R)\sin i$ is the maximum projected rotational velocity of the annulus with Keplerian rotational velocity v_K observed at an inclination angle i. The observed line profile at a given wavelength is the sum of profiles over all annuli, each with a different rotational velocity, and weighted according to the area and flux of each annulus. Only a finite range of radii in the disk contribute significant continuum emission at a given wavelength; for example, as shown in Figure 7.5, disk annuli with $R \gtrsim 6R_i$ do not contribute significant emission at optical wavelengths, while only regions of the disk with $3R_i \gtrsim R \gtrsim 10R_i$ contribute to the wavelength region near the 2.2 μm CO absorption bands. Thus, the effect of differential rotation is to smooth the profile (7.3) but not eliminate its essential 'double-peaked' shape. Figure 7.7 shows optical spectra of FU Ori objects which clearly demonstrate this doubling of many absorption lines.

The optical spectra shown in Figure 7.7 exhibit large rotational velocity broadening, ranging from about $45\,\mathrm{km\,s^{-1}}$ for V1057 Cyg to $\sim 110\,\mathrm{km\,s^{-1}}$ in Z CMa. Rapid rotation is also observed in the high-resolution near-infrared spectra of the first-overtone CO absorption in FU Ori and V1057 Cyg (Figure 7.8), as can be seen from both the widths of individual resolved features and the shape of the bandhead. This

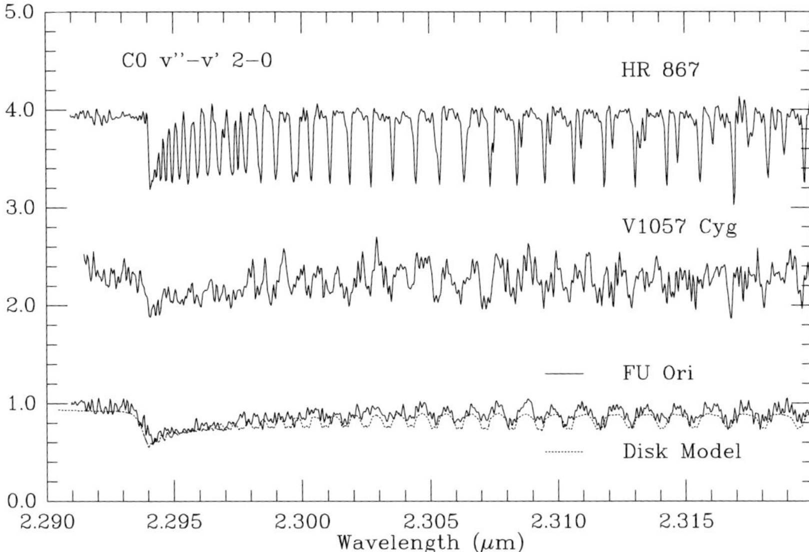

Fig. 7.8. Near-infrared first overtone ($v'' - v' = 2 - 0$) CO absorption bands in two FU Ori objects. The CO lines are strongly rotationally broadened compared with the bands in the M giant HR 867; the amount of the broadening is consistent with the FU Ori disk model (dashed line; see text). From Hartmann & Kenyon (1996).

large rotational velocity broadening is by itself a demonstration that these objects are not normal stars; though the main-sequence progenitors of supergiants may have been rapidly rotating, by the time they have expanded during evolution off the main sequence they become slow rotators.

The measurement of rotational velocities over a wide range of optical and near-infrared wavelengths may be exploited to provide another fairly conclusive test of the disk hypothesis. The disk regions responsible for the optical spectrum are hotter and lie at smaller radii than the regions producing the near-infrared spectrum (Figure 7.8). Since the inner disk regions must rotate more rapidly than the outer disk regions, it follows that if the FU Ori objects are disks, the differential rotation should manifest itself as a decreasing rotational velocity with increasing wavelength of observation.

It proves to be convenient to analyze the rotational line broadening using cross-correlation analysis. For objects rotating much more rapidly than the template star, the cross-correlation peak represents the line profile broadening averaged over many lines. Figure 7.9 compares optical and infrared cross-correlation peaks for FU Ori and V1057 Cyg, demonstrating that in both objects the broadening observed at 0.6 μm is roughly 1.5 times that observed at 2.2 μm. The observations clearly demonstrate large differential rotation inconsistent with stellar rotation. The only model so far which can account for this is a differentially-rotating disk.

The disk model also predicts that the variation of rotation with wavelength should be continuous. Welty *et al.* (1992) presented evidence of such a continuous variation of rotational velocity with wavelength in the optical spectrum $\sim$ 5000–8000 Å of V1057 Cyg and Z CMa. The effect is subtle ($\sim$ 10%) and uncertain, but consistent with Keplerian rotation (Welty *et al.* 1992).

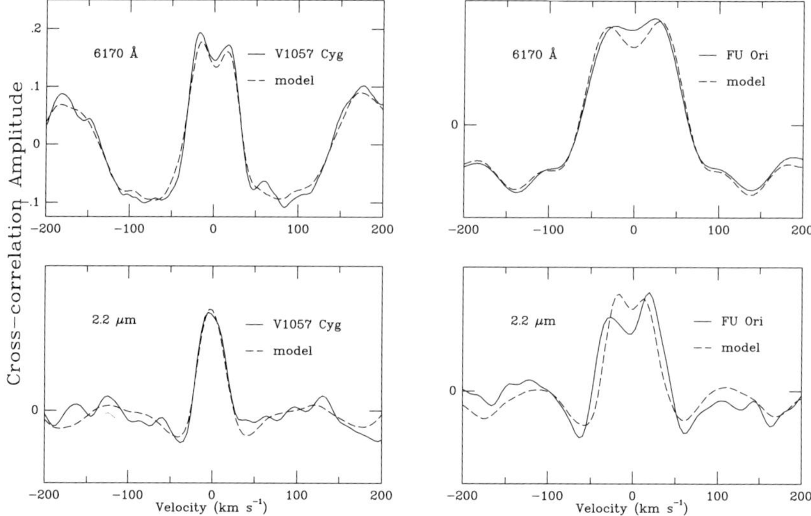

Fig. 7.9. Differential rotation in FU Ori and V1057 Cyg. Cross-correlation peaks show that the intrinsic line widths are larger at optical wavelengths than observed in the $v'' - v' = 2 - 0$ CO absorption bands at 2.2 μm. Cross-correlation of synthetic disk model spectra with the same templates (dashed lines) demonstrates that the difference in optical and infrared line widths is consistent with the assumption of Keplerian rotation (see text). From Kenyon, Hartmann, & Hewett (1988).

To provide a quantitative test of Keplerian rotation, synthetic disk spectra can be computed for the appropriate wavelength regions and then cross-correlated with the same template spectra used to produce the cross-correlations of the real FU Ori objects. The disk model line profiles can be synthesized by convolving the spectrum of each annulus with the appropriate rotational velocity broadening function (7.3), and summing the emission from the individual annuli. Figure 7.9 compares the results of synthetic disk models for FU Ori and V1057 Cyg, scaled to match the observed optical line widths (Kenyon *et al.* 1988). The overall agreement is good, although the predicted differential rotation is perhaps $\sim 25\%$ larger than observed. One should keep in mind that there are several uncertainties in the calculation of the model spectrum. In particular, the temperature at which the dust opacity begins to dominate the continuous spectrum (e.g., Figure 7.5) affects how much the outer disk regions contribute to the CO absorption features, and thus introduces uncertainty in how much these more slowly-rotating regions affect the overall line profile widths. Some theoretical models also suggest that the disks of FU Ori objects are so hot that they do not exhibit precisely Keplerian rotation, and are partially supported by radial gas pressure gradients (cf. §7.6).

Note that the predicted double-peaked *shapes* of the line profiles are in fairly good agreement with observations of FU Ori at both optical and infrared wavelengths, as shown by the cross-correlation peak shapes. (The instrumental resolution is not large enough to show clearly the infrared line doubling in the cross-correlation peak of V1057 Cyg in Figure 7.9.)

7.3 Disk properties

The theory of steady disk emission can be used to estimate physical properties of the FU Ori objects. The observed flux at the Earth is related to the disk accretion luminosity by

$$F = \frac{L_{acc} \cos i}{2\pi d^2},$$ (7.4)

where d is the distance to the Earth and i is the inclination of the disk to the line of sight. In practice the dominant uncertainties are the inclination, about which we have only guesses, and the extinction corrections, since the colors of FU Ori objects may well not be those of 'normal' stars. Since most of the radiation of FU Ori variables is emitted at optical wavelengths (Figure 7.4), and visual extinctions are typically $A_V \sim 2$–3 mag for the *lightly-reddened* objects, uncertainties of a factor of two in the luminosity are to be expected.

It is not certain whether the standard disk–boundary layer theory is applicable to FU Ori objects, since they exhibit no evidence of boundary layer emission (§7.6). Nevertheless, let us adopt the standard disk equations for simplicity. Then the accretion luminosity constrains the product of the mass and the accretion rate divided by the inner disk radius,

$$L_{acc} = \frac{GM_* \dot{M}}{2R_i}.$$ (7.5)

Consider a 'typical' FU Ori object with $L_{acc} = 250 \, L_\odot$. Fitting of extinction-corrected SEDs and optical line spectra by steady disk models suggest that the maximum temperature of the disk is $T_{max} \sim 7000$ K, where

$$T_{max} = 0.488 \left(\frac{3GM\dot{M}}{8\pi\sigma R_i^3} \right)^{1/4},$$ (7.6)

(equation (5.74)). Combining the luminosity with the maximum temperature results in an estimate of the inner disk radius, $R_i \sim 4.5 \, R_\odot$, and constrains the product $M\dot{M} \sim 0.75 \times 10^{-4} \, M_\odot^2 \, \mathrm{yr}^{-1}$. These inner disk radii are somewhat larger than typical radii of T Tauri stars ~ 2–3 $R_\odot$. The agreement is reasonably satisfactory considering the uncertainty in the boundary layer luminosity; it is possible that the disk advects or carries large amounts of accretion energy into the central star (§7.6), causing it to expand (Chapter 9).

The optical rotational velocities can be used to estimate the central mass. Steady disk models indicate that the effective emitting radius at $0.6 \, \mu$m for $T_{max} = 7000$ K is about 2.5 times the inner radius. Using this result, the apparent optical rotational velocity is

$$v \sin i \, (0.6 \, \mu\mathrm{m}) \sim \left(\frac{GM}{2.5R_i} \right)^{1/2} \sin i \sim 92 \, M_{0.5}^{1/2} \sin i \, \mathrm{km \, s}^{-1},$$ (7.7)

where $M_{0.5}$ is the central mass in units of one-half solar mass, and we have adopted a typical inner radius $R_i = 4.5 \, R_\odot$ (Kenyon *et al.* 1988; §7.6).

The observed optical rotational velocities of the four best-studied FU Ori objects are: Z CMa, $v \sin i \sim 110 \, \mathrm{km \, s}^{-1}$; FU Ori, $v \sin i \sim 65 \, \mathrm{km \, s}^{-1}$; V1057 Cyg, $v \sin i \sim 40 \, \mathrm{km \, s}^{-1}$; and V1515 Cyg, $v \sin i \sim 20 \, \mathrm{km \, s}^{-1}$ (cf. Figures 7.7, 7.9). It is evident that the rotational velocities of Z CMa and FU Ori are consistent with central stars of roughly one solar mass. The slow rotation of V1057 Cyg and V1515 Cyg require extremely small central masses unless these objects are observed at a very small inclination. Goodrich (1987) suggested that the circular reflection arcs seen around V1057 Cyg and V1515 Cyg (Figures 7.2, 7.3) are ovoidal cavities seen nearly along their axis. If these cavities are produced by outflow cavities perpendicular to the disk, parallel to the rotation axis (Chapter 8), then these accretion disks may be observed at low inclinations, and the projected rotational velocities can be consistent with masses typical of T Tauri stars.

Assuming that the typical central masses of FU Ori objects are about $0.5\,M_\odot$, maximum disk accretion rates must be $\sim 10^{-4}\,M_\odot\,\mathrm{yr}^{-1}$. One can then derive a minimum disk mass (prior to outburst) from the length of time of the rapid accretion event. In FU Ori and V1515 Cyg, the estimated outburst lengths ~ 100 yr indicate that the accreted disk mass is $\sim 10^{-2}\,M_\odot$ in one outburst, comparable to the minimum mass solar nebula. For V1057 Cyg the accreted mass during its present outburst will probably be closer to $10^{-3}\,M_\odot$. These values are comparable to the disk masses estimated for T Tauri stars from mm-wave emission (§6.3, Figure 6.5); however, it must be emphasized that these are merely lower limits to the disk masses, since this is only the material that is accreted over a very short period of time, not what is left behind.

To proceed further one requires some estimate of the disk viscosity. Such estimates are obviously uncertain, but the exercise of computing a disk structure adopting an 'α' viscosity yields some suggestive results. We use the steady α disk result (5.78),

$$\nu\Sigma \;=\; \alpha c_s H\Sigma \;=\; \frac{\dot{M}}{3\pi}, \tag{7.8}$$

and the steady disk temperature distribution (5.73),

$$T_d^4 \;=\; \frac{3GM\dot{M}}{8\pi\sigma R^3}. \tag{7.9}$$

Here we have set the factor $[1 - (R_i/R)^{1/2}] \sim 1$; this approximation makes little difference except at the innermost disk radii, and in any event the inner boundary condition is not well understood (§7.6). We also employ the supplemental equation (5.64),

$$H \;=\; c_s(R/v_\phi) \;=\; c_s/\Omega, \tag{7.10}$$

where $\Omega = (GM/R^3)^{1/2}$. Finally, we require a relation between the surface effective temperature T_d and the central temperature T at each radius R. Vertically averaging (in z) the opacity, and employing the diffusion approximation,

$$T^4 \;\simeq\; T_d^4(3/4)\tau_R \;=\; (3/4)T_d^4 k_R\Sigma, \tag{7.11}$$

where the optical depth τ_R through the disk is determined by the Rosseland mean opacity k_R (see Frank *et al.* (1992) for details).

With these approximations one can solve for the (vertically-averaged) disk structure. Bell & Lin (1994) provide convenient analytic approximations of the opacity for various regimes; in particular, for disk (midplane) temperatures $\gtrsim 10^4$ K,

$$k_R \;\simeq\; 1.5 \times 10^{20}\rho\,T^{-5/2}\,\mathrm{cm}^2\,\mathrm{g}^{-1}. \tag{7.12}$$

Using the vertical average $\rho = \Sigma/H$, one can solve for the disk structure for a fixed value of α. After some tedious algebra, the temperature distribution can be written as

$$T \simeq 1.3 \times 10^5 \, \alpha_{-2}^{-2/9} \, \mu_{0.6}^{5/18} \, \dot{M}_{-4}^{1/3} \, M_{0.5}^{5/18} \, R_{10}^{-5/6} \text{ K}, \tag{7.13}$$

where $\dot{M}_{-4}$ is the mass accretion rate in units of $10^{-4} \, M_\odot \, \text{yr}^{-1}$, α_{-2} is the alpha viscosity parameter in units of 10^{-2}, $M_{0.5}$ is once again the mass in units of one-half solar mass, and R_{10} is the cylindrical radius in units of $10 \, R_\odot$. Finally, we have set the mean molecular weight $\mu = 0.6$, roughly characteristic of ionized atomic gas.

These results suggest that the innermost disk must be extremely hot in its interior during outburst, far hotter than the surface temperature ~ 7000 K, as a result of the large optical depths which effectively trap the heat generated by accretion. The internal disk temperature is so large that in the innermost regions the thickness of the disk may not be negligible. For the fiducial parameters at $R = 10 \, R_\odot, H/R \sim 0.4$; therefore, the thin disk approximations may begin to break down in the inner disk regions. We return to this point when discussing the 'missing' boundary layer radiation in §7.6.

The surface density under these assumptions is

$$\Sigma \simeq 5.7 \times 10^4 \, \alpha_{-2}^{-7/9} \, \mu_{0.6}^{13/18} \, \dot{M}_{-4}^{2/3} \, M_{0.5}^{2/9} \, R_{10}^{-2/3} \text{ g cm}^{-2}. \tag{7.14}$$

This particular disk solution is applicable for temperatures $\gtrsim 10^4$ K. At lower temperatures, the opacity changes character dramatically in a way which may help produce outbursts (§7.5). With the fiducial parameters, $T \sim 10^4$ K occurs at $R_{10} \sim 20 \sim 0.87$ AU. The 'hot' ($T \gtrsim 10^4$ K) disk mass interior to this radius R_{out} is

$$M_d(hot) \sim 3.4 \times 10^{-3} \, \dot{M}_{-4}^{2/3} \, \alpha_{-2}^{-7/9} \, M_{0.5}^{2/9} \, (R_{out}/0.87 \text{ AU})^{4/3} \text{ M}_\odot. \tag{7.15}$$

This result suggests that, if $\alpha \sim 10^{-2}$, the disk may store roughly enough mass to fuel one FU Ori outburst within a radius of about 1 AU.

In Chapter 5 we showed that viscous disks evolve on the timescale

$$t_v \simeq R^2/v_v. \tag{7.16}$$

With the fiducial parameters, at R_{out} the viscous timescale is approximately $t_v \sim 70 \, \alpha_{-2}^{-7/9}$ yr, suggesting that the 'hot' region of the disk within about 1 AU can in principle account for both the amount of mass accreted and the observed decay timescales of FU Ori accretion events. However, the rise times of some FU Ori outbursts are ~ 1 yr, much shorter than this viscous time. As discussed in §7.5, there is at least one other, shorter, timescale for disk evolution – the *thermal* timescale – which offers a possibility for explaining rapid rises to maximum light.

7.4 Time variability and circumstellar envelopes

Although there is little spectral information available on the rise to maximum light in FU Ori objects, V1057 Cyg has faded substantially since the early 1970s, and its color evolution provides an important clue to physical conditions. After the outburst, the optical and near-infrared spectrum became redder as V1057 Cyg faded, consistent with optical spectra which indicate that V1057 Cyg evolved from an A spectral type near maximum ($T_{eff} \sim 8000$ K (Herbig 1977b)) to a mid-G spectral type ($T_{eff} \sim 6500$ K) at the current epoch. A series of steady accretion disk models with fixed inner radii but decreasing mass accretion rates matches the color evolution with decreasing

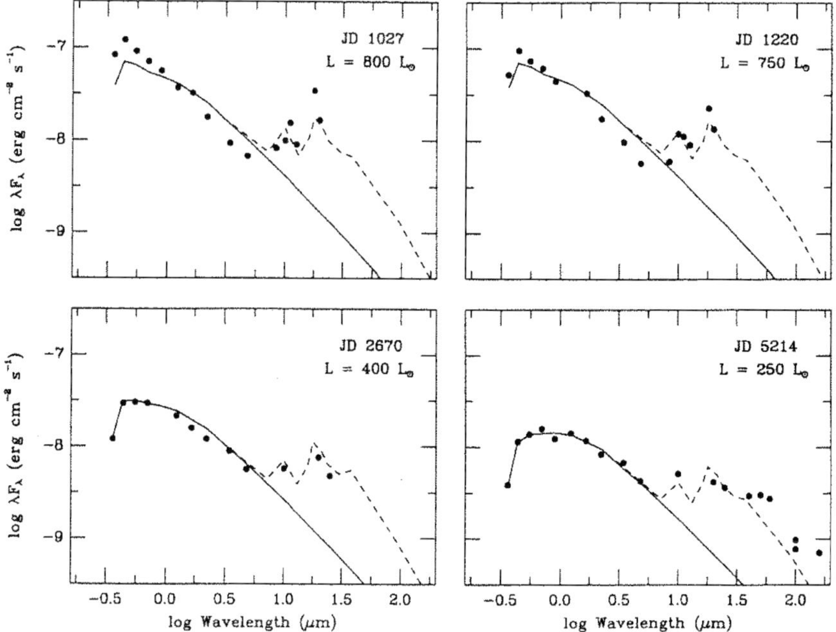

Fig. 7.10. Spectral evolution of V1057 Cyg during decay from maximum light. The central disk becomes redder, and can be fitted by a disk model with lower temperatures (solid line) as it fades from maximum. The emission at wavelengths $\lambda \gtrsim 10\,\mu$m decays rapidly, in proportion to the decay of overall system luminosity, consistent with its originating in a circumstellar dust envelope whose emission is powered by absorbing light from the central disk. The dashed line denotes a dusty infalling envelope model, as described in the text. From Kenyon & Hartmann (1991).

accretion luminosity fairly well from the optical to wavelengths $\sim 5\,\mu$m, reproducing the observed decrease in amplitude of decay at increasing wavelengths (Kenyon & Hartmann 1991; Figure 7.10).

One observation that steady disk models cannot explain is the observed variation in rotation velocity of V1057 Cyg. Herbig (1989) found that the optical rotational velocity appeared to decrease from about $70\,\mathrm{km\,s^{-1}}$ near peak light to about $45\,\mathrm{km\,s^{-1}}$ in the late 1980s. This is not explicable in terms of steady disk theory; more luminous steady disks have higher temperatures at larger radii, and therefore the optical spectrum should be produced at larger, more slowly-rotating regions during outburst. However, time-dependent models for outburst can show the opposite effect, simply because the temperature distribution can deviate substantially from that of a steady disk (Bell *et al.* 1995). While the time-dependent models are suggestive, they are difficult to test in detail using the evolution of the SED, because the disk spectrum is not very much different from that of a single-temperature blackbody over the limited wavelength range observed. Differences in the temperature structure due to time-dependent accretion therefore produce only second-order effects.

The wavelength dependence of the decay in light observed for V1057 Cyg has important implications for understanding the origin of the far-infrared excess emission. As shown in Figure 7.10, the amplitude of variation during decay decreases with increasing wavelength from the optical through the near-infrared. This can easily be understood in the context of the disk model. The optical wavelengths lie on the blue edge or 'Wien side' of the emission from the innermost disk regions, and therefore are the most sensitive to temperature variations. The timescales of variation should also lengthen at longer wavelengths, because the outer regions producing this emission have longer viscous timescales. However, the decay in the fluxes at $\lambda = 10\,\mu$m and $20\,\mu$m is much faster than observed in the near-infrared, but is consistent with the overall change in the system luminosity. Thus, the excess emission at $\lambda \gtrsim 10\,\mu$m is most likely produced by dust absorption of radiated accretion luminosity produced in the inner regions, rather than by local changes in the viscous energy dissipation.

The change in the spectral slope in the SED at 10 μm (Figure 7.4) similarly argues for another dust emission source at long wavelengths. The amount of flux emitted in long wavelength radiation is too large to be accounted for by a flat disk absorbing radiated accretion luminosity and reradiating this energy at longer wavelengths. One possibility is that the disk is flared at these wavelengths; however, it is difficult to make the disk sufficiently flared to account for the observations in detail (Kenyon & Hartmann 1991).

An alternative explanation for the far-infrared emission is a circumstellar envelope. Kenyon & Hartmann (1991) showed that an infalling envelope with a mass infall rate $\sim 5 \times 10^{-6}\,M_\odot\,\mathrm{yr}^{-1}$ typical of protostellar sources (Chapter 4) could explain the mid-IR observations of V1057 Cyg during its optical decline. Reproducing the mid-IR excess requires an envelope with $A_V \sim 50$ mag covering about half of solid angle, and extending in to ~ 5 AU from the central star (Figure 7.10). The envelope must have a cavity to allow a relatively clear line of sight to the inner disk. This model is consistent with interpretations of the morphology of the scattered light nebulae of V1057 and V1515 Cyg discussed above, and is reminiscent of the flattened collapse model results of §3.3 (Figures 3.3, 4.10).

If one includes objects for which no outburst has been clearly observed, but which have the appropriate spectroscopic signatures, roughly half of the FU Ori objects are heavily extincted. These statistics are consistent with the envelope solid angles inferred for V1057 Cyg and V1515 Cyg. The FU Ori candidate L1551 IRS 5 is one of the standard objects modelled with infalling protostellar envelopes (Adams *et al.* 1987; Butner *et al.* 1991; cf. §4.3, 4.5, Figure 4.7). Thus, FU Ori objects generally may have infalling material landing on their outer disks.

7.5   Outburst mechanisms

One suggested explanation of FU Ori outbursts is the passage of a companion star near the disk. The gravitational forces of the companion star may perturb the disk sufficiently to transfer substantial angular momentum from disk to star and thus enhance accretion. This idea was first suggested to the author by Toomre (1985, personal communication), but the only calculation directly applied to this problem to date is that of Bonnell & Bastien (1992). As part of a series of studies of binary formation, Bonnell & Bastien found that disks forming around each member of a binary system were strongly affected when the pair plunge close to each other on an eccentric orbit. Bonnell & Bastien found that the disks could be perturbed near periastron to high accretion rates comparable to that required for FU Ori systems; moreover, the evolution of one simulation suggested that disks might survive the collision process sufficiently well to undergo outbursts during three or four successive periastron passages.

This mechanism is attractive in that most stars are members of multiple systems, and theoretical arguments suggest that at least some outbursts need to be externally triggered (see below). On the other hand, no apparent radial velocity shifts have been observed in the brightest, best-studied FU Ori objects, to a detection limit of a few $km\,s^{-1}$ (Herbig 1977b; Hartmann & Kenyon 1987). The interaction must strongly perturb the disk at 1–10 AU to produce short rise times; to explain the event statistics, it is necessary to assume that the binary distribution was originally much more concentrated at small perihelion distances than presently observed in the field (e.g., Duquennoy & Mayor 1991), and that interactions, perhaps between stars and disks (Clarke & Pringle 1991), caused orbital evolution to the present-day distribution.

The mechanism investigated in the most detail for FU Ori outbursts is one involving thermal instability, which was originally developed to explain disk outbursts in accreting binary stars. The thermal instability model is attractive for a variety of reasons. Perhaps most important is that the thermal instability timescale is much shorter than the viscous timescale, $t_{th} \sim (H/R)^2 t_v$ (e.g., Pringle 1981), and therefore is a possible candidate to explain the short rise times of (some) outbursts.

The basic idea behind the thermal instability mechanism has been discussed in detail in many sources (see, e.g., Pringle (1981) and Frank *et al.* (1992)); here we provide only an outline. At a radial distance R in the disk, the viscous energy generation, F_{vis}, must be balanced by the radiative losses of the disk, F_{rad}, in thermal equilibrium. Integrating the energy equations over the vertical structure of the disk at a fixed radial distance R, and adopting the usual α viscosity treatment, equations (5.23) and (5.77) can be combined to yield

$$F_{vis} = \frac{9}{4}\alpha\Omega\Sigma c_s^2, \tag{7.17}$$

while

$$F_{rad} = 2\sigma T_d^4. \tag{7.18}$$

We also require the relation (7.11) between the internal disk temperature T and the surface temperature as derived before,

$$T^4 \approx \tau_R T_d^4, \tag{7.19}$$

Combining equations (7.17)–(7.19),

$$F_{rad}/F_{vis} \propto T^3 k_R(T,P)^{-1}\Sigma^{-2}\alpha^{-1} \tag{7.20}$$

(Kawazoe & Mineshige 1993; D'Alessio 1996).

Now consider a disk annulus initially in thermal equilibrium which is perturbed to a slightly higher internal temperature T. It can be shown that the thermal timescales for transferring energy are generally much shorter than the viscous timescales for transferring matter (Pringle 1981), so it suffices to take $\Sigma \approx$ constant. Then equation (7.20) indicates that the radiative cooling will exceed the viscous heating as long as the opacity does not increase more rapidly with temperature than T^3, and will therefore drive the disk back toward the thermal equilibrium. If, however, the opacity increases faster than T^3, heat is efficiently trapped within the disk, and the surface

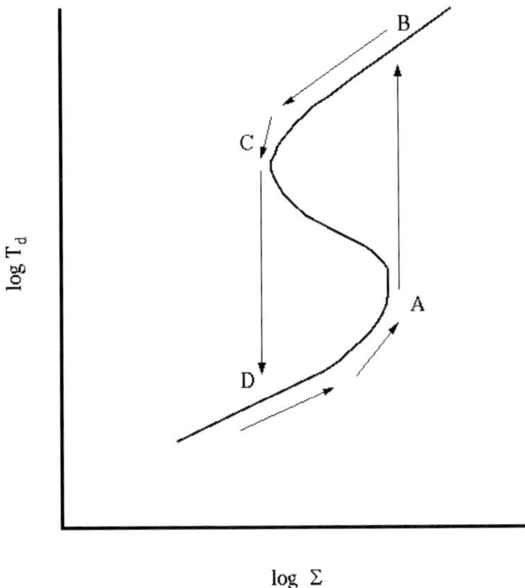

Fig. 7.11. Schematic 'S curve' for disk thermal instability, as discussed in the text.

cooling cannot keep pace with the increasing viscous heating. This leads to a thermal runaway, with the disk temperature increasing rapidly until the character of the opacity changes. Calculations generally indicate that the gas opacity should increase very steeply with increasing temperature for temperatures $\sim$ 5000–10000 K largely because of hydrogen ionization, with the precise range depending somewhat on density. This steep dependence on temperature ($\propto T^{10}$ (Bell & Lin 1994)) makes it possible to have thermal instabilities in this temperature range.

To explore the implications of this opacity dependence further, we consider the so-called 'S curves' resulting from these assumptions (cf. Frank *et al.* 1992). The thermal equilibrium equations for a steady-state accretion disk can be recast in terms of the surface (or effective) disk temperature (or mass accretion rates; cf. equation (5.73) and the surface density. Then the loci of thermal equilibrium in the T_d vs Σ plane for a specific disk annulus form an 'S'-shaped curve, with a 'kink' where the opacity causes thermal instability.

Figure 7.11 shows a schematic S curve typical of the results of detailed calculations. The region above and to the left of the curve corresponds to faster cooling than heating; the region below and to the right of the curve faster heating than cooling. In the limit cycle theory of outbursts, a disk annulus begins in a low accretion rate state at low Σ, lying on the lower branch of an S curve. A perturbation which increases the central temperature tends to drive the disk annulus vertically upward into a region where cooling exceeds heating, so the disk is stable. However, as material piles up in the disk, moving the equilibrium point upwards along the curve, eventually the disk annulus reaches an inflection point (A) in the S curve, at which point a positive temperature perturbation pushes the annulus vertically upward in the log T_d vs log Σ plane, where the disk annulus is thermally unstable. The annulus must jump up to the upper branch of the S curve at (B), where the surface (or effective) disk temperature and mass accretion rate are much higher. Eventually, the enhanced accretion onto the central star causes disk material to drain away, dropping Σ to the point (C) beyond which the only stable solution is on the low-accretion rate branch (D).

Thermal instability models are attractive as an explanation of FU Ori outbursts because: (1) the short rise times observed can be achieved (thermal timescales for the outburst); (2) the predicted inner disk mass fluxes in outburst are $10^{-4} \, \mathrm{M_\odot \, yr^{-1}}$ and the peak surface disk temperature of the disk at this point is just below 10^4 K, in good agreement with observations; and (3) the process can repeat, resulting in multiple events as required by outburst statistics, if there is a mass source which replenishes the disk material accreted during outburst. There is also some appeal to trying to find a 'unified' explanation for the outbursts of both close binary system disks as well as protostellar disks.

A key requirement of the thermal instability model is that of a moderately high accretion rate in the outer disk. At the low temperature end of the unstable regime, the gas opacity is quite low, resulting in an internal disk temperature that is not very much larger than the surface temperature. Thus, we can use the effective temperature at which the disk emits as a useful guide to the regions which can become thermally unstable. Bell & Lin (1994) showed that the critical effective temperature at the transition between the stable and unstable portions of the S curve is roughly ~ 2000 K. If we impose a fixed mass accretion rate $\dot{M}_{od}$ from the outer disk, we can then use this temperature constraint along with equation (5.73) to derive an approximate outer limit radius for the region of thermal instability (Bell & Lin 1994)

$$R_{max} \approx 20 \left(\frac{\dot{M}_{od}}{3 \times 10^{-6} \, \mathrm{M_\odot \, yr^{-1}}} \right)^{1/3} \left(\frac{M_*}{\mathrm{M_\odot}} \right)^{1/3} \left(\frac{T_{eff}}{2000 \, \mathrm{K}} \right)^{-4/3} \mathrm{R_\odot} . \qquad (7.21)$$

Disk regions at radii larger than R_{max} will never become thermally unstable because they cannot become hot enough to enter the required opacity regime. Thus, if there is to be a large amount of mass dumped onto the central star during an outburst, the outer limiting radius of the unstable region must be large; in turn, this tends to require a large outer disk (background) accretion rate.

Lin & Papaloizou (1985) first applied thermal instability models to FU Ori objects and generated optical outburst amplitudes similar to those observed. The eruptions were short-lived, however, and did not produce large infrared outbursts. The reason for this can be seen from equation (7.21). Lin & Papaloizou assumed background mass accretion rates of $\dot{M}_{od} \sim 10^{-7} \, \mathrm{M_\odot \, yr^{-1}}$, similar to the mass accretion rates of T Tauri stars, but then only the innermost disk regions became thermally unstable, so that very little mass was involved in the eruption; therefore the outburst decayed very quickly. Clarke, Lin, & Pringle (1990) managed to achieve longer-lasting outbursts with the same low background accretion rate by introducing a very large ($50\times$) perturbation in the surface density. This had the effect of producing a much higher mass flux into the unstable regions, and so triggered an outburst which lasted for ~ 30 yr, qualitatively resembling the decay of V1057 Cyg.

If FU Ori disks are acquiring mass from remnant infalling envelopes at $\sim 10^{-6}$– $10^{-5} \, \mathrm{M_\odot \, yr^{-1}}$ (§7.4), this mass source would tend to drive the mass accretion rates of the outer disk closer to values needed to make the thermal instability operate more easily. Kawazoe & Mineshige (1993) first argued that FU Ori outbursts are produced by thermally-unstable disks with high mass input rates, and noted the possible correspondence with infall rates. Bell & Lin (1994) and Bell *et al.* (1995) calculated similar models in more detail, and showed that many properties of the observed light curves can be reproduced with appropriate model parameters. Bell & Lin (1994) found that models simply fed at the appropriate accretion rate from the outer disk tended to become unstable at inner disk radii first; the instability propagates slowly outward to R_{max} (see also discussion in Smak (1984)). Under these circumstances, Bell *et al.* (1995) were able to reproduce the slow rise of V1515 Cyg, but not the rapid rise times of FU Ori and V1057 Cyg, which require external perturbations.

Unfortunately, many aspects of the thermal instability mechanism depend upon the unknown value(s) of α, which determines the timescales for evolution. Relatively small values of α are needed to obtain sufficiently long evolutionary timescales, and sufficiently large disk masses, to explain FU Ori outbursts. Bell & Lin (1994) and Bell *et al.* (1995) required a variable α, with values in the high accretion rate state about a factor of ten higher than in the low accretion state to obtain a sufficiently large outburst. Bell *et al.* found that they needed $\alpha_{low} \sim 10^{-4}$, which implies a relatively massive disk; inside ~ 0.5 AU, the disk mass approaches $0.1\,M_\odot$ in this model, without including outer regions where most of the disk mass presumably resides.

An idea which incorporates the thermal instability as part of the picture is the suggestion by Clarke & Syer (1996) that protoplanets or protostellar companions orbiting in the disk might help to dam up material in the disk. As discussed in §6.5, the general effect of companion gravitating bodies orbiting the central star is to force more rapid accretion of material inside the companion orbit and to push out material outside the orbit. Whether a gap develops in the disk as a result of this forcing depends upon the rate at which the disk tries to accrete. In Clarke & Syer's picture, in the low state the companion can produce a gap, helping to dam up material at a point outside its orbit; however, if the material becomes thermally unstable, the disk material can overwhelm the companion and cross the gap. In this way the companion can have the effect of producing a rapid outburst rise time through the perturbations of a small body on the inside of the disk, rather than postulating the interaction of a large body on the outer disk (e.g., Bonnell & Bastien 1992).

One interesting consequence of the thermal instability picture is that, as infall to the disk ceases, and therefore $\dot{M}_{in}$ drops below $10^{-6}\,M_\odot\,\mathrm{yr}^{-1}$ to approach typical T Tauri disk accretion rates $\sim 10^{-7}\,M_\odot\,\mathrm{yr}^{-1}$, one might expect to observe shorter and smaller amplitude outbursts, since only the innermost disk regions can become thermally unstable (equation (7.21)). Herbig (1977b) drew attention to small optical outbursts observed in pre-main-sequence stars which might constitute a separate class of 'EXor' variables, and it is tempting to speculate that the EXor outbursts are also produced by thermal disk instabilities.

These considerations emphasize the importance of understanding disk viscosity and/or angular momentum transport mechanisms in more detail, without which it is difficult to make definite theoretical predictions. For example, if the disk viscosity is magnetic (§5.3), the finite and variable conductivity predicted for protostellar disks indicates that the use of a constant α parameter for all parts of the disk is questionable. Gammie (1996a) has suggested that 'dead zones' of decreased accretion may occur where the disk ionization state is so low that magnetic viscosity is ineffective. Such dead zones might dam up disk material until other mechanisms permit accretion, providing a possible alternative explanation for FU Ori outbursts.

7.6   The boundary layer problem

The standard boundary layer model (§5.6) predicts large ultraviolet fluxes that are not observed in FU Ori objects (Kenyon *et al.* 1989). The ultraviolet fluxes and spectra of FU Ori and Z CMa show no evidence for appreciable excess emission above that predicted by the disk model. The absence of strong emission lines also points to a lack of boundary layer emission, which, if hotter than 3×10^4 K, and radiating half of the accretion luminosity, should produce enough extreme-ultraviolet photons to ionize a substantial fraction of an FU Ori object's wind; this is inconsistent with observations (Chapter 8).

As mentioned in the previous chapter, and discussed in detail in Chapter 8, the standard boundary layer model probably does not apply to T Tauri stars either, because the inner disks are disrupted by stellar magnetic fields. At the high accretion rates of FU Ori objects, one might expect that the dynamic pressure of the disk is sufficient to crush the stellar magnetic fields back up against the outer stellar layers (Shu *et al.* 1994); however, there should be boundary layer emission unless the star is rapidly rotating (equation (5.86)), unlike typical T Tauri stars.

The most likely explanation for the absence of boundary layer radiation is that FU Ori disks become quite hot internally at small radii, which causes departures from standard thin disk physics. Returning to the 'hot α disk' solution of §7.3, we may combine equations (7.10) and (7.13) to solve for the disk scale height,

$$\frac{H}{R} = \frac{c_s}{v_\phi} \simeq 0.43 \, \alpha_{-2}^{-1/9} \, \mu_{0.6}^{-13/36} \, \dot{M}_{-4}^{1/6} \, M_{0.5}^{-13/36} \, R_{10}^{1/12} \, . \tag{7.22}$$

Clarke *et al.* (1990) were the first to suggest that FU Ori disks may be so vertically thick that any radiation generated in the boundary layer is absorbed and diffused in the disk over a distance comparable to $H \sim R$. In this way the heat generated in any boundary layer will be spread over a large area – comparable to the region where most of the viscous disk radiation is generally produced – and therefore no hot emission component will be evident.

More generally, Popham *et al.* (1993, 1996) and Popham (1995) have argued that there may not be a well-defined boundary layer in FU Ori outbursting disks. Keplerian rotation is an adequate approximation for the thin disk only when the gas pressure term in the radial force balance equation (5.87),

$$\frac{v_\phi^2}{R} \approx \frac{1}{\rho} \frac{dP}{dR} + \frac{GM}{R^2} \, , \tag{7.23}$$

can be ignored. This is only true when

$$c_s^2 \ll \frac{GM}{R} \, , \tag{7.24}$$

which implies $(H/R)^2 \ll 1$. Since this is probably not true in inner FU Ori disks during outburst, gas pressure forces may become so large that the disk rotates at velocities substantially below the Keplerian velocity because of the pressure support term in equation (7.23). In such a situation the disk rotates more slowly than the local Keplerian angular velocity, and velocity gradients are not steep because of the gas pressure support. The result is that the dissipation of rotational energy is both reduced and less spatially concentrated than the conventional boundary layer.

This model makes the additional interesting prediction that the differential rotation of the disk with radius would be slower than Keplerian. As discussed in §7.2, the observations suggest slightly less differential rotation than Keplerian. While the observations provide some weak support for this picture (Popham *et al.* 1996), the uncertainties in modelling the near-infrared spectra preclude a definite test at this point. Another interesting consequence of this picture is that, even in the absence of magnetospheric interactions or stellar wind spindown, stars may be formed with equatorial rotational velocities less than Keplerian or break-up velocities (Popham 1995).

7.7 Outburst statistics and evolutionary significance

Since FU Ori objects are so rare, their evolutionary significance is somewhat hard to evaluate. Estimates of the star formation rate in the solar neighborhood (Miller & Scalo 1979) suggest that about 2×10^{-2} stars are born per year within 1 kpc of the Sun, where most of the known FU Ori objects are located. If the number of FU Ori outbursts in the same volume is ~ 5 over the last 60 years, and if every (low-mass) star experiences outbursts, then there must be ~ 4 outbursts per star. This is clearly a lower limit to the outburst frequency, since many of the more recent FU Oris have been discovered as heavily-extincted infrared sources.

If each star experiences a minimum of 4 outbursts, and each outburst accounts for the accretion of $\sim 10^{-2} \, \mathrm{M_\odot}$, then the total mass accreted is $\sim 4 \times 10^{-2} \, \mathrm{M_\odot} \, \mathrm{yr^{-1}}$, or about 10% of the mass of a typical T Tauri star. Another way of approaching this problem is to note that there are currently ~ 10 rapidly accreting FU Ori objects within about 1 kpc distance. Not all of these are at maximum light, or as luminous as FU Ori; still, probably $5 \times 10^{-4} \, \mathrm{M_\odot} \, \mathrm{yr^{-1}}$ or more is being accreted in FU Ori objects within 1 kpc, compared with a total mass rate of formation of low-mass stars $M \leq 1 \, \mathrm{M_\odot}$ of about $6.6 \times 10^{-3} \, \mathrm{M_\odot} \, \mathrm{yr^{-1}}$ (Miller & Scalo 1979). This again suggests that ~ 5–10% of the mass of low-mass stars is accreted (on average) in FU Ori eruptions. In comparison, typical parameters for T Tauri stars of $\dot{M} \sim 10^{-8} \, \mathrm{M_\odot} \, \mathrm{yr^{-1}}$ over a lifetime $\sim 10^6$ yr (§6.4) suggest that less mass is accreted from the disk during the T Tauri phase.

An independent reason for thinking that FU Ori outbursts may play a substantial role in the formation of stars comes from the apparent 'luminosity problem' for protostar candidates in the Taurus molecular cloud (§4.1, Figure 4.2). The luminosities of the protostellar or 'Class I' sources in Taurus are roughly an order of magnitude smaller than would be predicted if the mass added by infalling envelopes is accreted steadily onto the central star. One way around this problem is to assume that material is piling up in the disk because the disk accretion rate is usually lower than the infall rate; then the bulk of the disk material would be accreted in brief high-luminosity FU Ori events. Near-infrared spectroscopy limiting stellar luminosities and disk infrared emission is needed to determine accretion rates onto central stars and thus test this general picture.

8

Disk winds and magnetospheric accretion

The discovery of powerful highly-collimated mass ejection from pre-main-sequence stars (e.g., Figure 1.4) was a great surprise. Bipolar outflows are observed starting in the earliest stages of star formation, from objects heavily extincted by their natal clouds. It is fairly certain that accretion disks are responsible for this mass loss. A large fraction of the total accretion energy is required to power these outflows, suggesting that winds may play an important role in the energetics and angular momentum transport of inner disk regions. At present, the most favored mechanism to produce outflows invokes magnetic fields rotating with the disk which accelerate and collimate the ejected material into jets. The precise way in which this acceleration and collimation takes place is not clear, mostly because the magnetic field structure in the inner disk is not known.

Magnetic fields also play an important role in accretion onto pre-main-sequence stars. The magnetic fields of T Tauri stars are apparently strong enough to hold off disks from the stellar surface; the accreting gas deviates from the disk plane as it falls in along the stellar magnetic field lines, eventually shocking at the stellar surface and producing the observed hot continuum radiation (Figure 6.1). Some of the clearest evidence for magnetospheric accretion comes from the broad, asymmetric emission lines of classical T Tauri stars, which mostly originate in the infalling magnetospheric gas; the infall pattern explains both the velocity widths and the asymmetries of most line profiles. The torques produced by the interaction of the stellar magnetic field with the circumstellar disk can modify the rotation of the central star. This magnetic disk interaction can in principle explain the slow rotation of CTTS relative to WTTS of similar ages. However, the details of how this actually occurs are not well-understood at present.

This chapter provides an introduction to the complex physical processes involved in disk wind acceleration and magnetospheric accretion.

8.1 Outflows and jets

Many young stars eject powerful, highly-collimated, bipolar winds or jets in their early evolution. An enormous literature exists on the ways in which outflows propagate through and interact with the circumstellar medium. One of the earliest indications of outflow activity came from the recognition that the Herbig–Haro objects (Herbig 1951, 1952; Haro 1952, 1953) were produced by high-velocity material shocking with the ambient interstellar medium (Schwartz 1975). The discovery of highly-collimated emission jets (Dopita, Schwartz, & Evans 1982; Mundt & Fried 1983) has led to intensive research which has still not resolved many questions of jet structure, acceleration, and interaction with the interstellar medium. Many reviews of jets and HH objects have been presented (e.g., Mundt 1985; Dyson 1987; Reipurth 1989, 1991, 1997; Böhm 1990; Raga 1993; Edwards, Mundt, & Ray 1993); several reviews of molecular outflows are also available (Lada 1985; Fukui *et al.* 1993; Bachiller 1996). Outflows have been the subject of several conferences (e.g., Lizano & Torrelles 1995; Reipurth & Bertout 1997). Here we briefly outline some observations most relevant for understanding the physics of pre-main-sequence mass ejection.

Some of the earliest evidence for rapid mass loss from young objects came from observations of high-velocity molecular gas in star-forming regions (e.g., Snell, Loren, & Plambeck 1980). It now appears that most of this molecular gas is ambient material that has been swept up by the higher-velocity ejecta of young stars (Masson & Chernin 1993 and references therein). Most of the kinetic energy of the outflows is quickly dissipated and radiated away as the high-velocity material shocks with the ambient molecular medium. The time-integrated input momentum inferred from the molecular gas motions indicates that outflows from young stars carry away much more momentum than is contained in the radiation field of the central source (Lada 1985).

Of particular interest for understanding the processes of mass loss are direct observations of the initial ejecta of young stars. Highly-collimated optical jets from young stellar objects have been detected in forbidden emission lines, principally [OI] $0.6300, 0.6363\,\mu$m and [SII] $0.6717, 0.6731\,\mu$m. These jets may extend for distances of 10^3–10^4 AU or more from their source (e.g., Figure 1.4), and a few have been traced to distances of a pc from the source (Bally & Devine 1994). Both spectroscopic observations of radial velocities and proper motions of 'knots' or bright emission spots in the jets show that the optical emission lines are produced by material flowing outward at $\sim$ 100–300 km s^{-1}. In many cases the optical outflows are bipolar. In sources where only one jet (blueshifted) is seen, it is generally assumed that the (redshifted) counterjet is also present but is hidden at optical wavelengths by dust in the circumstellar environment. This interpretation is supported by the observation that most molecular outflows (which are observed at radio wavelengths, and therefore not affected by dust extinction) are bipolar (Bachiller 1996).

The optical emission lines indicate that the jet gas is being heated to temperatures of several thousand degrees or more, probably by shock waves (Schwartz 1975). The detailed motions within jets which can produce shock waves and individual emission knots are complex (Raga 1995; Masson & Chernin 1993); it is likely that time-variability of the ejection is at least partly responsible for some of the jet shocks and knots (Reipurth 1989; Raga 1995). The Herbig–Haro objects may be emission knots in jet material, or emission in the wind as it shocks with the interstellar medium, or emission from interstellar gas in the shock wave (often a bow-shock) which propagates ahead of the jet (cf. Hartigan, Raymond, & Hartmann 1987).

One fundamental uncertainty in interpreting jet observations is whether the observed optical emission delineates the entire wind. Models have been proposed in which a widely-diverging, cold wind is also present in addition to the jet (Shu *et al.* 1995). Since a cold wind does not radiate effectively in forbidden optical transitions and other emission lines, it would be difficult to detect. Part of the motivation for invoking cold diverging winds was the observation of molecular outflows which are generally much less collimated than jets, but which presumably are driven by the same source. The suggestion of mass ejection in addition to what was observed in optical emission lines was supported by early estimates of mass loss rates which suggested that the optical jets could not power molecular outflows.

The case for wide-angle cold winds has weakened substantially in recent years for several reasons. Multiple jet sources have now been discovered in some molecular outflow regions, suggesting that the individual poorly-aligned jets could produce a less-collimated total molecular outflow. Theories also have been developed which may in principle allow collimated jets to produce more diverging molecular outflows (see Raga (1995) for a discussion). These theoretical developments have taken on additional importance with improved estimates of jet momentum fluxes, which suggest that the observed jets can power associated molecular outflows (Cabrit & Bertout 1992; Edwards, Mundt, & Ray 1993; Hartigan *et al.* 1994; Chernin & Masson 1995). Driving by a collimated jet more easily explains many observational features of molecular outflows than wide-angle winds (Masson & Chernin 1993). Finally, and perhaps most importantly, recent high-resolution mm-wave observations have detected highly-collimated molecular outflows from some heavily-extincted (and therefore presumably extremely young) objects (Bachiller (1996) and references therein). Unlike the optical jet observations, which are sensitive to the presence of shock heating, observations of swept-up molecular material should be sensitive to a cold wide-angle wind. These results suggest that molecular outflows from individual sources probably are highly collimated, and that the outflows of older souces are less collimated because of the expansion of swept-up material with time.

8.2   P Cygni profiles

The production and acceleration of outflows rather than their interaction with the interstellar medium are the important issues from the point of view of accretion processes. To address the ejection process in detail, it is necessary to consider observations of outflows very near their sources.

The observational study of wind acceleration relies on the interpretation of velocity shifts and asymmetries in absorption and emission lines. Figure 8.1 illustrates some basic ways in which velocity fields and outflow geometry produce the so-called 'P Cygni' line profiles. The particular situation illustrated is that of spherically-symmetric outflow from a star. The expanding high-velocity outflow produces a broad line profile. Region (1) tends to contribute excess emission toward the observer, either because photons are created in the line, or because stellar photons are scattered into the line of sight. This region produces both blueshifted and redshifted emission. Region (2) would contribute emission at the highest redshifts, but is occulted by the stellar disk. In the case of an optically-thick line illustrated in the bottom left-hand panel of Figure 8.1, material in region (3) absorbs light from the star, producing a blueshifted absorption feature. The net result is an asymmetric line profile, with redshifted emission and blueshifted absorption – the canonical P Cygni profile observed in many hot stars.

Variations on this situation can be imagined which change the profile shape. For example, suppose that the star is surrounded by an opaque disk (right-hand panel of Figure 8.1). Then the region of occultation (2) becomes much larger. One can see that the effect of a disk at most inclination angles will be to eliminate the emission preferentially that arises from redshifted material; in the limit where the opaque inner disk edge joins the star, and the disk lies in the plane of the sky as seen by the observer,

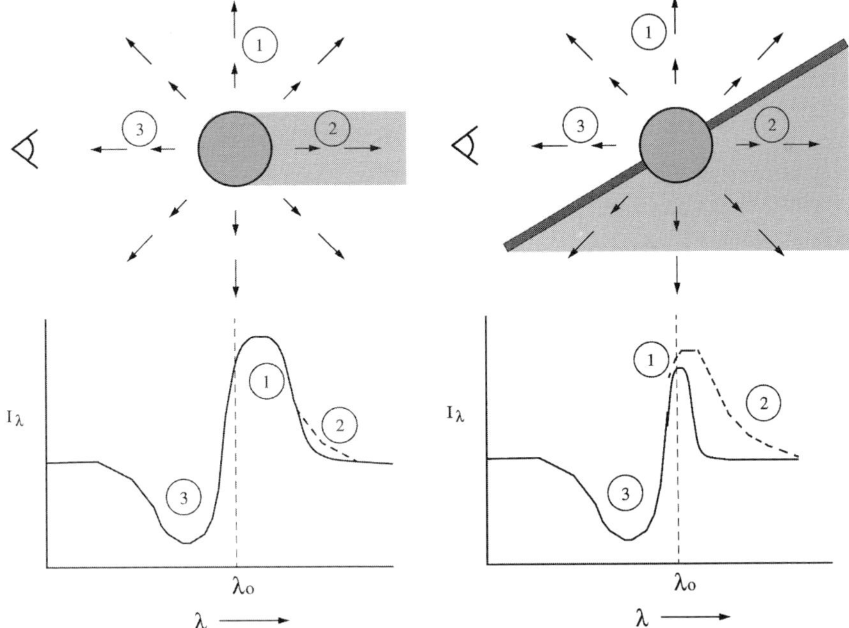

Fig. 8.1. Schematic diagram of geometry responsible for producing P Cygni wind line profiles (see text).

all of the redshifted emission will be occulted, and all that will be left is blueshifted absorption. If the spectral line is optically thin, then the absorption of star light will be negligible. Depending upon the physics of line excitation, region (3) may produce blueshifted *emission* rather than absorption; again, the disk will hide the redshifted material. The FU Ori objects and the T Tauri stars both show evidence for this disk occultation. In the case of FU Ori systems, the blueshifted spectral features are seen mainly in absorption in permitted atomic transitions, while in the T Tauri stars the winds are most easily detected in blueshifted emission from optical forbidden lines.

Figure 8.2 illustrates Balmer Hβ and Na I resonance line profiles in four FU Ori objects. Broad blueshifted absorption is observed, indicating mass ejection at several hundred km s^{-1}. There is no evidence for redshifted emission or absorption, indicating that unless the Earth is in some privileged place of observation, occultation of the redshifted outflow must be occurring. This is consistent with the FU Ori objects being optically-thick accretion disks.

In contrast, the Hα profiles of typical T Tauri stars illustrated in Figure 6.7 exhibit qualitatively different profiles.* Although there is some evidence for blueshifted absorption in some T Tauri stars, it is very much weaker than in the FU Ori objects. Moreover, in objects like BP Tau, DK Tau, and DN Tau (Figure 6.7), the *blueshifted* emission is either stronger or extends to higher velocity shifts than the redshifted emission. The simplest way of producing such asymmetry is to reverse the direction of the flow from that of Figure 8.1, producing the emission in *infalling*, not outflowing, gas (§8.12).

---

* The Hα and Hβ profiles of the FU Ori objects are quite similar.

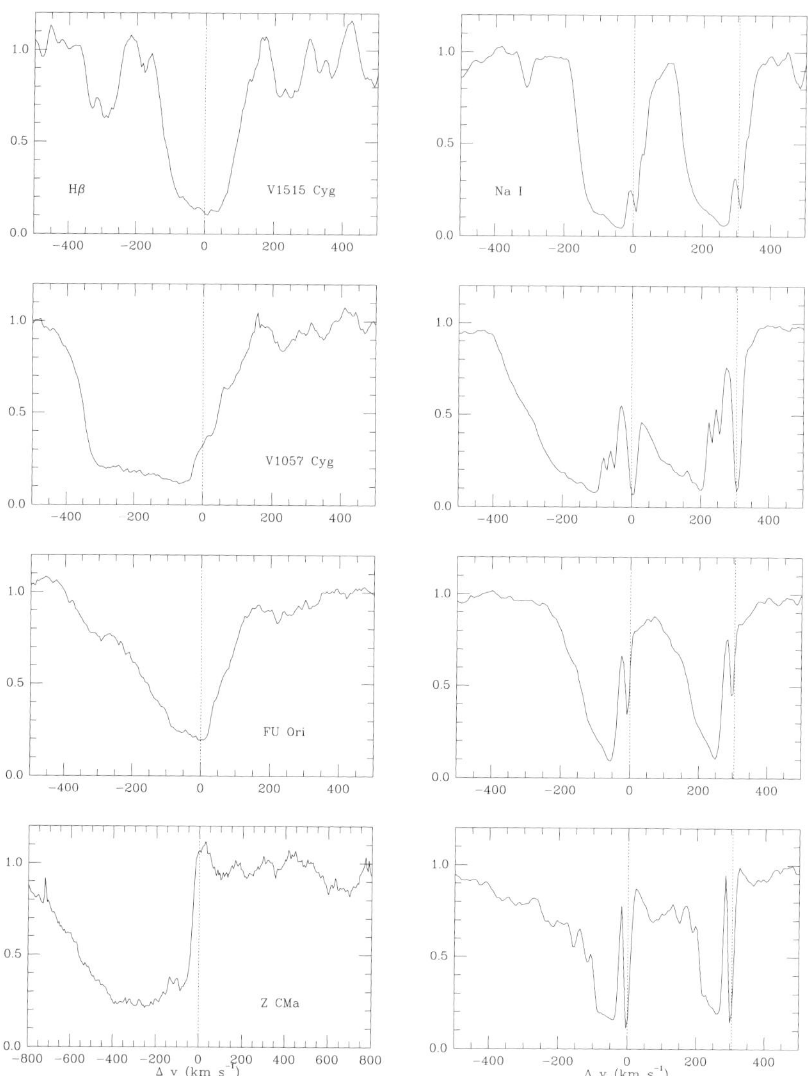

Fig. 8.2. Hβ (left-hand panels) and Na I (right-hand panels) line profiles observed in four FU Orionis objects. The dotted lines indicate the rest velocities for Hβ and the two Na resonance lines. The narrow absorption components of the Na I lines at rest velocities correspond to absorption in the adjacent cold, relatively slowly-moving interstellar medium. From Hartmann & Calvet (1995).

Similarly, the Na I wind absorption of T Tauri stars is very much weaker than in FU Ori objects, as shown in Figure 8.3. Na I blueshifted absorption is not detectable in many CTTS, and is not seen in any WTTS. In some CTTS, redshifted Na I absorption dominates (e.g., BP Tau in Figure 8.3; also Ulrich & Knapp (1979)), which is now attributed to magnetospheric infall (§8.12).

Thus FU Ori objects, with very rapid accretion rates, show the signatures of much higher mass loss rates than the more slowly-accreting CTTS. Moreover,

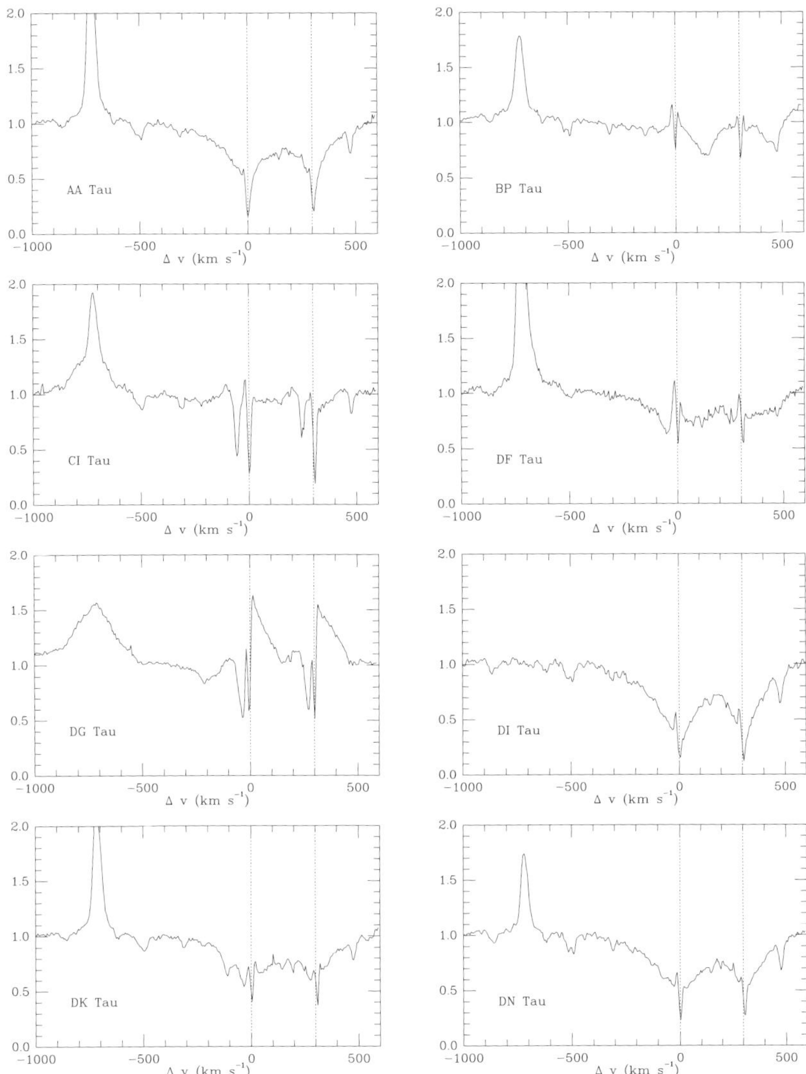

Fig. 8.3. Na I resonance line profiles for typical T Tauri stars. The dashed lines mark the rest velocities of the doublet; the data near these regions are affected by narrow interstellar absorption and night-sky emission components. The WTTS DI Tau exhibits the typical photospheric Na I absorption wings in the absence of any accretion processes. While blueshifted absorption may be seen in CI Tau, DF Tau, DG Tau, and DK Tau, it is very much weaker than observed in the more rapidly-accreting FU Ori objects (Figure 8.2). BP Tau exhibits redshifted absorption produced by magnetospheric accretion (§8.12). The strong emission line at wavelengths shorter than the Na I doublet is He I 5976.

the non-accreting WTTS do not show optical mass loss signatures. These results qualitatively indicate that disk accretion is responsible for powering outflows. In the following sections we explore the evidence for this conclusion in quantitative detail.

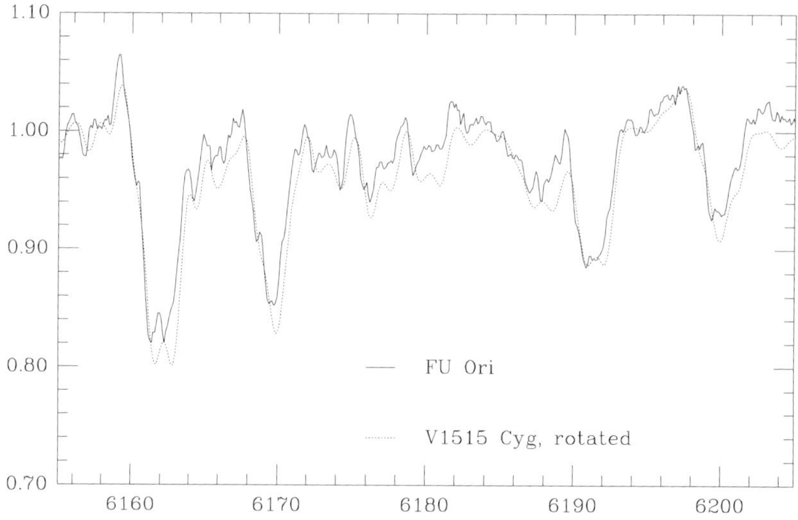

Fig. 8.4. A comparison between the observed spectrum of FU Ori near 6200 Å and a disk model spectrum (dotted line) synthesized by rotationally broadening the V1515 Cyg spectrum (see text). From Hartmann & Calvet (1995).

8.3   FU Ori disk winds

Observations demonstrate that the wind of FU Ori itself arises from the surface of the accretion disk. This is possible because wind velocity shifts can be detected in many weak absorption lines formed in the outer disk atmosphere. Because there are many available lines, with widely varying strengths, it is possible to follow the wind as it accelerates away from the disk photosphere.

Figure 8.4 shows a segment of the optical spectrum of FU Ori at high spectral resolution. For comparison the spectrum of another FU Ori object (V1515 Cyg), artificially broadened to match the rotational velocity of FU Ori, is included. Although the absorption features match up reasonably well, the strong lines and blends near 6162, 6170, and 6191 Å are blueshifted in FU Ori relative to the V1515 Cyg spectrum. Apparently, the wind of FU Ori is so strong that the upper photosphere is actually moving outward at detectable rates. Even with this expansion, the blend near 6162 Å is *double-peaked*; this implies that the material is *rotating* like the disk (cf. §7.2) as it expands outward.

Consider what might be observed from the differentially-expanding atmosphere of an FU Ori accretion disk. Weak absorption lines, formed near the disk photosphere, will exhibit the 'doubled' line profiles produced by disk rotation (§7.2; Figure 7.7), as shown by the uppermost profile (dotted line) shown in the upper panel of Figure 8.5. Stronger lines will be formed further up in the disk atmosphere, farther from the disk midplane. The atmosphere is cooler in these outer layers, and therefore the absorption profiles of these lines will be deeper, as in a normal stellar atmosphere. Since the gas density is lower at higher levels in the atmosphere, conservation of mass for a (roughly) steady wind implies a larger expansion velocity. The net result is that stronger lines are both deeper and more blueshifted. If the velocity shift due

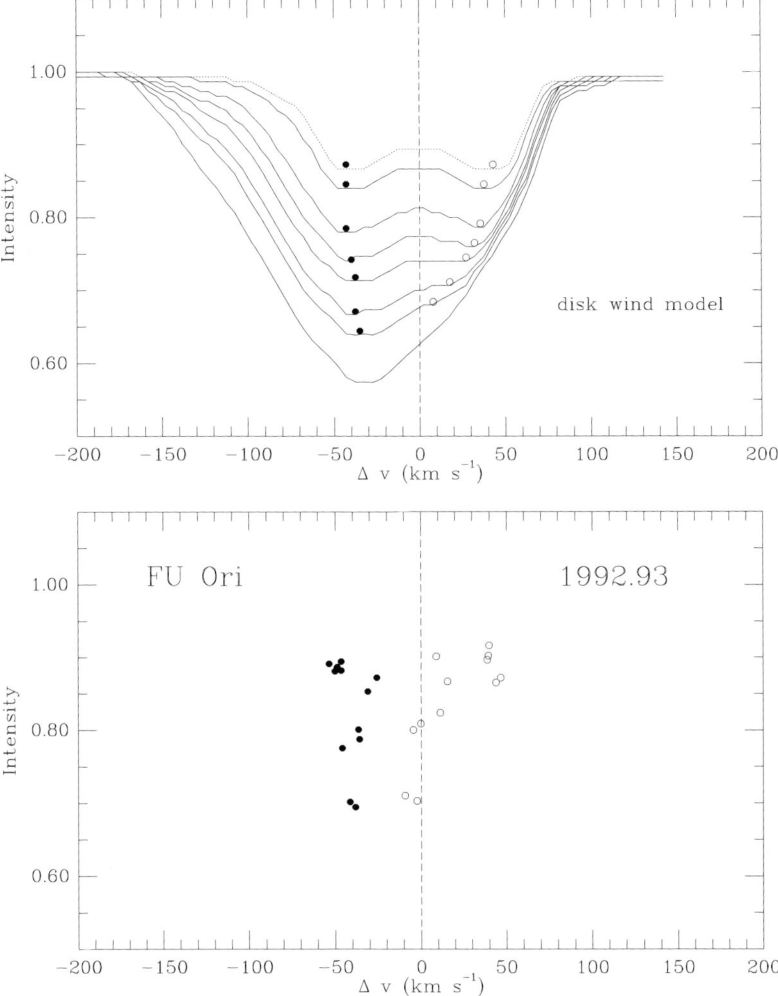

Fig. 8.5. Upper panel: Predicted disk wind profiles for a series of lines with different strengths, taken from Calvet *et al.* (1993). Filled and open circles mark the positions of the absorption dips for the blue and redshifted components, respectively. Lower panel: velocity shifts of absorption components in FU Ori, from Hartmann & Calvet (1995). See text.

to expansion is small compared with the rotational velocity, the combination of both motions will result in a profile that is dominated by rotation. If the lines are strong enough that the level of line formation in the disk atmosphere is very high, and the velocity of expansion dominates the line profile, the result will be a P Cygni profile with little or no evidence of rotation (e.g., Figure 8.1). Therefore, absorption lines formed in a disk wind should show increasing blueshifts with increasing line strength.

To explore this predicted behavior in more detail, Calvet *et al.* (1993) modelled the FU Ori disk atmosphere as a series of disk annuli, each with its own independent atmosphere; this is justifiable in the plane-parallel limit. The temperature structure of the atmosphere for each annulus was calculated for vertical radiative equilibrium with the appropriate surface flux at each radius. To make the results more realistic, the effects of ionization balance were included. Profiles were calculated for a series of typical Fe I lines of varying oscillator strengths assuming local thermodynamic equilibrium or LTE (Mihalas 1978), and adopting a linear acceleration of the expanding disk atmosphere combined with Keplerian rotation (see Calvet *et al.* (1993) for details).

The results of these calculations are shown by the sequence of line profiles in the upper panel of Figure 8.5. This sequence quantitatively demonstrates the profile evolution described above. Atomic spectral lines of progressively greater strength compose a sequence of line profiles, with increasing line depth accompanied by an increasing blueshift of the absorption. An interesting feature of the calculation in the upper panel of Figure 8.5 is that the absorption component on the red side of the line becomes increasingly blueshifted, as would be expected intuitively, but the blue absorption component does not appear to shift much in velocity. This is because the blue component is not an actual feature in the wind velocity profile, but instead is due to the convolution of the rotational broadening profile with the wind expansion profile (Calvet *et al.* 1993).

One simple way of characterizing this evolution of line profiles with increasing line strength is to consider the positions of the two absorption components in the line profiles, marked in the upper panel of Figure 8.5 by the open circles for the redshifted component and filled circles for the blueshifted component. As the line strength increases, the line becomes deeper and so the two absorption components appear at lower residual intensities; the line becomes more blueshifted, and the two absorption features move together. Eventually, for lines strong enough to be formed at a sufficiently high atmospheric level where the expansion velocity dominates the rotation, the two absorption components merge into one blueshifted feature.

This qualitative behavior of line profiles has been observed in FU Ori. The most detailed results were presented by Petrov & Herbig (1992), who measured the profiles of a large number of relatively unblended lines. Petrov & Herbig showed that as the depth of the line increased, the red absorption feature moved to more negative velocities, the blue absorption component remained at almost fixed radial velocity, and the centroid of the line profile became blueshifted, in qualitative agreement with the disk wind model prediction. The lower panel in Figure 8.5 shows confirming measurements of the absorption dip positions (Hartmann & Calvet 1995). The correspondence between the model predictions and the observed dip positions is quite good. The 'tomographic' view of the wind obtained by isolating lines of different strengths clearly demonstrates the evolution of the flow from pure Keplerian rotation to outflow, and represents the *only* situation at present where the wind can definitely be shown to arise from the accretion disk. (This effect is not observed in other FU Ori objects such as V1515 Cyg, probably because they have lower mass loss rates; Hartmann & Calvet (1995).)

8.4   T Tauri winds

Classical T Tauri stars show evidence for mass loss in the blue-shifted absorption seen the P Cygni profiles of optically-thick permitted spectral lines (Figures 6.7, 8.3). Although numerous efforts have been made to study the properties of these line profiles, and infer mass loss rates from them (e.g., Kuhi 1964; Kuan 1975; Herbig 1977a;

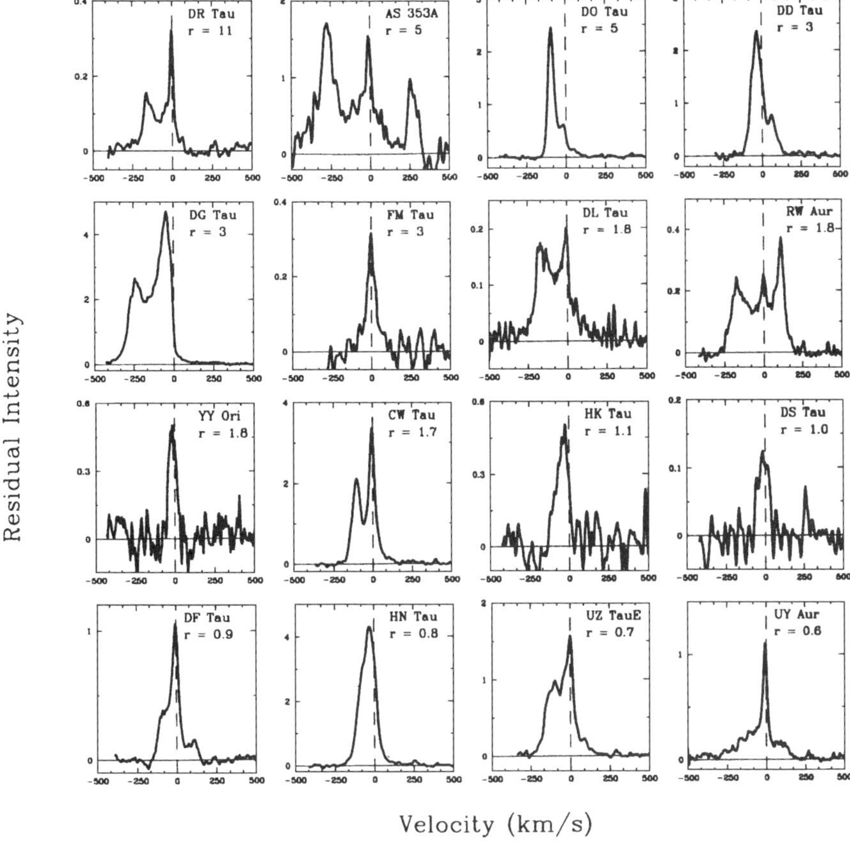

Velocity (km/s)

Fig. 8.6. [O I] λ 6300 Å emission of representative T Tauri stars. In most cases the emission is blueshifted with respect to the stellar velocity, denoted by the dashed vertical line. From Hartigan *et al.* (1995).

DeCampli 1981; Lago 1984; Mundt 1984; Natta & Giovanardi 1990; Hartmann *et al.* 1990; Reipurth, Pedrosa, & Lago 1996), mass loss rates remain very uncertain for two main reasons. First, in many CTTS, the absorption line components are quite weak or absent in optical tracers, making mass loss determinations difficult (§8.5). Second, it appears that much or most of the *emission* portion of the line profile is produced by *infall* (§8.12), while the blueshifted absorption comes from a distinct *outflow* region.

Some T Tauri stars exhibit spatially-resolved jets (cf. Figure 1.4) which can be observed in optical emission lines such as Hα, and particularly in forbidden [O I] 6300, 6363 Å and [S II] 6717, 6731 Å emission lines. Optical observations of T Tauri stars have usually been made with spatial resolutions comparable to typical seeing disks $\sim$ 1 arcsec, and so also include emission from the near environments of these stars ($\sim$ 140 AU in the nearest star-forming regions). At this spatial resolution, many T Tauri stars exhibit optical forbidden emission lines with velocity widths $\sim$ 100–300 km s^{-1}, comparable to the velocities seen in spatially-resolved jets; these forbidden emission lines are almost always blueshifted with respect to the star (Appenzeller *et al.* 1984; Edwards *et al.* 1987; Figure 8.6). Unless these objects are mostly beaming their outflows toward earth, some obscuration of the receding flow must be present, such as

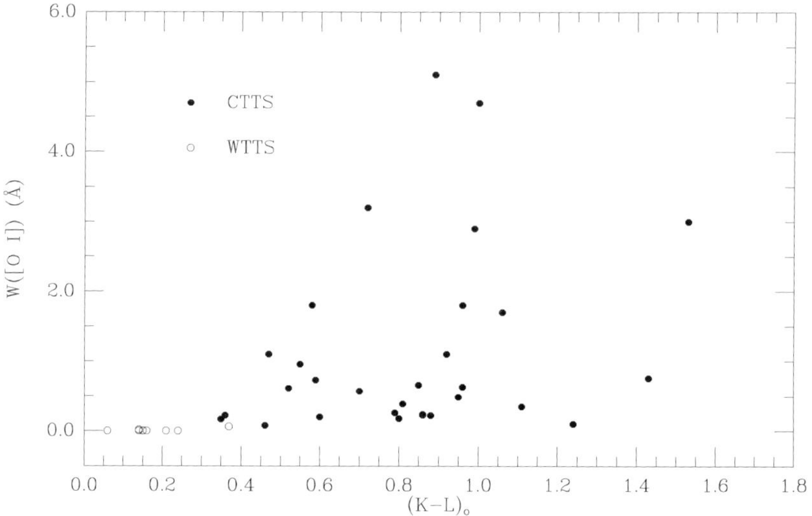

Fig. 8.7. The equivalent width of forbidden [O I] 6300 Å emission vs. $K - L$ color (see Figures 6.9 and 6.10). Only objects with near-infrared excess emission, as indicated by a color $(K - L)_o \gtrsim 0.3$ exhibit forbidden-line emission produced by mass loss (see text, Figure 8.6). Data taken from Hartigan *et al.* (1995).

would be provided by a dusty circumstellar disk (Figure 8.1). Since redshifted emission is rarely observed, the observations suggest that the jet is already collimated on size scales of 100 AU or less even if a disk is present; otherwise, at typical inclination angles, one would observe some portion of the receding flow (Figure 8.1). *HST* images of a few objects also suggest that at least the emitting material in the outflow is collimated on size scales of $\lesssim$ 30 AU (Kepner *et al.* 1993; Burrows *et al.* 1996; Ray *et al.* 1996).

Optically-detectable CTTS mass loss is *always* associated with evidence for an inner circumstellar disk, and nearly always with detectable disk accretion (Hartigan *et al.* 1995). As shown in Figure 8.7, only those T Tauri stars with near-infrared excess emission attributable to a disk (CTTS) exhibit detectable blueshifted forbidden line emission (Edwards 1995); the WTTS do not show evidence for strong winds. There is also a good correlation between evidence for the hot continuum emission, thought to be produced by accretion onto the central star, and the blueshifted jet/wind emission. The theoretical explanation of this behavior is that jet/wind is produced in some way by accretion disks, and taps a significant fraction of the energy released by accretion.

8.5   Mass loss rates from permitted absorption lines

Mass loss rates have been estimated using the profiles of optically-thick permitted spectral lines, with special emphasis on analyzing the blueshifted absorption produced by expansion (Figures 8.1–8.3). Usually it suffices to adopt the Sobolev approximation (§4.7) to calculate the optical depth in terms of physical quantities. In this approximation it is assumed that the velocity gradients in the expanding atmosphere are large, and therefore only a narrow volume in the wind is responsible for the line absorption and emission at a given velocity shift. Rybicki & Hummer (1978) provide a clear discussion of the Sobolev treatment; here we outline some relevant results.

The observed flux F_v at a given frequency v from line center is proportional to

$$F_v = \int d\Omega \left\{ I_c \exp(-\tau_v) + S[1 - \exp(-\tau_v)] \right\}, \qquad (8.1)$$

where the integral is taken over solid angle as seen by the observer, I_c is the background continuum source intensity, and S and τ_v are the line source function and optical depth, respectively, evaluated along the line of sight to the observer in regions where the velocity shifts the spectral line of central frequency v_o to v, i.e. where $v = v_o(1 - v/c)$, where v is the line-of-sight velocity. The term in curly brackets is the formal solution to the transfer equation (Appendix 3) in the limit where only a small part of the atmosphere is at the correct velocity shift, and therefore one can assume that S remains constant through this small volume along the line of sight.

To illustrate the essential aspects of the problem with a minimum of complication, consider the case in which the line source function is negligible and the background continuum is constant and uniform over area da, and the wind seen projected against this continuum is also uniform across this area. Then, for an object at distance d, the blueshifted absorption portion of the line profile is

$$F_v \sim \frac{da}{d^2} I_c \exp(-\tau_v). \qquad (8.2)$$

Suppose a spectral line of a species A has a Gaussian line profile with intrinsic velocity width v_{th} (due to thermal Doppler motions and any other velocity broadening). The line absorption per atom in the absorbing state is (Mihalas 1978)

$$\alpha_v = \frac{\pi e^2}{mc} \frac{f}{\pi^{1/2} \Delta v_D} \exp[-(\Delta v/\Delta v_D)^2], \qquad (8.3)$$

where $\Delta v_D = v_o v_{th}/c$ and the oscillator strength is related to the spontaneous emission probability by

$$f = 1.5 \times 10^{-8} \lambda_\mu^2 (g_2/g_1) A_{21}, \qquad (8.4)$$

where λ_μ is the wavelength of the transition in microns.

The absorption profile can be written in terms of a velocity shift (related to the frequency shift from line center),

$$\alpha_v = \frac{\pi e^2}{mc} f \frac{c}{\pi^{1/2} v_o v_{th}} \exp[-(\Delta v/v_{th})^2]. \qquad (8.5)$$

where

$$\Delta v = c(v - v_o)/v_o. \qquad (8.6)$$

If we expand the velocity field in the line-of-sight direction z about the point z_o where the expansion shifts the absorption line to the velocity of interest, then

$$\Delta v = v - v_o = \frac{dv}{dz}(z - z_o). \qquad (8.7)$$

Assuming that the population N_1 remains constant over the (narrow) region of interest, we estimate the Sobolev optical depth as

$$\tau_v \simeq \int dz\, \alpha_v N_1 = \frac{\pi e^2}{mc} f \frac{c}{v_o} \frac{N_1(z_o)}{|dv/dz|_{z_o}}. \qquad (8.8)$$

Note that the intrinsic line width v_{th} drops out in this approximation. This can be understood as follows. Only atmospheric regions with velocity shifts $\sim \pm v_{th}/2$ of the central velocity contribute to the optical depth integral along a given line of sight; therefore any increase in thermal velocity width, which reduces the opacity per unit distance, correspondingly increases the distance over which the line can effectively contribute to the absorption, and thus the optical depth at v is unaffected (§4.7).

An estimate of the wind absorbing optical depth τ_v from analysis of the line profile then yields an estimate of the wind mass loss rate,

$$\dot{M}_w \sim (2) \times da\, v\, \mu m_H\, N_H(v),\tag{8.9}$$

where

$$N_H(v) \propto \tau_v\, |dv/dz|\, (N_H/N_1).\tag{8.10}$$

The factor of (2) is intended to account for the bipolar nature of the flow (other treatments often assume spherical wind expansion, so that $2da = 4\pi R^2$).

To illustrate the basic parameter dependence of wind absorption features, we evaluate the jet or wind optical depth at a (near-escape) velocity v_w, which we assume is attained at a distance of about $R \sim 2R_*$. We further estimate $|dv/dz| \sim v_w/R$, and take the wind to flow through an area $da = \pi R^2$. Then the line optical depth is

$$\tau_v \sim 3 \times 10^9 f\, \lambda_{0.5\mu}\, (N_1/N_H)\, \dot{M}_{-8} R_4^{-1} v_{100}^{-2}\, |dv_{100}/dz_4|^{-1},\tag{8.11}$$

where $\lambda_{0.5\mu}$ is the wavelength of the transition in units of $0.5\,\mu$m, $\dot{M}_{-8}$ is the mass loss rate in units of $10^{-8}\,\mathrm{M}_\odot\,\mathrm{yr}^{-1}$, R_4 is the adopted radius in units of $4\,\mathrm{R}_\odot$, v_{100} is the wind velocity in units of $100\,\mathrm{km\,s}^{-1}$, and $|dv_{100}/dz_4|$ is the velocity gradient in the same distance and velocity units.

Ideally, one would like to estimate $\dot{M}_w$ in cases where the fractional abundance N_1/N_H is not sensitive to assumed or poorly-known parameters. For example, at the (low) wind temperatures estimated in T Tauri winds, most of the Ca atoms in the wind should be singly ionized, and therefore the population of the ground state of the Ca II resonance lines at $0.39\,\mu$m relative to the total hydrogen number density is given simply by the cosmic abundance of Ca, $N_1 \sim N(Ca) \sim 2 \times 10^{-6} N_H$. Unfortunately, with this abundance and the oscillator strengths of order unity for the resonance lines, the predicted wind optical depths are $\sim 3 \times 10^3$ for the fiducial parameters, which serve as a reasonable order-of-magnitude estimate for most T Tauri winds. Such large optical depths are far too large to be estimated with any accuracy from the depths of the blueshifted absorption features (Kuhi 1964; Mundt 1984).

To use blueshifted wind absorption to estimate mass loss rates, it is generally necessary to analyze lines from ions that are not the dominant stage of ionization of a particular atom, such as the resonance lines of Na I at $0.59\,\mu$m, or lines whose lower levels are highly excited states and therefore strongly depopulated relative to the ground state, such as the Balmer lines. However, the mass loss rates derived from such lines are therefore very sensitive to the adopted temperature and/or ionization structure in the wind. Natta & Giovanardi (1990) tried to circumvent these problems by simultaneously analyzing the Na I resonance absorption lines and infrared Brackett hydrogen emission lines of T Tauri stars. Over a moderately wide range of conditions, the Na I ground-state populations are more sensitive to the local density than to the temperature. In contrast, the hydrogen lines are quite sensitive to temperature, and only moderately sensitive to density. Combining results for the two species, one can estimate the wind temperature and density independently, and therefore determine the mass loss rate given some assumption about the wind velocity as a function of radius. Unfortunately, the ionization fraction of Na I is so low that blueshifted $0.59\,\mu$m resonance line absorption is observed in only a small minority of T Tauri stars, which presumably have the highest mass loss rates. Natta & Giovanardi (1990) derived mass loss rates for a few T Tauri stars in the range of $\sim 10^{-8}$–10^{-7} M$_\odot$ yr^{-1}, strongly based on Na I absorption estimates. These results may be biased somewhat because of the use of the emission flux of the infrared hydrogen lines to constrain the wind temperature; it now appears that this emission is produced in the accreting magnetospheric flow, not the wind (Najita, Carr, & Tokunaga 1996; Folha, Emerson, & Calvet 1997; §8.12).

A revised version of this type of analysis adopting typical sizes of wind regions and velocity gradients, but using only absorption strengths (or limits on absorption strengths) is shown in Figure 8.8. Many CTTS show some evidence for blueshifted Hα absorption, but such absorption is much less frequently seen in the higher Balmer lines; this suggests that the high Balmer series lines are optically thin in the wind, and therefore τ(Hα) cannot be much larger than unity. Observations at high spectral resolution generally show no evidence for blueshifted absorption in Brγ (Najita *et al.* 1996) or Paβ (Folha *et al.* 1997). Similarly, most CTTS do not show blueshifted Na I absorption (cf. Figure 8.3); for the modest number of strong-emission stars which do show such absoption, τ(Na I) $\sim$ 1–10 (Natta & Giovanardi 1990). From these rough constraints Figure 8.8 suggests that the mass loss rates of most CTTS are $< 10^{-8}$ M$_\odot$ yr^{-1}; mass ejection rates of the strongest emission stars may be in the vicinity of several $\times 10^{-8}$ M$_\odot$ yr^{-1} (Calvet 1997).

The FU Ori objects, with disk accretion rates $\sim 10^2$–10^3 larger than T Tauri stars, have correspondingly much larger mass loss rates. As shown in Figure 8.2, it is evident from the absorption blueshifts that both the Balmer and Na I lines are formed in outflows. The differing temperature- and density-sensitivity of hydrogen lines on the one hand and the Na I resonance doublet on the other provides a means for simultaneously estimating wind temperatures and densities, and therefore determining mass loss rates. The analysis of Croswell, Hartmann, & Avrett (1987) found a mass loss rate of FU Ori is $\sim 10^{-5}$ M$_\odot$ yr^{-1}, i.e. an order of magnitude smaller than the disk accretion rate, and a wind temperature $\sim$ 5000–6000 K. The precise values derived depend upon the assumed velocity gradient (equation (8.10)). The results also depend somewhat upon the geometry of the outflow; Croswell *et al.* obtained line profiles in better agreement with observations if the wind is collimated rather than spherically symmetric.

By analyzing specific metal lines with a detailed disk atmosphere model, Calvet, Hartmann, & Kenyon (1993) were able to estimate a mass loss rate of $\sim 10^{-5}$ M$_\odot$ yr^{-1} for FU Ori, similar to the estimate of Croswell *et al.* (1987). Although the mass loss rate derived in this way also depends inversely on the assumed velocity gradient of the outflow, the use of many more individual lines formed in the low-velocity portion of the outflow where the gas temperatures are more closely constrained to photospheric values, suggests that this estimate for FU Ori is relatively robust. Photospheric line profile shifts are not apparent in the other well-studied FU Ori objects, suggesting that the mass loss rates in these systems are substantially lower, $\lesssim 10^{-6}$ M$_\odot$ yr^{-1}. In

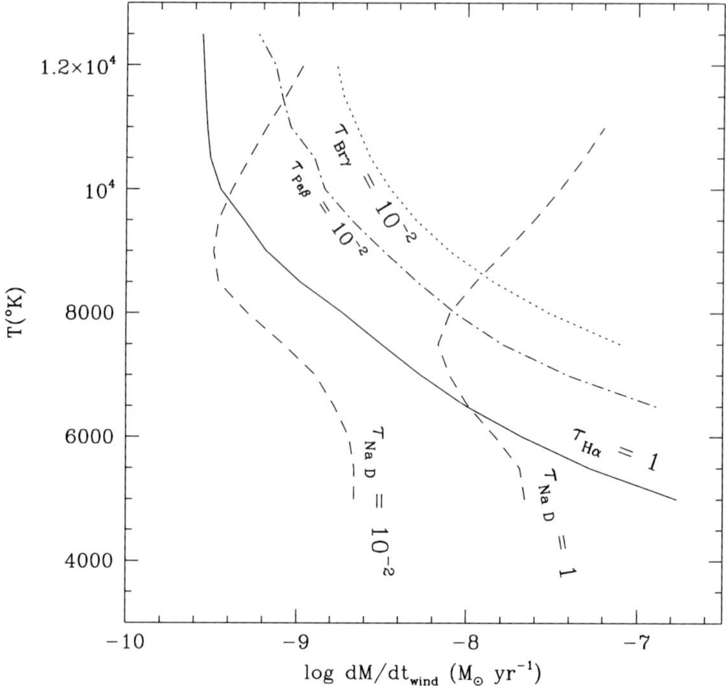

Fig. 8.8. Predicted optical depths for various permitted lines in T Tauri winds as a function of wind temperature and mass loss rate (see text). From Calvet (1997).

general, the estimated mass loss rates from FU Ori objects are typically 10^{-1}–10^{-2} of the mass accretion rates.

8.6   Mass loss rates from forbidden emission lines

As shown in Figure 8.6, many T Tauri stars exhibit high-velocity blueshifted forbidden-line emission, which in principle can be analyzed to estimate mass ejection rates. The main uncertainty involved is the temperature structure of the outflowing gas.

The forbidden lines are optically thin and so the total luminosity is proportional to the number of emitting atoms in the observed volume,

$$L_{12} = \int dV N_2(A) A_{21} E, \tag{8.12}$$

where the subscripts 1 and 2 refer to the lower and upper atomic levels of the spectral line being observed in volume V, $N_2(A)$ is the number density of atoms of species A in the upper level 2, A_{21} is the Einstein spontaneous emission probability for the transition, and

$$E = h\nu_{21} \tag{8.13}$$

is the energy of the observed spectral line. If observations are limited to a region of radial extent R_w, and the expansion velocity v_w is estimated from the radial velocity shifts, the (one-sided) mass loss rate is

$$\dot{M}_w \;\sim\; M(A)\frac{v_w}{R_w} \;\sim\; \frac{v_w}{R_w}\int dV\, N_2(A)\,(N_A/N_2)(N_H/N_A)\mu m_H \,, \tag{8.14}$$

where $M(A)$ is the mass of species A in the volume under observation, N_A/N_H is the abundance of atom A relative to hydrogen, and μ is the mean molecular weight. The mass loss rate can then be determined from observations of the forbidden-line luminosity, if an estimate of the fractional abundance N_2/N_A can be made.

It is often possible to assume that the only processes involved in producing the line emission are collisional excitations between the lower level 1 and the upper level 2, along with spontaneous radiative decay. This is especially true if level 1 is the ground state in which most of the ions reside at any instant. Under these assumptions the equation of statistical equilibrium can be written as

$$N_1 N_e C_{12} \;=\; N_2(A_{21} + N_e C_{21}), \tag{8.15}$$

where the C_{ij} are the collision rates, and N_e is the density of electrons responsible for inducing the collisional transitions (Spitzer 1978). Because the distribution function of the colliding electrons can generally be taken to be Maxwellian, the collision rates are related by the principle of detailed balance (Mihalas 1978),

$$C_{12} \;=\; (N_2/N_1)^* C_{21} \;=\; (g_2/g_1)\,\exp(-E/kT)C_{21}\,, \tag{8.16}$$

where the starred quantities refer to the LTE ratios, and the g_i are the statistical weights of the atomic levels. With this result one can write

$$L_{12} \;=\; \int dV\, \frac{N_1 N_e\, C_{21}\,(g_2/g_1)\,\exp(-E/kT)}{1 + N_e C_{21}/A_{21}}\, E \,. \tag{8.17}$$

The collision rate is a function of temperature; usually it is written

$$C_{21} \;=\; \frac{8.63 \times 10^{-6}\,\Omega_{21}}{g_2\, T^{1/2}}\ \mathrm{cm^3\,s^{-1}}\,, \tag{8.18}$$

where Ω_{21} is usually a slowly-varying function of temperature.

To determine mass loss rates we require the number of ions in state 1 relative to the total atomic abundance, N_1/N_A, the gas temperature, and the electron density. Since level 1 is often the ground state of ion A for strong forbidden lines, it is sometimes possible to assume that most of the atoms of species A are in the ground state, i.e. $N_1 \sim N_A$, *if* this is the dominant stage of ionization for A. For example, [O I] and [S II] are likely to be dominant stages of ionization provided the gas temperature is not $>> 10^4$ K. This conclusion must be modified for the time-dependent ionization structure of shocks, in which the ionization states are not unique functions of the local density and temperature but depend upon initial shock conditions as well (cf. Hartigan *et al.* 1994). The observed emission in these forbidden lines is likely to be dominated by only those regions where the conditions can produce $N_1 \sim N_A$. This in turn tends to restrict the temperature range relevant to the emission accordingly, although there is no guarantee that *all* of the wind posesses such temperatures.

The line emission (equation (8.17)) generally depends upon the electron density. In some cases it is possible to make an independent estimate of the electron density by comparing the emission from different lines. For many T Tauri stars, the ratio of

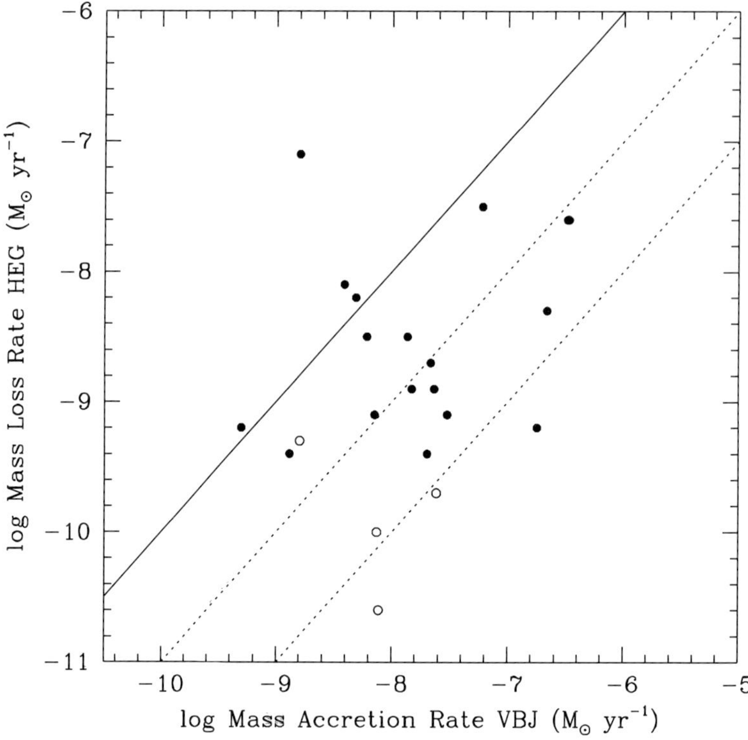

Fig. 8.9. T Tauri mass loss rates estimated from (high-velocity) forbidden-line emission plotted as a function of mass accretion rates determined from hot continuum emission (§6.4). Open circles are upper limits to the [O I] mass ejection rate. The mass loss rates are taken from Hartigan *et al.* (1995), while the mass accretion rates are taken from Valenti *et al.* (1993) (see text).

the [S II] 6717, 6731 Å forbidden lines indicates emission in the high-density limit $N_e C_{21} \gg A_{21}$, in which case the dependence on electron density is eliminated,

$$L_{12} \sim \int dV N_1 A_{21}(g_2/g_1) \exp(-E/kT) E. \tag{8.19}$$

Figure 8.9 shows T Tauri mass loss rates estimated from the [O I] 6300 Å and 6363 Å emission lines, using equations (8.14) and (8.17), and including constraints on electron densities from the [S II] lines (Hartigan *et al.* 1995). To derive these mass loss rates, Hartigan *et al.* adopted an emitting temperature of ~ 8000 K for the collision rates, based on results from shock models for jets which suggest that this is the characteristic (post shock) temperature where most of the emission is radiated. Since $E/k \sim 2.2 \times 10^4$ K for the [O I] lines, the derived mass loss rates are not highly sensitive to the temperature unless it is very much lower than 8000 K.

Shock models actually exhibit a range of temperatures in the cooling gas, and this can be taken into account in detailed calculations (Hollenbach 1985; Hartigan *et al.* 1994). The Hartigan *et al.* (1995) results do not take this into account, mainly because without more information (spatial resolution, etc.) it is difficult to know just how the forbidden-line emitting region is heated, how many shocks there are, etc. These uncertainties plus the likely time-variability of both mass ejection and accretion could account for most of the scatter in Figure 8.9.

The general trend in Figure 8.9 suggests mass ejection rates $\sim 10^{-1}$ of mass accretion rates (Chapter 6). Because of the possible existence of cold, non-emitting gas, the forbidden-line mass loss rates generally are lower limits to the true values. It is difficult to say how much larger the total mass ejection rates are, but in a few T Tauri stars independent mass loss rate estimates are generally no more than an order of magnitude larger (Natta & Giovanardi 1990).

The optically-bright FU Ori objects do not show the forbidden-line jet emission, although the heavily-extincted objects V346 Nor and L1551 IRS 5 do show spatially-resolved jet emission. (Z CMa has an optical jet, but it is not clear if it is associated with the FU Ori object or the more luminous infrared companion.) Spatially-resolved emission may not be detected if insufficient time has elapsed since the outburst; at typical distances of 500–1000 pc, the jet must be $\gtrsim 10^3$ AU long to be resolved from ground-based observations, and it would take about 50 years for the jet to expand this far at $300\,\mathrm{km\,s^{-1}}$. The reasons why the inner wind does not emit strongly in forbidden lines are less clear, but it may be that the densities are so high in the wind that the forbidden lines are effectively collisionally deexcited (8.15), and the radiative cooling required by any shock heating is accomplished through other (permitted) transitions.

In one case where the forbidden-line jet has been studied in detail, L1551 IRS 5, Stocke *et al.* (1988) made an estimate for the mass loss rate of $\lesssim 10^{-7}\,\mathrm{M_\odot\,yr^{-1}}$, whereas Cohen, Bieging, & Schwartz (1982) estimated $\dot{M}_w \lesssim 3.5 \times 10^{-7}\,\mathrm{M_\odot\,yr^{-1}}$ from an analysis of the radio continuum emission from the ionized jet. It is suggestive that IRS 5, which at $\hat{L} \sim 20\,\mathrm{L_\odot}$ has an apparent luminosity intermediate between most FU Ori objects and T Tauri stars, appears to have an intermediate rate of mass loss. Radio continuum observations of some other FU Ori objects (Rodriguez, Hartmann, & Chavira 1990; Rodriguez & Hartmann 1992) suggest *ionized* mass loss rates $\sim 10^{-7}\,\mathrm{M_\odot\,yr^{-1}}$, but these are only lower limits since it is highly likely that FU Ori winds are mostly neutral.

In summary, mass loss rates for T Tauri winds/jets are not very well constrained. Most T Tauri stars probably have mass loss rates $\lesssim 10^{-8}\,\mathrm{M_\odot\,yr^{-1}}$, based on the absence of blueshifted Na I absorption and weak forbidden-line emission. In the few cases where mass loss estimates can be made by more than one technique, the agreement is at the order of magnitude level. Overall, the mass ejection rates of T Tauri stars appear to be $\sim 10^{-1}$ of disk mass accretion rates.

8.7   Magnetocentrifugal acceleration and collimation

The results of §§8.5 and 8.6 suggest that wind mass loss rates in both accreting T Tauri systems and FU Ori objects are roughly 10^{-1} of the mass accretion rates. The significance of this estimate can best be seen in terms of energy fluxes. Because typical wind or jet velocities are approximately a few times the Keplerian velocity of the inner disk, the above ratio of mass loss rates to accretion rates implies similar amounts of energy are released in accretion and in wind ejection. The process which accelerates the outflows of YSOs must be highly efficient in utilizing accretion power.

What causes the outflows and jets observed from T Tauri stars and FU Ori objects? In stars like the Sun, the outer coronal atmosphere is sufficiently hot that winds may be thermally driven (perhaps augmented by magnetic wave pressure). In hot stars,

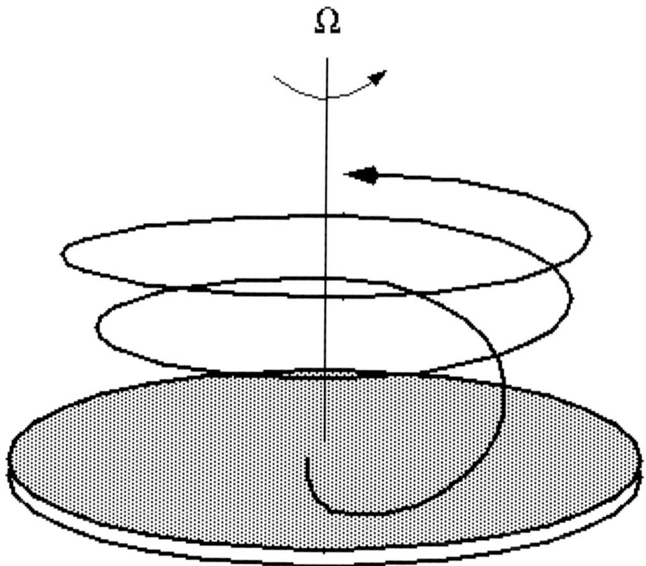

Fig. 8.10. Schematic drawing of a magnetic field line for a magnetocentrifugally-driven disk wind, illustrating the azimuthal winding-up of the magnetic field which can lead to collimation around the rotation axis (see text).

radiation pressure is sufficiently large to produce massive winds. Neither of these mechanisms appears applicable to T Tauri stars or FU Ori objects. Thermal pressure is not important when the gas sound speed is very much less than typical escape velocities (see next section). The temperatures of FU Ori and T Tauri winds are fairly well constrained to be $\lesssim 10^4$ K, and the associated sound speeds $c_s \lesssim 10 \, \mathrm{km \, s^{-1}}$ are very much smaller than typical escape velocities 200–400 $\mathrm{km \, s^{-1}}$. Similarly, radiation pressure is unlikely to be important because of the low energy densities in the low-temperature radiation fields of these objects. Radiation pressure driving of winds tends to be most effective for objects with high luminosity to mass ratios like O stars, unlike FU Ori objects and T Tauri stars. Moreover, the momentum fluxes of molecular outflows are generally much larger than the photon momentum flux L/c. Although multiple scattering can produce larger momentum fluxes in principle, in practice this is very difficult to arrange (Lada 1985). Finally, neither thermal acceleration nor radiation pressure forces naturally account for the highly collimated nature of outflows from young stars.

It is clear that the circumstellar disk plays a vital role in producing strong mass ejection. The T Tauri stars without inner disks (the WTTS) do not exhibit the same signatures of outflows; moreover, in FU Ori the wind can be shown to be emerging directly from the rotating disk surface. The favored mechanism for producing pre-main-sequence outflows is magnetocentrifugal acceleration from the circumstellar disk. Magnetic fields can effectively fling material outwards at high rates if they are rotating locally at speeds near the Keplerian velocity (Hartmann & MacGregor 1982). The magnetic fields also collimate the outflow along the rotation axis, an effect initially found by Suess & Nerney (1975) in solar wind models. It seems surprising at first that a mechanism which relies on radially outward motion for the initial acceleration ends up focussing the ejection in the direction perpendicular to the rotation axis, but the collimation results from the toroidal field that is eventually built up in the rotating flow, as illustrated in Figure 8.10. Material is launched from the rotating disk along a magnetic field line away from the rotation axis. Initially the magnetic field may be strong enough to enforce near co-rotation of the gas with the disk, but as the outflow expands the *poloidal* magnetic field eventually is no longer strong relative to the gas pressure forces. In this weak-magnetic field region, the angular velocity of the gas decreases as it tries to expand away from the rotation axis by conservation of angular momentum. This decreasing angular velocity causes the magnetic field to become wound up azimuthally. The azimuthal field exerts a curvature force toward the rotation axis which collimates the flow. Although the details of this situation are complex, this general type of model appears to provide the most likely explanation for collimated pre-main-sequence outflows.

8.8 Magnetohydrodynamic flows

To illustrate some of the basic issues involved in magnetocentrifugal wind accelera-tion, we briefly outline some basic magnetohydodynamic (MHD) results. We assume axisymmetric steady flow and infinite conductivity for simplicity, although it may be important to eliminate the latter constraint for the cold material involved in protostel-lar outflows (Königl 1989). The treatment below follows Mestel's development (1961, 1968). Important discussions have been given by Blandford & Payne (1982), Pudritz & Norman (1983), Lovelace, Wang, & Sulkanen (1987), Königl (1989), Pelletier & Pudritz (1992), Lovelace, Romanova, & Contopoulos (1993), Safier (1993), and Shu *et al.* (1994). Reviews of the subject have been presented by Pudritz, Pelletier, & Gomez de Castro (1991), Königl & Ruden (1993), and Spruit (1996).

We take cylindrical coordinates R, ϕ, z, where the total distance from the coordinate center is $r = (R^2 + z^2)^{1/2}$. Splitting the magnetic field and velocity into poloidal and toroidal components, we may write

$$\mathbf{B} = \mathbf{B}_p + \mathbf{B}_\phi, \tag{8.20}$$

$$\mathbf{v} = \mathbf{v}_p + R\Omega\hat{\phi}, \tag{8.21}$$

where Ω is the angular velocity at R. In the infinite conductivity MHD limit, the electric field in the frame moving with the fluid must vanish, and so the induction equation yields (e.g., Priest 1984)

$$\nabla \times (\mathbf{v}/c \times \mathbf{B}) = \mathbf{0}. \tag{8.22}$$

Using this result and Stokes' theorem,

$$\int \nabla \times (\mathbf{v}_p \times \mathbf{B}_p) \cdot d\mathbf{A} = \oint d\mathbf{S} \cdot (\mathbf{v}_p \times \mathbf{B}_p) = 0, \tag{8.23}$$

where the integral on the left-hand side is taken over the surface area bounded by the closed curve S. In particular, the line integral on the right-hand side may be taken to be a circle centered on the $\hat{z}$ axis; if the flow is axisymmetric, $\mathbf{v}_p \times \mathbf{B}_p$ must be constant on any such circle. Therefore $\mathbf{v}_p \times \mathbf{B}_p = 0$, and the poloidal velocity can be written as

$$\mathbf{v}_p = \kappa \mathbf{B}_p, \tag{8.24}$$

where κ is a scalar quantity. The poloidal velocity is parallel to the poloidal magnetic field as a consequence of the assumed infinite conductivity.

Using the Maxwell equation $\nabla \cdot \mathbf{B} = 0$, the toroidal component of equation (8.22) can be written after some manipulation as

$$(\mathbf{B} \cdot \nabla)(\Omega - \kappa B_\phi / R) = 0. \tag{8.25}$$

This equation can be interpreted as the change in the second bracketed quantity along a magnetic field line. Therefore, following any individual field line, we have

$$\Omega - \kappa B_\phi / R = \Omega_\circ, \tag{8.26}$$

where $\Omega_\circ$ is a constant along the given field line; this is the angular velocity of the *pattern* of the magnetic field in steady flow.

The continuity equation is

$$\nabla \cdot (\rho \mathbf{v}) = 0 = \nabla \cdot (\rho \kappa \mathbf{B}) = (\mathbf{B} \cdot \nabla) \rho \kappa. \tag{8.27}$$

Again, this defines a constant along a field line,

$$\rho \kappa = \frac{\rho v_p}{B_p} \equiv \eta. \tag{8.28}$$

The continuity equation in this form simply states that the mass flux density per unit poloidal magnetic flux density is a constant along a given field line. Equivalently, one may say that the poloidal mass flow is bounded by poloidal magnetic field lines, because the magnetic flux is 'frozen in' to the flow.

The momentum equation can be written using the first Maxwell equation as

$$\rho(\mathbf{v} \cdot \nabla)\mathbf{v} = -\rho \nabla \phi - \nabla P + (\nabla \times \mathbf{B}) \times \mathbf{B}/4\pi. \tag{8.29}$$

In (rotating) cylindrical coordinates, the toroidal component of the equation of motion is

$$\rho(\mathbf{v} \cdot \nabla)R^2 \Omega = R \left[\frac{(\nabla \times \mathbf{B}) \times \mathbf{B}}{4\pi} \right]_\phi. \tag{8.30}$$

After some additional manipulation, this equation can be written as

$$(\mathbf{B} \cdot \nabla)(\rho \kappa R^2 \Omega - R B_\phi / 4\pi) = 0. \tag{8.31}$$

Thus we have yet another quantity conserved along magnetic field lines. Since $\rho \kappa \equiv \eta$ is a constant along a streamline, this constraint can also be written as

$$R^2 \Omega - R B_\phi / 4\pi \eta = l, \tag{8.32}$$

where l is another constant along the streamline.

The first term on the left-hand side of equation (8.32) is the angular momentum per unit mass carried by the gas. The second term is the angular momentum carried by the magnetic field per unit mass. It is non-zero only if $B_\phi \neq 0$, i.e. the magnetic field must curve in the direction of rotation to exert a torque. We identify l as the total specific angular momentum carried by the flow (Mestel 1968). Then the angular momentum flux along a streamline is

$$\rho v_p l \;=\; \rho v_p R^2 \Omega \;-\; R B_p B_\phi / 4\pi . \tag{8.33}$$

These results may be combined to find an equation for the angular velocity of the gas (Mestel 1968),

$$\Omega \;=\; \frac{\Omega_\circ \;-\; 4\pi\eta^2 l/\rho R^2}{1 \;-\; 4\pi\eta^2/\rho} . \tag{8.34}$$

One critical parameter controlling the azimuthal motion is

$$\frac{4\pi\eta^2}{\rho} \;=\; \frac{4\pi\rho v_p^2}{B_p^2} , \tag{8.35}$$

which represents the ratio of the poloidal ram pressure of the flow to the pressure of the poloidal magnetic field. When the magnetic field is strong, it forces the gas to rotate with the same angular velocity, $\Omega \to \Omega_\circ$; if the magnetic pressure is much lower than the poloidal gas ram pressure, then $\Omega \to l/R^2$, i.e. the gas conserves its angular momentum along the field line.

It is evident that the magnetic field must be strong in the region where the outflow is accelerated to produce strong mass loss. In general the wind density decreases as the gas moves outward, and therefore the flow must pass through a point where equation (8.34) becomes singular; this requires

$$\frac{4\pi\eta^2}{\rho} \;=\; \frac{4\pi\rho v_p^2}{B_p^2} \;=\; 1 . \tag{8.36}$$

In general, singular points appear in steady flow problems where the flow speed matches the speed of a backward-propagating wave. In this particular case, the singular point is called the Alfvén critical point, because the poloidal velocity matches the Alfvén velocity,

$$V_{a,p}^2 \;=\; \frac{B_p^2}{4\pi\rho} , \tag{8.37}$$

which is the velocity of purely transverse waves along the magnetic field. There are two other pure MHD wave modes – the so-called 'fast' and 'slow' magnetosonic waves (e.g., Jackson 1962) – and in general the flow must pass through critical points associated with these modes as well (e.g., Weber & Davis 1967).

For smooth flow the numerator of equation (8.34) must also vanish at the Alfvén point, and therefore

$$l \;=\; R_A^2 \Omega_\circ . \tag{8.38}$$

The total angular momentum carried by the flow along a streamline is therefore equal to the value that the gas would have if it rotated at the same angular velocity as the base of the flow out to a cylindrical radius R_A. Although the actual flow does not precisely co-rotate out to the Alfvén radius (or, considering all the field lines in space, the Alfvén surface), and the magnetic field carries some of the angular momentum of the flow (equation (8.32)), this picture provides a convenient way of roughly categorizing the outflow into two reasonably distinct regions. Interior to the Alfvén surface, the flow is dominated by the magnetic field, and forced nearly into solid-body rotation; exterior to the Alfvén surface, the poloidal magnetic field becomes less important (although the same is not true of the azimuthal field), and the magnetic field lines become much more wound up (equation (8.34)).

The momentum equation may also be used to derive an energy constant of the motion in the absence of dissipation. Taking the dot product of the momentum equation with the velocity, and assuming isothermal flow for simplicity,

$$\rho \mathbf{v} \cdot \left[\nabla \frac{v^2}{2} + -\nabla \frac{GM_*}{(R^2 + z^2)^{1/2}} + c_s^2 \nabla \ln \rho \right] = \mathbf{v} \cdot \frac{(\nabla \times \mathbf{B}) \times \mathbf{B}}{4\pi}, \qquad (8.39)$$

where c_s is the sound speed. After some manipulation, this can be written as

$$\rho \mathbf{v} \cdot \nabla \left[\frac{v^2}{2} + c_s^2 \ln \rho - \frac{GM_*}{(R^2 + z^2)^{1/2}} - R^2 \Omega_\circ \Omega \right] = 0. \qquad (8.40)$$

The term in square brackets,

$$E = \frac{v^2}{2} + c_s^2 \ln \rho - \frac{GM_*}{(R^2 + z^2)^{1/2}} - R^2 \Omega_\circ \Omega, \qquad (8.41)$$

is an energy constant along the direction of motion, i.e. along a flow line, and is usually called the Bernoulli constant.

The final term in the energy constant is due to the effects of the magnetic field. To illustrate this more clearly, it can also be written as

$$\nabla \cdot (\rho \mathbf{v} (R^2 \Omega_\circ \Omega)) = \nabla \cdot (\rho \kappa \mathbf{B}_p R^2 \Omega_\circ \Omega) = \nabla \cdot \left(\frac{\Omega_\circ R B_\phi \mathbf{B}_p}{4\pi} \right)$$

$$= \nabla \cdot \left(\frac{1}{4\pi} (\mathbf{v} \times \mathbf{B}) \times \mathbf{B} \right) = -\nabla \cdot \left(\frac{c}{4\pi} \mathbf{E} \times \mathbf{B} \right). \qquad (8.42)$$

Thus, this energy term corresponds to an electromagnetic Poynting flux, and is responsible for accelerating the outflow in the absence of thermal gas pressure.

8.9   MHD disk winds

To illustrate the essential effects involved in launching a magnetocentrifugal wind, we first assume that the magnetic field in the innermost wind regions is sufficiently strong to enforce corotation in the region under consideration (cf. eq.(8.34)). Then the Bernoulli constant (8.41) can be approximated as

$$E = \frac{v_p^2}{2} + c_s^2 \ln \rho - \frac{R^2 \Omega_\circ^2}{2} - \frac{GM_*}{(R^2 + z^2)^{1/2}} = \frac{v_p^2}{2} + c_s^2 \ln \rho - \Phi, \qquad (8.43)$$

where Φ is an effective potential term including the effects of rotation and magnetic fields. The behavior of the flow depends upon the form of Φ, which in turn depends upon the geometry of the flow.

To start with a particularly simple case, suppose that the outflow starts from a ring rotating at $\Omega_\circ$ and is confined to the equatorial plane. Suppose further that the flow is confined to a flux tube whose cross-sectional area A varies as R^2. Then using the continuity equation for the mass flow rate

$$\dot{m}_w = \rho A v_p = \text{constant}, \qquad (8.44)$$

and differentiating the Bernoulli constant (8.43) with respect to R, one can substitute for the $\ln \rho$ term using (8.44), resulting in

$$(v_p^2 - c_s^2)\frac{dv_p}{dR} = \frac{v_p}{R}\left(2c_s^2 + R^2\Omega_\circ^2 - GM_*/R\right).\tag{8.45}$$

(Note that for this flow, $v_p = v_r = v_R$, i.e. the poloidal motion is purely radial.) If $\Omega_\circ = 0$ this equation is the familiar steady flow equation for the spherical thermally-driven wind (Parker 1963); in this case equation (8.45) is singular at the sonic point $v_p = v = c_s$; the radius at which the sonic point occurs is then $R_s = GM/2c_s^2$. The rotational term accounts for the effect of the magnetic field; it has the same sign as the thermal pressure term, and therefore helps accelerate the outflow. This singular point is actually where the flow speed matches the magnetospheric slow mode velocity, but by assuming co-rotation we have effectively assumed very large magnetic field strength and therefore large Alfvén velocity; in this limit the magnetospheric slow mode velocity approaches the sound speed (for radial propagation; see also Blandford & Payne (1982)).

Once the density at the sonic (slow mode) point is found by combining the location of the sonic point from equation (8.45) with the Bernoulli constant, the sonic (or slow mode) point determines the mass flow rate,

$$\dot{m}_w = \rho_s A_s c_s.\tag{8.46}$$

To illustrate this in more detail, consider first the simple thermally-driven case, assuming rotational (and magnetic) effects are negligible. Then the sonic point is the so-called Parker point,

$$R_P = \frac{GM_*}{2c_s^2},\tag{8.47}$$

and the density at the sonic point can be evaluated in terms of the density $\rho_\circ$ at the starting radius $R_\circ$,

$$\ln\frac{\rho_s}{\rho_\circ} = -\frac{1}{2} - \frac{GM_*}{R_\circ c_s^2}\left(1 - \frac{R_\circ}{R_P}\right).\tag{8.48}$$

If $R_P \gg R_\circ$, it follows that $GM/R_\circ c_s^2 \gg 1$. This means that the density at the sonic point must be very much less than the density at the reference level $R_\circ$, and thus the mass loss rate $\dot{m}_w \propto \rho_\circ \exp[-(1/2) - GM/R_\circ c_s^2]$ will be very low.

Returning to the case with magnetocentrifugal acceleration, the sonic point is given by the cubic equation

$$R_s^3\Omega_\circ^2 + 2c_s^2 R_s - GM_* = 0,\tag{8.49}$$

and the density at the sonic point becomes

$$\ln\frac{\rho_s}{\rho_\circ} = -\frac{1}{2} + \frac{-\Phi(R_\circ) + \Phi(R_s)}{c_s^2}.\tag{8.50}$$

The effect of the rotating magnetic field is to move the sonic point closer in to the initial radius and therefore generally increase the mass loss rate. T Tauri and FU Ori winds are generally quite cold, and so the effect of thermal pressure is rather small, i.e. the Parker radius R_p is large (and large in comparison with the sonic point (slow mode) radius). If the rotation at $R_\circ$ is written as a fraction f of the local Keplerian velocity,

$$\Omega_\circ = f(GM_*/R_\circ^3)^{1/2}, \tag{8.51}$$

and assuming that the Parker sonic point is at a much larger radius than the magnetosonic point, equation (8.49) yields approximately

$$R_s^3 \approx R_\circ^3/f^2. \tag{8.52}$$

Thus as $f \to 1$, the sonic point moves inwards to the initial or reference radius $R_\circ$. In the limit that the ring is rotating at the local Keplerian velocity, the sonic point lies right at the ring surface, where the density is large, and therefore large mass fluxes result. Outflow from a disk obviously does not lie precisely in the equatorial plane; however, the above discussion outlines the basic physics of magnetocentrifugal flows.

Consider now the opposite case in which the magnetic field lines are completely vertical, i.e. perpendicular to the disk plane. We suppose that the footpoint in the disk is rotating at the local Keplerian velocity. Now the Bernoulli constant is

$$E = \frac{v_p^2}{2} + c_s^2 \ln \rho - \frac{GM_*}{2R_\circ} - \frac{GM_*}{(R_\circ^2 + z^2)^{1/2}}. \tag{8.53}$$

It is evident that the effective potential term does not vary except on distance scales $\sim R_\circ$, and thus the sonic point of a cold flow will be far above the disk, and the mass loss rate will be vanishingly small. The vertical field configuration does not take advantage of the magnetocentrifugal acceleration which takes place when the field line is tilted with respect to the rotation axis.

From these two limits one can see that there must be some critical angle of the magnetic field line to the rotation axis which allows the sonic point to lie close to the disk surface and thus produce a high mass loss rate. The easiest way to determine this condition is to take the strong-field limit and assume that the sound speed is extremely small, so that thermal effects can be completely neglected. Then the Bernoulli equation simplifies to

$$E = \frac{1}{2}v_p^2 + \Phi, \tag{8.54}$$

where the 'effective' potential is

$$\Phi = -\frac{GM_*}{R_\circ}\left[\frac{1}{2}\frac{R^2}{R_\circ^2} + \frac{R_\circ}{(R^2 + z^2)^{1/2}}\right]. \tag{8.55}$$

Consider now a small displacement along the field line, with a coordinate given by s, and

$$ds^2 = dR^2 + dz^2. \tag{8.56}$$

At the base of the flow, the disk material is rotating at the local Keplerian velocity. This is an equilibrium state, because $d\Phi/ds = 0$ at $z = 0$. However, if $d^2\Phi/ds^2 < 0$, this equilibrium is *unstable*; any small perturbation along the field line will result in an increased (outward) poloidal velocity from equation (8.54). If θ is the angle between the field line and the disk plane, the critical stability criterion

$$\frac{\partial^2\Phi}{\partial s^2} = 0 \quad (R = R_\circ, z = 0) \tag{8.57}$$

requires $\tan^2\theta_c = 3$, or $\theta_c = 60°$ (Blandford & Payne 1982). Disk magnetic field lines which are tipped away from the rotation axis by an angle greater than $30°$ result in an unstable equilibrium, and rapid outflow will commence at the disk. For smaller tilt angles from the rotation axis, launching of an outflow requires thermal pressure to initiate the motion, and so the mass loss rates will be correspondingly smaller. If the flow is very cold, values of $\theta \geq 60°$ will result in essentially no mass ejection.

A complete analysis of the flow requires consideration of wind behavior at large distances, where the approximation of co-rotation is not valid. The outflow may or may not pass through a fast-mode critical point (Blandford & Payne 1982; Spruit 1996), depending upon how 'magnetic' the solution is at large distances, i.e. how much of the energy flow is accounted for by the kinetic energy of the gas vs how much is contained in the Poynting flux at large distances (equation (8.42)). We will not consider this problem further; although this is of interest for understanding the collimation of jets, it is not essential for understanding the effects of the wind ejection on the disk.

The disk wind carries away both angular momentum and energy from the disk, and therefore may affect disk accretion. To explore this most simply, consider the case where thermal pressure can be neglected. Using (8.40) along a flux tube with area dA,

$$\rho v_p dA \left[\frac{v^2}{2} - \frac{GM_*}{(R^2 + z^2)^{1/2}} - R^2\Omega_\circ\Omega \right] = \text{constant}, \tag{8.58}$$

where v_p is the poloidal velocity. This can be written using (8.32) as

$$\rho v_p dA \left[\frac{v^2}{2} - \frac{GM_*}{(R^2 + z^2)^{1/2}} - \Omega_\circ \left(l + \frac{RB_\phi}{4\pi\eta} \right) \right]. \tag{8.59}$$

Because both $\Omega_\circ$ and l are constants of the motion,

$$\rho v_p dA \left[\frac{v^2}{2} - \frac{GM_*}{(R^2 + z^2)^{1/2}} + \Omega_\circ l_B \right] = \text{constant} = \dot{e}_w, \tag{8.60}$$

where we identify $\dot{e}_w$ as the energy flux along the flux tube, and

$$l_B = -\frac{RB_\phi}{4\pi\eta} \tag{8.61}$$

is the (non-constant) angular momentum carried by the magnetic field.

The outflow angular momentum and energy are carried both by the gas and by the magnetic field. However, it is the magnetic field coupling the outflow and disk which is responsible for extracting the energy and angular momentum from the disk needed to drive the outflow. Just above the disk surface, the outflowing gas contains essentially the same angular momentum that it had in the disk, and so its removal does not change the angular momentum per unit mass of the remaining disk material. The magnetic field produces a torque on the disk as it carries away angular momentum; the angular momentum flux carried away by the wind can be evaluated at the base of the flow as

$$\frac{dj_w}{dt} = \rho_\circ v_{po} dA_\circ l_{Bo}. \tag{8.62}$$

Suppose now that a magnetic flow tube spans an annular area corresponding to a radial range on the disk of dR at $R_\circ$. In steady state, the flow cannot extract more angular momentum than corresponds to the accretion rate through the disk. Away from the inner disk edge, the angular momentum lost as material accretes through dR at $R_\circ$ is

$$\frac{dj_{acc}}{dt} \; = \; \dot{M}\frac{d}{dR}\Omega_K R^2 dR \; = \; \frac{1}{2}\dot{M}\Omega_\circ R_\circ^2\frac{dR}{R_\circ} \, . \tag{8.63}$$

If the angular momentum carried away by the wind flux tube spanning dR is exactly equal to the angular momentum transport needed for accretion, the energy carried away by the wind is

$$\dot{e}_w \; = \; \Omega_\circ l_{Bo}\dot{m}_w \; = \; \frac{1}{2}\frac{GM_*\dot{M}}{R_\circ}\frac{dR}{R_\circ} \, . \tag{8.64}$$

Therefore, a wind which removes all of the angular momentum needed for disk accretion carries off all of the accretion energy as well. An accretion disk whose angular momentum transport is entirely the result of magnetocentrifugal mass loss is generally not self-luminous, since all of the accretion energy is put into the wind, leaving nothing to radiate.[*]

An analysis of the solution at large distances is needed to derive the precise collimation of the flow and the asymptotic flow speed. However, a crude estimate of the terminal velocity of the latter can be made using the Bernouilli equation (8.41) and the assumptions that most of the angular momentum of the flow is initially carried by the magnetic field, and that the angular momentum asymptotically carried by the magnetic field is negligible; then

$$\frac{v_\infty^2}{2} \; \sim \; -\frac{GM_*}{2R_\circ} + \Omega_\circ l_{Bo} \; \sim \; -\frac{GM_*}{2R_\circ} + \Omega_\circ^2 R_A^2 \, . \tag{8.65}$$

Thus the terminal velocity of the flow is typically of the order of the azimuthal velocity at the Alfvén point.

8.10   Applications of MHD disk wind theory

Following the initial self-similar disk wind models investigated for active galactic nuclei by Blandford & Payne (1982), magnetocentrifugal disk wind models for pre-main-sequence objects have been constructed by Pudritz & Norman (1983, 1986), Lovelace *et al.* (1987), Königl (1989), Pelletier & Pudritz (1992), Wardle & Königl (1993), Safier (1993), and Paatz & Camenzind (1996). An extreme limit of the disk wind, where only the innermost edge of the disk contributes to the mass loss, has been developed by Shu and collaborators (Shu *et al.* 1994; Najita & Shu 1994).

MHD disk wind models can successfully account for a number of observed features of outflows from YSOs. First, magnetic acceleration is effective for cold flows; even when the gas is only lightly ionized, so that the magnetic field is imperfectly coupled to the gas through collisions between ions and neutrals, massive flows can still be driven (Königl 1989). Second, *if* the magnetic field has the appropriate geometry of

---

[*] This conclusion depends upon the assumption that the magnetic field is anchored in disk material rotating at the local Keplerian velocity. This may not always be the case. For example, in non-ideal MHD models of disks, field lines attached to ions may rotate below Keplerian velocities, with resultant energy dissipation and heating in the disk (Königl 1989; Wardle & Königl 1993). Usually this sub-Keplerian rotation is small, and so the fraction of accretion energy deposited in the disk is also small. If the disk is rotating at rates well below Keplerian, then gas must be ejected with substantial velocities from the disk surface, or there must be strong thermal pressure acceleration (i.e. a hot disk corona), to produce sizable mass ejection (Shu 1991).

open field lines, the wind can tap into a large fraction – potentially all – of the energy released by disk accretion (equation (8.65). Since the accretion energy is often a significant fraction of the total luminosity, it follows that the outflow momentum flux $\dot{M}v_\infty$ can greatly exceed L/c. Third, the outflow velocities can be explained easily. Asymptotic flow speeds are of order the azimuthal velocities at the Alfvén surface (see above); these velocities must be larger than footpoint velocities, which are hundreds of km s^{-1} in the innermost disk regions. Finally, the winding-up of toroidal magnetic field in these models provides a natural explanation for the highly-collimated nature of jets.

Note that the outflow energy is taken from the disk. Therefore, flows which carry away a large fraction of the accretion energy must be produced in the inner disk, because this is where most of the accretion energy is produced (Pringle 1989).

The crucial unknowns are the disk magnetic field strength and its geometry. The required field strengths can be estimated from the Alfvén point (surface), where both rotational (and poloidal) velocities must be an appreciable fraction of the observed wind terminal velocity. Thus, from the relation (8.36),

$$B_A^2 \sim 4\pi\rho_A v_A^2, \tag{8.66}$$

we estimate

$$B_A \sim (\dot{M}v_\infty)^{1/2} R_A^{-1} \sim 200\,(\dot{M}_{-5}\,v_{300})^{1/2} R_{10}^{-1}\,\mathrm{G}, \tag{8.67}$$

where the mass loss rate $\dot{M}_{-5}$ is measured in units of $10^{-5}\,\mathrm{M}_\odot\,\mathrm{yr}^{-1}$, the wind velocity v is measured in units of $300\,\mathrm{km\,s}^{-1}$, and the Alfvén radius is in units of $10\,\mathrm{R}_\odot$.

For MHD disk winds to explain FU Ori outflows, the magnetic fields in the wind acceleration region must be hundreds of gauss, and probably stronger at the disk surface. Although these magnetic field strengths seem quite large, physical conditions in FU Ori disks may be sufficient to contain these fields by gas pressure forces. If the innermost regions of FU Ori disks have surface densities $\Sigma \sim 6 \times 10^4\,\mathrm{g\,cm}^{-2}$ and scale heights $\sim 2 \times 10^{11}\,\mathrm{cm}$ (§7.3), the central densities may be as high as $\rho_c \sim 3 \times 10^{-7}\,\mathrm{g\,cm}^{-3}$; then with central temperatures $\sim 10^5\,\mathrm{K}$, the midplane disk pressures may reach $\sim 5 \times 10^6\,\mathrm{dyne}$. The equipartition magnetic field which matches this gas pressure is $B_{eq} \sim 10^4\,\mathrm{G}$. Thus, one might imagine that $\sim \mathrm{kG}$ fields can be contained within FU Ori inner disks, which might be sufficiently 'weak' for the Balbus–Hawley instability to operate, and yet are still strong enough to drive the observed outflows from the disk surfaces. Unfortunately, even such large field strengths are too small to detect with conventional Zeeman splitting techniques, especially in the presence of rapid disk rotation.

Similarly, one may make an estimate of the maximum magnetic field strengths allowed in T Tauri disks. Using the relations (5.64) and (5.78),

$$v\Sigma = \alpha c_s H\Sigma = \alpha c_s^2 \Sigma\Omega^{-1} = \dot{M}/3\pi, \tag{8.68}$$

the disk midplane pressure is

$$P_c = \dot{M}\Omega(3\pi\alpha H)^{-1}. \tag{8.69}$$

Evaluating this at the inner disk for $H/R \sim 0.1$, $R = 10\,\mathrm{R}_\odot$, we have

$$B_{eq} \sim 180\,\dot{M}_{-8}^{1/2}\,M_{0.5}^{1/4}\,R_{10}^{-5/4}\,\alpha_{-2}^{-1/2}\,\mathrm{G}, \tag{8.70}$$

where $\dot{M}_{-8}$ is the mass accretion rate in typical T Tauri units of $10^{-8}\,\mathrm{M_\odot\,yr^{-1}}$, $M_{0.5}$ is the central stellar mass in units of $0.5\,\mathrm{M_\odot}$, R_{10} is the radius in units of $10\,\mathrm{R_\odot}$, and α_{-2} is the viscosity parameter in units of 10^{-2}. If typical T Tauri mass loss rates are of order $10^{-9}\,\mathrm{M_\odot\,yr^{-1}}$ (Figure 8.9), the required magnetic field strengths for wind ejection estimated from (8.67) are ~ 10 G, so the above estimate indicates that T Tauri disks can contain the required fields.

Whether the magnetic field structures in disks are really suitable for generating strong winds is much less clear. The 'ideal' structure for mass loss is one in which the magnetic field is open on both sides of the disk (Lovelace *et al.* 1987), such as might occur if the magnetic field has been dragged in from the interstellar medium (Stepinski 1995). But this requires a complicated interaction so that (at least some) of the accreting disk material can pass through the magnetic fields in the disk; otherwise, accretion might stop, and accretion is the energy source ultimately powering the outflow. It is possible that the finite conductivity of protostellar disks allows the disk gas to diffuse sufficiently rapidly through the magnetic fields to allow accretion to proceed (König1 1991; Wardle & König1 1993), but it is not clear whether the actual ionization state of protostellar disks is consistent with these models. For example, the inner disks of FU Ori objects must be extremely hot and completely ionized, and thus have very high conductivities, yet winds and rapid disk accretion coexist. All this must occur in such a way that the magnetic field makes an angle $\gtrsim 30°$ with the rotation axis (cf. §8.9), and there is uncertainty whether these field angles can be maintained, at least in models assuming simple turbulent diffusivity in which the wind carries away all the angular momentum for accretion (Lubow, Papaloizou, & Pringle 1994).

Alternatively, the disk magnetic field may be self-generated through dynamo activity (e.g., Stepinski 1995). There is some evidence that the Balbus–Hawley instability may produce dynamo magnetic field generation (Hawley *et al.* 1996). In this case one might expect the disk field to be highly irregular, with both closed and open magnetic field lines at the disk surface, such as present in the Sun, and like the picture of disk fields originally sketched by Blandford & Payne (1982). Conventionally, this is not thought to be very efficient in producing disk winds; but such efficiency in ejecting mass may not be necessary. It was thought originally that the MHD disk wind might carry away all the disk angular momentum needed for accretion. In this limit, as discussed in §8.9, essentially all of the accretion energy is carried away by the wind. However, this leaves very little accretion energy to be radiated by the disk. Such a model is difficult to maintain for FU Ori objects. As discussed at the beginning of §8.7, the estimated kinetic energy flux in the wind of FU Ori is comparable to the *radiated* disk accretion luminosity. The observational evidence suggests that there must be another source of angular momentum transport in FU Ori disks other than magnetocentrifugal winds, since the amount of energy dissipated in the disk and subsequently radiated away is comparable to the wind luminosity.

Numerical results (Stone & Norman 1994) suggest that MHD wind angular momentum loss can be extremely rapid for favorable magnetic field geometries; thus, it may be necessary to assume that much of the magnetic field of the disk does *not* have a favorable geometry. The magnetic field strengths at the disk surface required to explain the observed mass loss may be much smaller than the central disk magnetic field strengths; this would allow for some cancellation of magnetic fields by small scale structure (e.g., Blandford & Payne 1982).

In any event, it seems much more plausible simply to assume that other sources of viscosity and angular momentum transport operate in FU Ori disks and dominate the transfer of angular momentum and thus control the accretion rate. This is likely to be especially applicable to the outer disk, where most of the disk angular momentum resides, but from which there is little evidence for mass loss.

8.11 Magnetospheric accretion in T Tauri stars

As discussed in §6.4, it has become increasingly apparent that the effects of the stellar magnetic field cannot be neglected in understanding accretion onto T Tauri stars. Some of the most direct evidence for magnetospheric accretion comes from the permitted emission line profiles of T Tauri stars. Although wind-produced blueshifted absorption can be observed at times, especially in Hα, the observed emission profiles generally look quite different than the predictions of simple wind models. Often there is no blueshifted absorption detected in the high Balmer lines, nor in the Brackett or Paschen lines (Najita, Carr, & Tokunaga 1996; Folha, Emerson, & Calvet 1997), whereas physically-consistent wind models tend to predict strong blueshifted wind absorption components in these lines (Hartmann *et al.* 1990). The Balmer emission lines frequently peak on the *blueshifted* side of line center; this is difficult to achieve with wind models, but it is a natural consequence of infall (§4.7; see following section).

It has been known for many years that a few T Tauri stars – the so-called 'YY Orionis' stars – show 'inverse P Cygni' profiles with high-velocity redshifted absorption (e.g., Walker 1972; Wolf *et al.* 1977). Using improved CCD detectors it has become clear that high-velocity infall is a common phenomenon in T Tauri stars (Edwards *et al.* 1994), as shown by faint but detectable *redshifted* absorption. This suggests a picture of T Tauri envelopes in which blue-shifted permitted-line absorption, along with blueshifted forbidden-line emission, are formed in a low-density, expanding wind, while much of the permitted-line emission is formed in dense, hot magnetospheric accretion columns (Figure 6.1).

8.12 Models of magnetospheric accretion

Magnetospheric models were originally developed to investigate disk accretion onto magnetized neutron stars (Ghosh & Lamb 1979) and have also been applied to the highly-magnetized white dwarf AM Her systems (see Frank *et al.* (1992) for a discussion). The way in which the stellar magnetic field interacts with the accretion disk is very complicated and not well understood (see Shapiro & Teukolsky (1983) for a discussion), and we only outline schematically some of the issues involved here; other treatments are given by Cameron & Campbell (1993), Ghosh (1995), Ostriker & Shu (1995), and Armitage & Clarke (1996).

The basic idea behind the magnetospheric truncation of the disk can be seen most easily in the context of spherical (free-fall) accretion. Suppose that the magnetic pressure balances the ram pressure of accretion,

$$B^2 = 4\pi\rho v^2 = \dot{M}v_{in}/r^2, \tag{8.71}$$

where v_{in} is the infall velocity. In such a situation, the (sufficiently-ionized) accreting gas cannot fall in freely; its motion must be restricted or even halted by the magnetic force. If the infall velocity v_{in} is set equal to its maximum value, the free-fall velocity from infinity, and the stellar magnetic field is a dipole, so that $B \propto r^{-3}$, then the radius at which the magnetic field is strong enough to hold off the infall is roughly

$$\frac{r_T}{R_*} = \frac{B_o^{4/7} R_*^{5/7}}{\dot{M}^{2/7}(2GM)^{1/7}} = 3.7 B_3^{4/7} \dot{M}_{-7}^{-2/7} M_{1/2}^{-1/7} R_2^{5/7}, \tag{8.72}$$

where B_3 is the surface magnetic field in kG, $\dot{M}_{-7}$ is the mass accretion rate in $10^{-7} M_\odot \, \mathrm{yr}^{-1}$, $M_{1/2}$ is the central mass in units of 0.5 $M_\odot$, and R_2 is the stellar radius in $2 R_\odot$. The argument is modified for disk accretion. The truncation radius r_T is in general smaller than given above, because disk ram pressures (dominated by the azimuthal motion) generally are larger than free-fall ram pressures for the same mass accretion rate. This is because the circular velocity is comparable to the free-fall velocity, whereas the disk radial velocities are very small and so the disk densities are large. Nevertheless, equation (8.72) illustrates the basic parameter dependence of the problem; the numerical factor can be modified suitably for the disk accretion case (e.g., Wang 1996). The results illustrate that the magnetic fields of order ~ 1 kG likely to be present on young stars (Basri *et al.* 1992) can significantly affect the accretion process (Königl 1991).

The discussion of MHD flows showed that magnetic fields can transport angular momentum efficiently. A basic parameter of the disk accretion problem is the co-rotation radius, where the Keplerian angular velocity of the disk equals the rotational angular velocity of the star. Magnetic field lines connecting the star with disk regions outside the corotation radius will tend to spin up the disk and spin down the star. This is exactly what is needed to make the CTTS rotate slowly. Unfortunately, *accretion* requires field lines to connect with the disk *inside* the corotation radius. The simplest way to see this is to consider the Bernoulli constant for strong radial magnetic fields (8.43). Taking the radial derivative of this energy constant near the disk truncation radius yields

$$v_p \frac{dv_p}{dR} = -\frac{GM_*}{R_T^2} + R_T \Omega_\circ^2 = -\frac{GM_*}{R_T^2}(1 - f^2), \tag{8.73}$$

where

$$f = \frac{\Omega_\circ}{\Omega_T} = \left(\frac{R_T}{R_{co}}\right)^{3/2}, \tag{8.74}$$

and R_{co} is the corotation disk radius at which the Keplerian angular velocity matches that of the star. It can be seen that if $f > 1$, i.e. $R_T > R_{co}$, the sign of the acceleration is positive and the gas will move outward, not inward. Only if $f < 1$ will material accrete along the magnetic field lines.

To produce an equilibrium situation in which accreting CTTS are slowly rotating, it is necessary to have some magnetic field lines which attach to the disk outside of corotation at the same time accretion is occuring along magnetic field lines which meet the disk interior to co-rotation (Figure 8.11). The accreting material and associated magnetic stresses add angular momentum to the star, which must be given back out to the disk by the magnetic field lines interacting with the disk outside of corotation. Ghosh & Lamb (1979) accomplish this by postulating a region of finite width through which the disk interacts with the magnetic field, ranging from inside to outside of corotation. Since the magnetic field lines have a constant pattern speed $\Omega_\circ$ which does not necessarily correspond to the angular velocity of the disk, Ghosh & Lamb assume

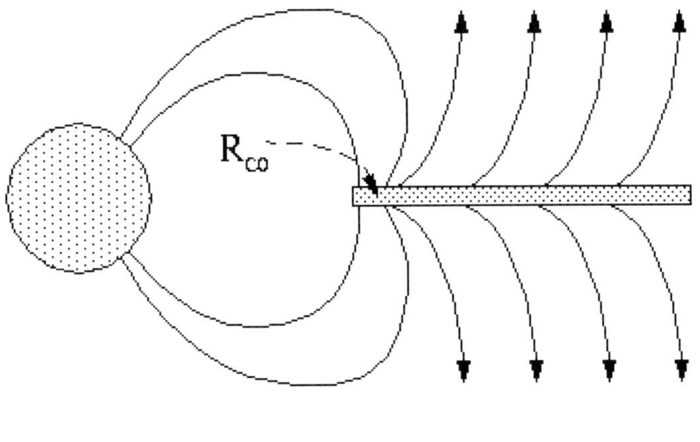

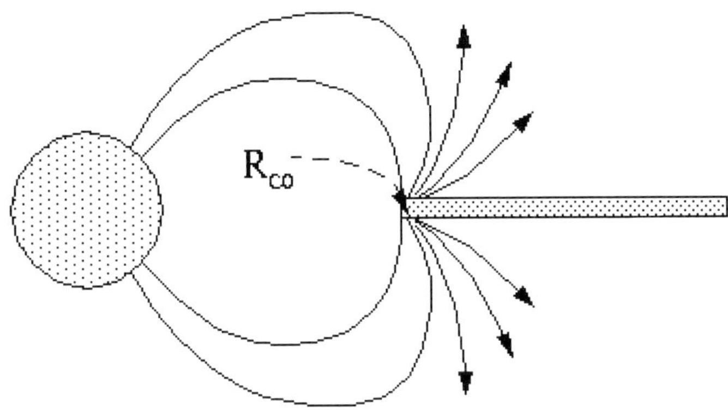

Fig. 8.11. Two versions of accretion disk–wind–stellar magnetosphere structure. In the upper drawing, the magnetic field penetrates the disk through a finite range of radii, as in the Ghosh & Lamb (1979) model, modified as in Camenzind (1990). Accreting material comes from disk regions interior to the co-rotation radius R_{co}, while the magnetic field lines linking the star to the disk outside of R_{co} provide a spindown torque opposing the incoming angular momentum flux from the accreting region. In the lower panel, the model of Shu *et al.* (1994) is indicated, in which all the stellar magnetic field lines which penetrate the disk do so exactly at R_{co}. See text.

that the magnetic fields pass through the disk in a steady fashion, being dragged or pulled to some extent. The details of the interaction regions are quite unclear, and it is far from obvious that the stellar magnetic fields pass smoothly through the disk. Wang (1995) argued that the magnetic field interaction must be confined to regions near the corotation radius. Even so, if the magnetic field were fixed to disk regions rotating even slightly differently than the star, after only a few rotations the magnetic field would be stretched out of all proportion (van Ballegooijen 1994), and would have to reconnect in some complicated way (Aly & Kuijpers 1990).

Shu *et al.* (1994) have tried to avoid this problem by postulating that the interaction of the stellar magnetic field with the disk occurs only at a very small region right at co-rotation, hoping in this way to avoid winding up the magnetic field. However, it seems highly unlikely that the stellar magnetic fields are completely regular, axisymmetric, and steady. Hot accretion cannot produce light variations modulated by stellar rotation unless there are departures from complete axisymmetry. Variations of line profiles suggest that departures from axisymmetry in the magnetosphere may affect both the mass accretion and ejection (Giampapa *et al.* 1993; Johns & Basri 1995b; Petrov *et al.* 1996). Moreover, disk accretion is certainly not completely steady (e.g., Herbst *et al.* 1994; Gahm 1994; Gullbring 1994; Gullbring *et al.* 1996; Johns & Basri 1995a; Johns-Krull & Basri 1997). It seems implausible to assume that the stellar magnetic field adjusts instantaneously so that the truncation radius in equation (8.72) is always exactly at corotation.

Rather than assume that fields easily slip through all regions of the disk at precisely the needed rate to avoid winding up, as in the Ghosh & Lamb model, or postulate that the stellar magnetic field only interacts with the disk precisely at co-rotation, it seems much more plausible to assume that magnetic fields penetrate finite regions of the inner disks of T Tauri stars, and are being constantly wound up, with occasional field reconnection which releases the stored-up magnetic energy, perhaps in high-energy flares and/or in heating the infalling magnetospheric gas (or the disk). In other words, a complex model of constantly time-varying magnetospheric accretion, along the lines of the picture treated in an idealized way by van Ballegooijen (1994), seems to be the most plausible.

Some observations suggest that the accretion rates of some T Tauri stars actually vary considerably over periods of a few years to decades (EXor events; Figure 1.7; §6.4; §7.5), and it is tempting to suggest that time-variability of the stellar magnetic field may either enhance or restrict the rate of accretion onto the central object; in general, this may result in either spinning the star up or down at various times (Clarke *et al.* 1995). This picture does not apply to the FU Ori objects, because at the enormous accretion rates during outburst the magnetosphere is probably crushed back against the central star (equation (8.72)), although in this case it is possible that the high central disk temperatures result in pressure-supported accretion at velocities below breakup (Popham 1995; §7.6).

Independently of the exact physics involved in star-disk magnetic interactions, observations of line profile asymmetries provide strong evidence for accretion flows out of the disk. To see how models might be constructed to compare with observations, note that the low densities and relatively high temperatures in the accretion flow are likely to lead to high electron densities, and therefore conductivities, so that ideal MHD should be a good approximation. Moreover, because the magnetic field must be strong enough to disrupt the dense disk, the lower-density accretion flow must be strongly dominated by the magnetic forces. This yields a particularly simple velocity field which can be used to predict observable quantities. In the limit where the magnetic field is sufficiently strong to enforce co-rotation of the gas in the magnetosphere, and neglecting gas pressure forces, the Bernoulli constant can be evaluated to find the velocity at radial distance r along a field line which meets the disk

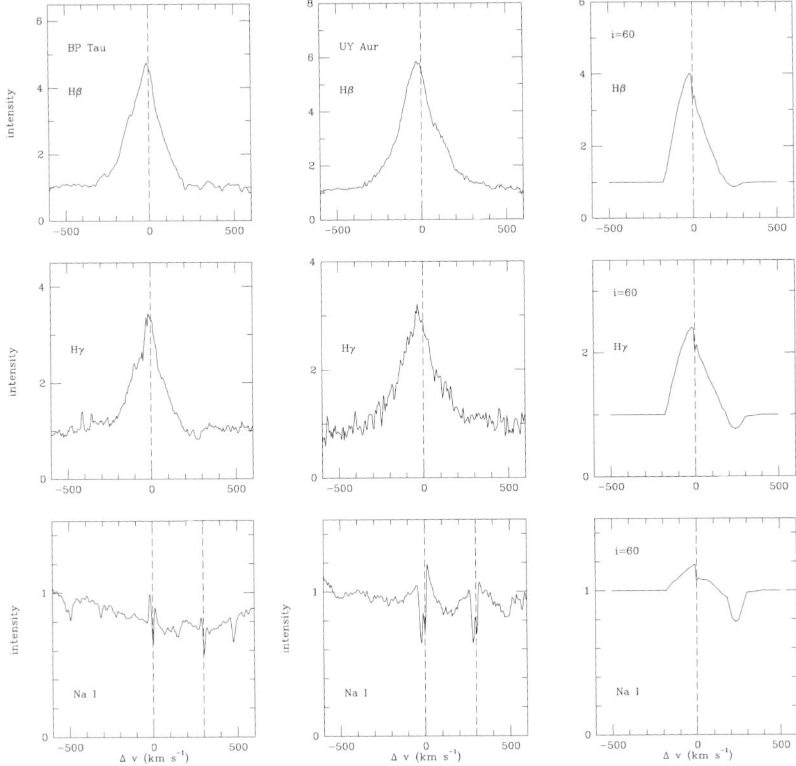

Fig. 8.12. Hβ, Hγ, and Na I line profiles of two accreting T Tauri stars, BP Tau and UY Aur, compared with line profiles calculated for magnetospheric infall models (Hartmann, Hewett, & Calvet 1994). Often the redshifted absorption is not very strong in the Balmer lines, but this is consistent with the model calculations; the overall asymmetry of the emission component, however, is very suggestive when compared with the infall model. See text; compare also with Figure 4.13.

near R_T,

$$v_P^2 \simeq \frac{2GM_*}{r}\left(1 - \frac{r}{R_T}\right) + \Omega_\circ^2(R^2 - R_T^2);\qquad(8.75)$$

if R_T is near the co-rotation radius R_{co}, then

$$v_P^2 \simeq \frac{2GM_*}{r}\left(1 - \frac{r}{R_T}\right) + \frac{GM_*}{R_T}\left(\frac{R^2}{R_T^2} - 1\right).\qquad(8.76)$$

Most accreting CTTS are slowly rotating, i.e. $R_{co} \gg R_*$, and so for a first approximation one can ignore the azimuthal velocity, in which case the poloidal velocity is simply that of free-fall. If the poloidal structure of the magnetic field is known, then the velocity vectors can also be determined, as well as the density along streamlines from the mass conservation equation.

Using this simple free-fall model, and assuming a non-rotating dipolar magnetosphere, Hartmann, Hewett, & Calvet (1994) found that it was possible to produce line profiles that look remarkably like observations, even of stars not initially considered to be YY Ori objects. The principal uncertainty in these calculations is the temperature structure; Hartmann, Hewett, & Calvet (1994) assumed a simple smooth temperature distribution that is nearly constant over the magnetosphere, but drops near the disk. Figure 8.12 compares the computed line profiles with observed profiles of two typical T Tauri stars. The infall asymmetry can be observed in the Balmer emission lines in both the model and observations; the emission peak is slightly blueward of line center for the same reason as discussed in §4.7 in the context of protostellar envelope collapse. The model indicates weak redshifted absorption in the Balmer lines, with stronger absorption in $H\gamma$ than in $H\beta$, consistent with the general results of Edwards *et al.* (1994) for accreting T Tauri stars.

The strength of the redshifted absorption was found to depend strongly upon the geometry (inclination) as well as the relative temperatures of the magnetospheric gas and the accretion shock. Thus, these calculations suggest that the YY Ori stars are simply CTTS for which parameters conspire to produce more noticable redshifted absorption. The detailed observational results of Edwards *et al.* (1994) support this interpretation; with high signal-to-noise spectra and careful subtraction of photospheric absorption features, weak redshifted absorption components are quite commonly observed in CTTS.

To explain the observed redshifted absorption, and the velocity widths of the underlying emission, it is necessary to invoke infall from $R \gtrsim 1.5R_*$; this is a conservative lower limit which ignores projection effects that make the line-of-sight velocities smaller than the absolute motions. It is very difficult to imagine that any standard boundary layer can form between the disk and star given such large magnetospheric radii, no matter what detailed geometrical form the stellar magnetic fields exhibit.

The line profile modelling suggests that much of the permitted-line emission of T Tauri stars comes from dense infalling material in the stellar magnetosphere, while the wind arises from outside the magnetosphere (Figure 8.11). Since the emission comes from the magnetosphere, the wind need not be very optically thick in the Balmer lines. $H\alpha$, with the largest oscillator strength, can exhibit detectable absorption while the higher Balmer series lines, with much smaller optical depths, can remain relatively transparent, explaining the absence of blueshifted absorption in the high Balmer lines (Edwards *et al.* 1994). These results are consistent with the forbidden-line estimates (§§8.5, 8.6) which suggest that T Tauri stars generally have winds with mass loss rates $\sim 10^{-9}\,M_\odot\,\mathrm{yr}^{-1}$, lower than typical accretion rates $\sim 10^{-8}\,M_\odot\,\mathrm{yr}^{-1}$ (§6.4) which produce the high densities of the infalling magnetospheric gas.

9

Disk accretion and early stellar evolution

The low-mass stars produced by molecular cloud collapse have central temperatures that are initially too low for hydrogen fusion to proceed. The energy lost by radiation from the stellar surface therefore must be balanced by the release of gravitational potential energy, causing the star to contract. With appropriate allowance for deuterium fusion, which typically can slow or halt pre-main-sequence contraction for $\lesssim 1$ Myr, the position of a young star in the HR diagram provides an indication of its 'age', and can be used as a clock by which to explore disk evolutionary timescales.

The (post infall) age derived for a T Tauri star depends in part on its 'initial' position in the HR diagram, i.e. its luminosity and effective temperature (or radius) at the end of protostellar accretion. Under certain conditions, low-mass protostars may finish their main accretion phase on or near the 'birthline' (Stahler 1983, 1988). The birthline is generally near the 'deuterium main sequence', where protostars first become hot enough to fuse deuterium rapidly. Whether the deuterium main sequence really defines an exact starting point for low-mass T Tauri stars depends on initial conditions, as well on as how much thermal energy is added to the central star during protostellar accretion. Although many of the youngest low-mass stars are observed to lie near the birthline in the HR diagram, observational uncertainties preclude a definitive test of the theory. The predicted birthline positions of intermediate mass stars are much less certain because deuterium fusion plays a minor role at high luminosities.

(Slow) disk accretion during most of the T Tauri phase probably does not affect stellar evolutionary tracks very much. Aside from the question of birthline positions, the principal uncertainties at present in absolute ages of T Tauri stars derive from possible errors in assigning stellar masses from pre-main-sequence evolutionary tracks.

Observational studies of the late stages of T Tauri disk evolution, particularly relevant for understanding planet formation, are hampered at present by a lack of good samples of stars with ages ~ 10 Myr. These older stars lie quite close to the hydrogen burning main sequence, and so age determinations demand accurate distance determinations – which, unless the stars are in clusters, are difficult to come by. Even with the limitations of samples and methods, however, it is possible to make an initial assessment of the time decay of disk accretion. Preliminary results for stars with ages of a few Myr or less suggest that accretion rates do decrease with age in roughly the manner predicted by simple models of disk evolution. Further research is needed to characterize better the evolution of disks at later times, closer to the expected epochs of planet formation.

9.1   Pre-main-sequence stellar evolutionary tracks

If a low-mass star is much larger than it would be on the hydrogen-burning main sequence, its central temperature will not be sufficiently high to fuse hydrogen in its interior. Under these circumstances, the only energy available to supply the stellar luminosity is the internal thermal energy. (Deuterium fusion will be considered separately below.) As energy is lost to space, the internal energy must decrease, and so the star must contract; this contraction continues until the central temperature rises sufficiently that hydrogen fusion begins.

Hayashi, Hoshi, & Sugimoto (1962) and Hayashi (1966) began to establish the modern foundation for the subject of pre-main-sequence evolution by showing that low-mass stars above the main sequence are likely to be nearly completely convective (see Shu (1991) for an overview). The interiors of convective stars are found to be nearly adiabatic over most of their mass, because the convection is so efficient at energy transport (Schwarzschild 1958). The polytropic equation describing the stellar interior is

$$P\rho^{-\gamma} = K = \text{constant} = P\rho^{-(1+1/n)} ; \qquad (9.1)$$

in terms of the polytropic index, $n = 1/(\gamma - 1)$. A convective star composed of a perfect gas has $\gamma = 5/3$ and a polytropic index $n = 3/2$ (Chandrasekhar 1967).

The equation of hydrostatic equilibrium (2.27) can be integrated using the polytropic assumption to construct simple models of stellar structure. The numerical results, plus the scaling used to transform into non-dimensional variables, provide several important results, as summarized by Chandrasekhar (1967). Without going into details, we note that for the $n = 3/2$ polytrope, the relation between the stellar radius and mass is

$$R_* M_*^{1/3} \propto K . \qquad (9.2)$$

The mass–radius relation requires the specification of the constant K, i.e. the adiabat corresponding to the stellar interior conditions. This constant cannot be found from analyzing only convective energy transport (Schwarzschild 1958). The radiative atmosphere provides the outer boundary condition necessary to specify K; it is joined to the essentially adiabatic interior by a region where both convective and radiative transport are important. To solve for the structure of this latter region, it is necessary to have an estimate of the efficiency of convection where radiative transport is important. This is conventionally modelled in terms of an efficiency parameter, the so-called 'mixing length' (Schwarzschild 1958). The resulting mass–radius relations will be dependent upon opacities, particularly in the radiative atmosphere, and upon the adopted mixing length or convection efficiency. Because it is quite difficult to calculate the needed molecular opacities, treating many millions of lines, and because the mixing length theory is heuristic rather than predictive, there is substantial uncertainty in predictions of HR diagram positions (stellar luminosity and effective temperature) for contracting pre-main-sequence stars. The situation may be clarified substantially when studies of T Tauri binaries provide stellar masses independent of position in the HR diagram; this will permit calibration of the combined effects of opacity and convection in outer stellar layers.

Hayashi *et al.* (1962) noted that for a pre-main-sequence star of a given (low) mass, it was not possible to find solutions to the stellar structure equations for which the effective temperature was less than a certain value. This occurs because the (gaseous) opacity decreases very rapidly below ~ 4000 K. The photosphere of the star must be optically thick to radiate the required luminosity out into space, but the low opacity at low temperatures makes it impossible to match an atmosphere to the interior solution for a given stellar mass below a certain temperature (cf. Shu 1991). This property of the opacity tends to make low-mass pre-main-sequence stars contract at *roughly* constant effective temperature.

The overall properties of pre-main-sequence contraction of convective stars can be illustrated by applying a global energy balance equation. The equation of hydrostatic equilibrium is (2.27)

$$\frac{1}{\rho}\frac{dP}{dr} = -\frac{GM_r}{r^2} = -\frac{d\Phi}{dr}, \tag{9.3}$$

where M_r is the mass interior to radius r and Φ is the gravitational potential. With the polytropic assumption (9.1), one can integrate equation (9.3) with radius to obtain the ratio of pressure to density at radius r,

$$(n + 1)P/\rho = \Phi(R_*) - \Phi(r). \tag{9.4}$$

The total gravitational potential energy of the star is

$$W = -G\int_0^{R_*} \frac{M_r\,dM_r}{r} = \frac{1}{2}\int_0^{R_*} \Phi\,dM_r, \tag{9.5}$$

where the final right-hand result can be found by integrating by parts twice (Chandrasekhar 1967). Substituting, one finds

$$W = \frac{1}{2}\int_0^{R_*} dM_r\left[-(n+1)\frac{P}{\rho} + \Phi(R_*)\right] = -\frac{n+1}{2}\int_0^{R_*} P\,dV - \frac{1}{2}\frac{GM_*^2}{R_*}, \tag{9.6}$$

where V is the volume and $dM_r = 4\pi r^2\rho dr$.

From the virial theorem for an unmagnetized, non-rotating equilibrium configuration (2.19),

$$\int dV\,3P = \int dV\,\rho r\nabla\Phi = \int dM_r GM_r/r = -W. \tag{9.7}$$

Combining these equations yields

$$W = -\frac{3}{(5-n)}\frac{GM_*^2}{R_*} = -\frac{6}{7}\frac{GM_*^2}{R_*} \tag{9.8}$$

for $n = 3/2$.

The internal thermal energy of the star is

$$U = \int_0^{R_*} c_V\,T\,dM_r = \frac{1}{\gamma-1}\int_0^{R_*} \frac{P}{\rho}\,dM_r = \frac{1}{\gamma-1}\int_0^{R_*} P\,dV, \tag{9.9}$$

where c_V is the specific heat at constant volume (Chandrasekhar 1967). Then, for a perfect gas,

$$U = -\frac{W}{3(\gamma-1)} = -\frac{W}{2}, \tag{9.10}$$

which is the result required by the virial theorem,

$$2U + W = 0. \tag{9.11}$$

Finally, we have the total energy of the polytropic star:

$$E = U + W = \frac{3}{7}\frac{GM_*^2}{R_*} - \frac{6}{7}\frac{GM_*^2}{R_*} = -\frac{3}{7}\frac{GM_*^2}{R_*}. \tag{9.12}$$

If the (pre-main-sequence) star does not have an internal fusion energy source, the energy radiated into space (the photospheric luminosity L_*) must cause a corresponding change in the energy of the star;

$$L_* = -\frac{d}{dt}E = \frac{d}{dt}\frac{3}{7}\frac{GM_*^2}{R_*}. \tag{9.13}$$

The negative sign enters because a decrease in the total stellar energy results in a positive luminosity. It is evident from equation (9.13) that, at fixed stellar mass M_*, the radiation loss L_* will cause the star to contract. The stellar contraction releases gravitational potential energy to replace the energy lost from the stellar surface. By the virial equilibrium equation (9.11), half of this gravitational energy is converted into thermal energy which is needed to replace the surface energy losses to maintain hydrostatic support.

We can obtain qualitative insight into Hayashi track evolution by making the simple approximation that the effective temperature is constant,

$$T_{eff} = (L_*/4\pi\sigma R_*^2)^{1/4} \approx \text{constant.} \tag{9.14}$$

For times t long after some (arbitrary) starting time, when the radius was much larger than the current radius at time t, combining the constant effective temperature approximation with (9.13) results in

$$L_* = L_\circ \left(3\frac{t}{\tau_{kh}}\right)^{-2/3}, \tag{9.15}$$

where

$$\tau_{kh} = \frac{3}{7}\frac{GM_*^2}{R_\circ L_\circ} \tag{9.16}$$

is called the Kelvin–Helmholtz timescale and $L_\circ$ is the luminosity when the star has a reference radius $R_\circ$. Thus, as the star ages, it contracts and becomes fainter. The rate of decrease in the stellar luminosity (and in the stellar radius) slows with increasing age.

Figure 1.2 shows some pre-main-sequence evolutionary tracks calculated by D'Antona and Mazzitelli (1994). The low-mass stars initially descend in the HR diagram nearly vertically; the evolutionary tracks for different masses are separated in effective temperature partly because the differing gravities affect the outer atmospheric opacities. The simple estimate (9.15) does a reasonably good job of explaining the rate of contraction shown in Figure 1.2. For example, for parameters $M_* = 0.8\,M_\odot$, $R_* = 2\,R_\odot$, and $L_* = 1\,L_\odot$, the Kelvin–Helmholtz time is $\tau_{kh} = 4.3 \times 10^6$ yr, so that the Hayashi track age should be about $t \sim 1.4 \times 10^6$ yr, in reasonable agreement with the tracks.

This simple scaling of contraction rates predicts that stars differing by a factor of ten in age should have luminosities differing by $\Delta \log L \sim 2/3$, approximately what is found in detailed calculations. The principal departure from the power-law decay of the luminosity with time occurs for low-mass stars between the ages of $\sim 10^5$ and

$\sim 3 \times 10^5$ yr, and is mostly due to deuterium fusion energy release, which occurs when the central stellar temperature reaches $\sim 10^6$ K (§9.5). This nuclear energy release slows the stellar contraction, because the star is no longer supplying the energy lost in surface radiation from its internal store of thermal energy; this causes the isochrones to approach each other in Figure 1.2. Deuterium fusion represents a modest energy supply, and therefore cannot halt the overall contraction for long. Once the deuterium is completely fused, gravitational contraction resumes unhindered until the star reaches the hydrogen-fusion main sequence.

Aside from the starting HR diagram position (§9.5), the ages of low-mass pre-main-sequence stars on convective tracks can be determined from the stellar mass, radius, and luminosity. The radius and luminosity of the T Tauri star can be determined fairly directly from observations, if the distance is known. Uncertainties in the masses assigned to some T Tauri stars on the basis of theoretical evolutionary tracks may be as large as a factor of two (Mazzitelli 1989); such a change in mass would correspond to a factor of four change in the Kelvin–Helmholtz timescale and thus in the age. This situation probably will improve in the near future as the masses of binary T Tauri systems become available.

For stars with masses $\gtrsim 0.8\,\mathrm{M_\odot}$, contraction at nearly constant effective temperature eventually stops when the star develops a radiative core. The star then proceeds to the main sequence along nearly horizontal paths in the HR diagram (Henyey, LeLevier, & Levee 1955; Figure 1.2). This change occurs because radiative stars of uniform gaseous composition exhibit a well-defined relation between mass and luminosity (Schwarzschild 1958). In a radiative star the transport of energy can be described by the diffusion equation (Appendix 3) through the interior. Therefore, using a rough scaling argument,

$$\frac{1}{\kappa\rho}\frac{dT^4}{dr} \sim \frac{1}{<\kappa\rho>}\frac{T^4}{R_*} \propto \frac{L_*}{R_*^2}. \tag{9.17}$$

If we set the mean interior temperature $T \propto M_*/R_*$ (Chapter 2; also see below), and take the mean density $<\rho> \propto M_*/R_*^3$, then we arrive at a crude mass–luminosity relationship

$$M_*^3 <\kappa>^{-1} \propto L_*. \tag{9.18}$$

For high-mass stars, where electron scattering dominates, the mean opacity is nearly constant, and $L_* \propto M_*^3$; for lower-mass stars, the opacity dependence changes the mass–luminosity relationship. As a pre-main-sequence star becomes increasingly radiative, this mass–luminosity scaling becomes more relevant, and contraction to the main sequence occurs at nearly constant luminosity.

9.2 Observational HR diagrams: Taurus

As an example of the application of evolutionary tracks to determine the ages of pre-main-sequence stars, we consider the Taurus-Auriga molecular cloud complex in some detail. As noted in Chapter 1, Taurus is not a typical star-forming region, but its relatively low extinction means that it is much easier to characterize the stellar population with current techniques than in other star-forming clouds.

Establishing the luminosity and effective temperature of a star generally involves the following steps. Ratios of the strengths of photospheric absorption lines are used to determine a spectral type for the star, which corresponds to some photospheric effective temperature (e.g., Cohen & Kuhi 1979). To calculate the luminosity, the observed fluxes must be corrected for interstellar extinction, which causes the star to become redder, i.e. short-wavelength light is more strongly absorbed and scattered than long-wavelength light (Figure 4.1). Efforts have been made to determine 'standard' interstellar extinction laws which relate the change in reddening, i.e. the change in the differential extinction as a function of wavelength, to absolute extinction (Mathis 1990). If the spectral type or effective temperature of the star is known, the observed colors or fluxes of the star as a function of wavelength can be compared with the properties of non-obscured standard stars of similar temperature/spectral type to establish the extinction corrections needed to convert the observed fluxes to intrinsic stellar fluxes. Finally, the observed fluxes can be converted to a luminosity if the distance is determined, usually by identifying main-sequence stars which can be associated with the cloud by either reflection nebulae or enhanced extinction.

These procedures are (relatively) easy to implement for WTTS, which basically exhibit the emission properties of normal stars. The determination of spectral types and luminosities is much more difficult for the accreting CTTS because of the need to disentangle the stellar photospheric emission from the hot accretion continuum (§6.4). The following discussion is based on Taurus CTTS luminosities estimated from observed fluxes at $1.25\,\mu m$ (the J band flux) along with a bolometric correction based on the optical spectral type and an extinction correction *neglecting* the effect of hot continuum radiation (Kenyon & Hartmann 1995). This simple procedure is advantageous because most stars in Taurus have spectral energy distributions which peak near $1\,\mu m$ (Figure 6.2); thus, using the J flux minimizes the contributions of the hot continuum (strongest at blue-ultraviolet wavelengths) and the disk (which is more important at longer wavelengths. In addition, near-infrared measurements are less affected by the necessary extinction corrections. The disadvantage of this procedure is that, by neglecting the effects of the hot continuum on optical colors, the extinction to the star is underestimated (e.g., Hartigan *et al.* 1995). It does not appear that neglecting the hot continuum greatly affects the luminosities derived for most T Tauri stars (Kenyon & Hartmann 1995).

Currently there are ~ 100 T Tauri stars in Taurus which can be placed in the HR diagram with some (modest) accuracy. Figure 1.2 shows the distribution of these Taurus pre-main-sequence stars in the HR diagram, superimposed on theoretical stellar evolutionary tracks and isochrones. The stars clearly lie above the main sequence, with a median age of $\sim 10^6$ yr suggested by comparison with the theoretical tracks. There are very few stars lying above the 10^5 yr isochrone; a sprinkling of 'WTTS' lie near the main sequence, and these objects will be discussed further below.

Figure 9.1 shows the stars of Figure 1.2 binned into age intervals, using the D'Antona & Mazzitelli (1994) tracks. The median age (luminosity) of the WTTS is slightly older (lower) than the corresponding distributions for the CTTS, although this difference is modest compared to the dispersion of the individual distributions. It should be emphasized that Figures 1.2 and 9.1 do not take into account the presence of close binary companions (e.g., Ghez *et al.* 1993; Leinert *et al.* 1993; Simon *et al.* 1995) which may bias results. Specifically, if the photometric or flux measurement includes more than one star, the result is an apparently overluminous star which will be assigned too low an age from the HR diagram. The maximum effect for a binary

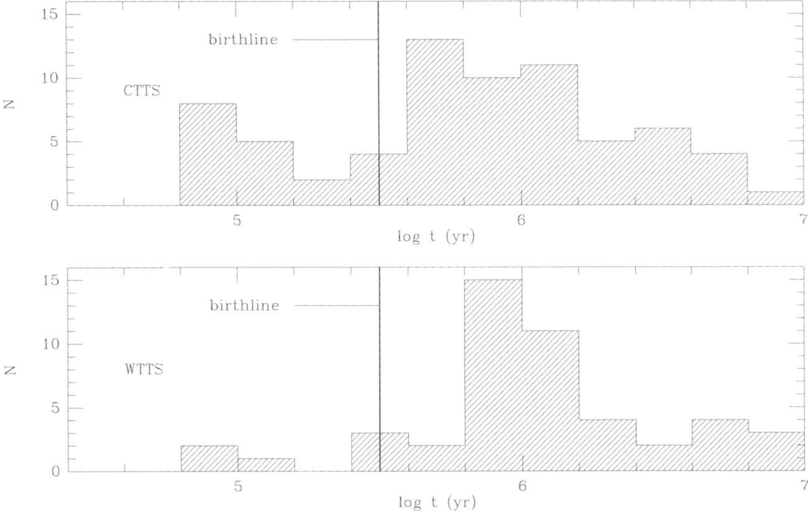

Fig. 9.1. Distribution of Hayashi track ages (see text) for CTTS and WTTS in Taurus. Stellar data was taken from Kenyon & Hartmann (1995), and the 'CMA' evolutionary tracks of D'Antona & Mazzitelli (1994) (cf. Figure 1.2) were used for the age calibrations. An age of $\sim 3 \times 10^5$ yr is used to approximate the position of the 'birthline' (see §9.5).

is to double the apparent stellar luminosity (if both stars are equal), in which case the age is underestimated by about a factor of three (equation (9.15)).

At present the significance of the age differences shown in Figure 9.1 is not clear. Uncertainties in the extinction corrections, distances, and spectral types might lead to an error of almost 0.2 in log L, corresponding to nearly a factor of two uncertainty in the age, not much smaller than the observed spread in the distributions of WTTS and CTTS (Kenyon & Hartmann 1995). The WTTS seem to be less numerous than the CTTS at the youngest ages; however, given that it is more difficult to place a CTTS in the HR diagram than a WTTS, this result should be viewed cautiously. There is a very small number of stars which appear to be (relatively) luminous and very young; however, we do not yet know whether there is some other observational error involved; perhaps these objects are multiple systems. Ongoing searches for close stellar companions among T Tauri stars should help clarify the situation in the next few years.

Figures 1.2 and 9.1 indicate the presence of some relatively old WTTS in Taurus, quite near the main sequence. There is a distinct gap in the HR diagram between the young pre-main-sequence stars and these objects, suggesting that they are part of a different population than the main body of Taurus stars. If these stars really have ages of $\gtrsim 30$ Myr, as suggested by their apparent positions in the HR diagram, they may have traveled distances $\gtrsim 60$ pc from their birthplaces at a typical (small) velocity dispersion for Taurus stars $\sim 2\,\mathrm{km\,s^{-1}}$ (Herbig 1977a; Jones & Herbig 1979). In other words, there is no guarantee that these stars 'belong' to Taurus at all (Briceño *et al.* 1997). If these 'older' stars are part of a more dispersed population, the age analysis is even more difficult, because the (unknown) distances of these objects may not be the same as the distance to Taurus; even small changes in distance can make very large changes in age estimates. Briceño *et al.* (1997) have shown that X-ray surveys are often biased toward finding young stars at distances < 140 pc; thus, main-sequence stars may be identified as 'pre-main-sequence' objects if the adopted distances are too large. This issue has important consequences for current attempts to understand the time evolution of circumstellar disks, because this spatially-distributed population of stars is often used to constrain the timescale over which disk infrared emission 'disappears' (Strom *et al.* 1989). Progress in this area requires the study of disk properties in stellar clusters, where distances and ages are better determined.

9.3   Formation of protostars

Pre-main-sequence evolutionary tracks such as the ones shown in Figure 1.2 assume, following the the original calculations of Hayashi, that low-mass stars are formed far up on convective tracks in the HR diagram. However, as shown in Figure 9.1, there are relatively few stars found far up on Hayashi tracks. Why is this so?

One possible reason is that stars evolve quickly through this region of the HR diagram. As is evident from the scaling law (9.15), stars high up on Hayashi tracks, which have large R_* and L_*, will contract quickly because they have relatively low energy content $\propto M_*^2/R_*$ to supply the large photospheric luminosity. Therefore one should observe relatively few luminous low-mass pre-main-sequence stars.

If star formation occurs at a constant rate, then the total number of stars with ages less than t is $N_T \propto t$. Since the luminosity of a star decending on a Hayashi track is $L_* \propto t^{-2/3}$ (equation (9.15)), the number of stars with luminosity greater than L should be $N_T \propto L^{-3/2}$. The distributions shown in Figure 9.1 seem to have relatively fewer stars at high luminosity than predicted by this distribution, but it is difficult to make a statistically significant argument for the small numbers of stars in Taurus. The analysis is complicated by the apparent difference between WTTS and CTTS distributions, which may result from binary companions (§9.2), errors in correcting for accretion, etc.

Another reason why stars might not be observed high on Hayashi tracks is that optical samples do not detect the youngest, most heavily-extincted stars. However, as discussed in §4.1, the median luminosity of Class I stars in Taurus is the same as that of the optically-visible T Tauri stars (Figure 4.2), providing no evidence for objects higher on their Hayashi tracks. This constraint is even stronger if allowance is made for some accretion luminosity in Class I objects; i.e. the Class I system luminosity must be an upper limit to the intrinsic stellar luminosity. The luminosity function of Taurus Class I sources suggests that protostars do not generally lie in the upper regions of the HR diagram.

It seems intuitively plausible that stars cannot be formed arbitrarily high on Hayashi tracks, since it takes a finite amount of time to accrete the necessary mass from the infalling envelope. Hayashi track evolution at an 'age' of, say, 10^3 yr seems unlikely to be relevant if it takes 10^5 yr to accumulate the stellar mass (Chapter 3). The first numerical calculations of low-mass protostars formed by collapsing envelopes, in particular those of Larson (1969a,b), indicated that envelopes which collapse on timescales of 10^6 yr form protostars at positions roughly near the 10^6 yr isochrones of Hayashi tracks. Although these results have been superseded by much more detailed calculations, it is remarkable how close Larson's results for initial stellar radii are to those of later, more sophisticated calculations (e.g., Stahler *et al.* (1980a,b); see Shu (1991) for an historical overview, and Boss (1995) for a more recent review).

Several lines of argument lead to the conclusion that the initial luminosities (or radii) of low-mass protostars are not likely to be large. One argument depends upon the outer boundary condition for the protostellar core. The accretion energy liberated under spherical non-rotating collapse to a protostar with radius R_p is

$$L_{acc} = GM_p\dot{M}/R_p,\qquad(9.19)$$

where M_p is the mass of the central core and $\dot{M}$ is the mass infall rate. The protostar's photosphere radiates energy into space at a rate

$$L_{phot} = 4\pi R_c^2 \sigma T_{eff}^4.\qquad(9.20)$$

Assuming that most of the accretion energy is radiated, Stahler *et al.* (1980a) argued that there is a maximum radius at which the energy balance can be maintained, resulting in a lower limit to T_{eff} given the rapid decrease of opacity with decreasing temperature (§9.1). They estimated that a solar-mass protostar accreting at expected rates in spherical collapse would have a radius only a few times that of the present-day Sun. This result assumes essentially free radiation of the accretion energy into space; although the accretion-produced radiation is absorbed in the surrounding dusty protostellar envelope (Chapter 4), the 'back-radiation' from the circumstellar dust shell can be shown to be relatively unimportant (Stahler *et al.* 1980a).

Another way of looking at this problem is to note that the development of a protostar in hydrostatic equilibrium requires a source of thermal energy to create sufficient gas pressure to balance gravity and the ram pressure of accretion. A non-rotating star in virial equilibrium must have a total energy (§9.1)

$$E = U + W = W/2.\qquad(9.21)$$

The initial star-forming cloud had a total energy of essentially zero in comparison with the final star. The production of an equilibrium (stellar) core mass thus depends upon the generation of sufficient internal energy to balance the increase in the gravitational potential energy during infall. There are two potential sources of this internal energy; the dissipation of gravitational energy during infall, releasing heat, and the possible fusion of deuterium as it is mixed into the star. The problem of fixing the size of the stellar core schematically reduces to calculating the amount of thermal energy produced and trapped in the protostar.

To produce a protostellar core in hydrostatic equilibrium, not only must there be a source of thermal energy to produce sufficient gas pressure; this thermal energy also must be trapped effectively within the core, otherwise rapid radiation losses will prevent core formation. This can be seen by using the diffusion approximation (9.17) once more, in the form

$$T \sim \left(\frac{L_c \tau_R}{4\pi R_c^2 \sigma}\right)^{1/4}.\qquad(9.22)$$

For hydrostatic equilibrium, it is necessary that

$$T \sim \frac{GM_c}{(k/\mu m_H)R_c},\qquad(9.23)$$

(cf. §9.4), where μ is the mean molecular weight and m_H is the mass of the hydrogen atom.

From these equations it is straightforward to show that the Rosseland mean optical depth τ_R must be very large for any reasonable parameters, and this demands a small compact core. Equations (9.22) and (9.23) can be combined, using a Kramer's opacity law

$$\kappa_R = 6.6 \times 10^{22} \rho T^{-7/2} \, \text{cm}^2 \, \text{g}^{-1} \tag{9.24}$$

(Frank *et al.* 1992) to yield the protostellar size given some estimate of the luminosity. If the protostellar core produces luminosity in its interior at a rate

$$L_p = f_p L_{acc}, \tag{9.25}$$

the core radius is approximately

$$R_c \sim 1.6 \, \mu_{0.6}^3 \, M_{0.5}^{11/5} \, \dot{M}_{-5}^{-2/5} \, f_p^{-2/5} \, \text{R}_\odot. \tag{9.26}$$

where $\mu_{0.6}$ is the mean molecular weight in units of 0.6, $M_{0.5}$ is the protostellar mass in units of $0.5\,\text{M}_\odot$, and $\dot{M}_{-5}$ is the mass accretion rate in units of $10^{-5}\,\text{M}_\odot\,\text{yr}^{-1}$. This crude estimate suggests that low-mass protostellar cores are likely to be relatively small for reasonable internal luminosities (see Boss (1995) for a more detailed description of numerical simulations of core formation).

Why should there be internally-generated protostellar luminosity? There are two possible reasons; readjustment of the material in the stellar interior, and fusion energy release. To examine the first possibility schematically, imagine that a thin shell of mass M_s and radius R_s has accreted from infinity, and the excess energy has been radiated away so that the shell is at a low (effectively zero) temperature. Then the total energy is simply the shell's gravitational potential energy,

$$W_s = -\frac{GM_s^2}{2R_s} = E_s. \tag{9.27}$$

If matter is brought from infinity to make this shell, an amount of energy $-W_s$ must be released; thus, the accretion luminosity is

$$L_{acc} = -\frac{dW_s}{dt} = \frac{GM_s(t)\,\dot{M}}{R_s}, \tag{9.28}$$

in agreement with the standard result. However, a cold thin shell with all the matter at the outer radius R_s is not a possible configuration for a pressure-supported star. The temperature and density structure of the star must readjust to maintain hydrostatic equilibrium, and in this process of redistribution of mass energy may be released (or absorbed) in the stellar interior.

The details of the stellar readjustment cannot be derived without detailed interior calculations (e.g., Stahler *et al.* 1980a,b). To see qualitatively what is involved, suppose that the protostar can be represented by an $n = 3$ radiative polytrope (§9.1). (Detailed calculations suggest that protostars may be radiative initially, since the accretion onto the outer surface provides a strong outside-in heating, which suppresses convection; Stahler *et al.* (1980a,b).) In this case the gravitational potential energy is

$$W_p = -(3/2)GM_p^2/R_p. \tag{9.29}$$

By the virial theorem half of this energy must go into thermal energy to support the star against gravity (§9.1), and so the total energy of the core is

$$E_p = -(3/4)GM_p^2/R_p. \tag{9.30}$$

If we set $M_p = M_s$ and $R_p \leq R_s$ initially, we see by comparing equations (9.30) and (9.27) that $E_p < E_s$. That is, the energy of the adjusted protostar is more negative than the energy of the initial shell. The only way this can occur is if some energy has been lost from the system, e.g. by radiation from the surface (photosphere). In this limit there is some internally-generated luminosity resulting from accretion which can help to stop collapse. Alternatively, if the matter is accreted so quickly that the surface photospheric luminosity is negligible, and energy is conserved, then $R_p = 3R_s/2$, i.e. the core must expand. In general both effects may occur and detailed calculations are required to obtain quantitative results, especially to include time-dependent hydrodynamic effects.

To illustrate the importance of the interior structure, suppose instead that the protostar is an $n = 3/2$ (convective) polytrope. In this case the total internal energy of the protostar, $E_p = -(3/7)GM_p^2/R_p$, is slightly *larger* than the energy of the cold shell (9.27). If there is no readjustment in radius, there is a *negative* settling luminosity. To rearrange material from a cold shell into a convective star without adding thermal energy, it is necessary for the star to *contract* (§9.5).

Alternatively, if the protostar is small enough and hot enough, deuterium fusion may proceed, representing an energy source to help heat the star and halt collapse. This possibility is considered in more detail in the following two sections.

9.4   The 'birthline'

For convenience, the formation of a protostar might be schematically divided into two phases; a 'core' formation phase, in which a central mass in hydrostatic equilibrium develops, followed by a longer phase of accretion onto the quasi-static protostar. The process of building a protostellar core is likely to be a complicated hydrodynamic event which may be dependent upon initial conditions varying from star to star. Moreover, the numerical difficulties in following this collapse are severe (see discussion in Tscharnuter (1991)). Once a hydrostatic core is established, however, a simplified treatment following the accretion of mass onto the protostar becomes possible. This procedure was used to follow evolution during this second, quasi-static phase of protostellar formation by Stahler *et al.* (1980a,b) and Stahler (1988) for the case of spherical accretion.

Stahler *et al.* (1980a,b) considered the evolution of protostellar cores starting from a small mass $M \sim 0.01\,M_\odot$ and modest radius $\sim 3.5\,R_\odot$. They found that such cores were initially radiative as they accreted from the infalling envelope. However, the protostars then contracted until the central temperature was high enough for deuterium fusion to proceed; at that point contraction halted and deuterium fusion energy release in the interior eventually made the protostars convective. SST found that the settling luminosity (§9.3) generally was small; (§II.b.i of Stahler *et al.* 1980b).

Expanding on this work, Stahler (1983, 1988) argued that, for a reasonable range of initial conditions and accretion rates, protostars should evolve in such a way that their radii were a strongly-constrained function of their mass, because of the strong sensitivity of deuterium fusion to temperature. Once spherical accretion ended,

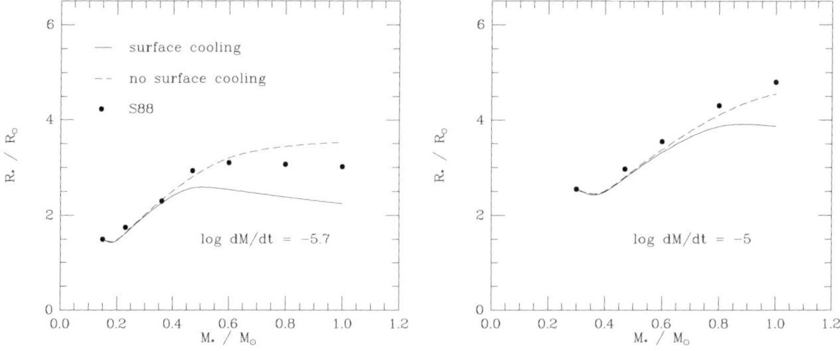

Fig. 9.2. Birthline evolutionary tracks for two different mass accretion rates. The dots correspond to the results of Stahler (1988) and the curves are calculated from the simple evolutionary model for cold disk accretion (§9.5). From Hartmann, Cassen, & Kenyon (1997).

the protostars of fixed $R_p(M_p)$ would appear along a well-defined locus in the HR diagram he called 'the stellar birthline'. Stahler argued that the predicted birthline was consistent with the positions of the *youngest* optically-visible stars, and thus it provides a reasonable starting point from which to time the contraction of convective T Tauri stars down their Hayashi tracks.

The essential basis of Stahler's (1988) argument is as follows. For sufficiently large mass accretion rates $\dot{M}$, the concentration of deuterium is kept high by the addition of freshly-accreted material even as the star fuses deuterium into heavier elements. In this limit, the central temperature of the protostar is forced to remain nearly constant by the thermostatic nature of deuterium fusion (produced by the high sensitivity of the deuterium fusion rate to temperature). If the central temperature falls slightly, the star will resume contracting ($T \propto M_p/R_p$) until the temperature is high enough for fusion; if the central temperature increases slightly, the resulting rapid increase in L_D (equation (9.38)) will cause the star to expand (equation (9.34)). This temperature sensitivity causes the star to try to maintain a constant M_p/R_p; as the star accretes, the radius grows in the same proportion. This 'deuterium main sequence' behavior leads to a reasonably well-defined $R_p(M_p)$ or protostellar 'birthline'.

Figure 9.2 shows evolutionary tracks calculated for two different mass accretion rates which might correspond to Class I sources (Chapters 3, 4). The solid dots represent the original calculations of Stahler (1988). There is an initial $M_* \propto R_*$ dependence, which extends to higher masses at higher accretion rates; this is the region where the concentration of deuterium remains high, and so the central stellar temperature must remain fixed (Figure 9.3). At higher masses the $R_p(M_p)$ relation flattens out when the available deuterium is mostly fused into heavier elements (Figure 9.3), allowing the star to heat up beyond the point of deuterium fusion.

Note that in the case of spherical accretion, protostars of given radius and mass do not have the same effective temperatures that normal low-mass pre-main-sequence stars would have. The accretion energy release makes the photospheric temperature much larger. If accretion stops rapidly, however, the post-protostar (now T Tauri

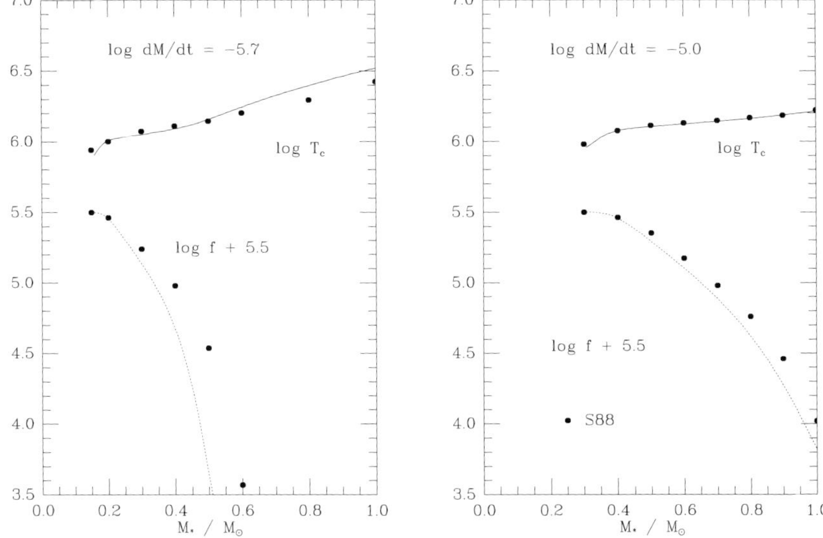

Fig. 9.3. Variation of central protostellar temperature and deuterium concentration f corresponding to the evolutionary tracks shown in Figure 9.2 (the 'surface cooling' case of Figure 9.2). Dots show the original results of Stahler (1988); curves illustrate the disk accretion cases of §9.5. From Hartmann *et al.* (1997).

star) would quickly lower its effective temperature at nearly constant R_* (Stahler *et al.* 1980b).

9.5   Stellar evolution with disk accretion

The results of Stahler (1988) were very important in emphasizing the importance of deuterium fusion in early stellar evolution, and in making quantitative estimates of protostellar properties. However, given the likely angular momentum of accreting material (Chapters 3), quasi-spherical infall to a central hydrostatic core probably occurs over a very short initial period, followed by the main phase of accretion through a disk. It is therefore necessary to consider how disk accretion might differ from spherical accretion. It turns out to be possible to address this problem in a simple way which (at least qualitatively) illustrates the general 'birthline' physics of protostellar evolution as well as the effects of disk accretion during the T Tauri phase (Hartmann *et al.* 1997; see also Palla & Stahler 1992).

Assume that disk accretion, whether through a boundary layer or the magneto-spheric accretion column (§§6.4, 8.11, 8.12), covers only a small fraction of the stellar photosphere. In this case, most of the stellar photosphere can radiate freely to space, unaffected by accretion. Under these conditions, we expect typical low-mass pre-main-sequence stars or stellar cores to be completely convective, and therefore adopt the $n = 3/2$ polytropic approximation. We further neglect the rotational energy of the star for simplicity; this should not be a bad approximation for most T Tauri stars, which are slowly rotating.

Suppose a small mass of gas Δm is added to the star, with gravitational potential energy

$$\Delta W = -\frac{GM_*\Delta m}{R} \tag{9.31}$$

and internal energy

$$\Delta U = \epsilon \frac{GM_*\Delta m}{R}, \tag{9.32}$$

where ϵ is introduced to parameterize the thermal energy content of the accreted material (e.g. Prialnik & Livio 1985). The star then readjusts to a polytropic configuration with mass $M_* + \Delta m$ and radius $R_* + \Delta R$. We assume that this mass is added over a time Δt, and that the surface luminosity of the star during this time is L_*. Then conservation of energy requires (cf. equation (9.13))

$$L_*\Delta t - \frac{3}{7}\frac{G(M_* + \Delta m)^2}{(R_* + \Delta R)} = -\frac{3}{7}\frac{GM_*^2}{R_*} - \frac{GM_*\Delta m}{R_*}(1 - \epsilon) + L_D\Delta t, \tag{9.33}$$

where L_D is the deuterium fusion luminosity (see also Palla & Stahler (1992)). Expanding this equation to first order, writing $\dot{M} = \Delta m/\Delta t$, and letting $\Delta t \to 0$, we have

$$L_* = -\frac{3}{7}\frac{GM_*^2}{R_*}\left[\left(\frac{1}{3} - \frac{7\epsilon}{3}\right)\frac{\dot{M}}{M_*} + \frac{\dot{R}_*}{R_*}\right] + L_D. \tag{9.34}$$

In the limit that $\dot{M} = 0$ and $L_D = 0$, the standard gravitational contraction result (9.13) is recovered.

The first two terms on the right-hand side of equation (9.34) account for change in the energy of the star due to the accreted matter. To isolate the basic effects, assume that the surface photospheric luminosity and the deuterium fusion luminosity are small (equivalently, assume that a finite mass and energy are added on a short enough timescale that surface energy losses and deuterium fusion energy release are negligible). First, suppose $\epsilon << 1/7$, so that the material is accreted 'cold'. Then the accretion of matter will cause the star to contract ($\dot{R}_* < 0$). The physical reason for this is that the cold accreted material does not support itself; the star must readjust to generate additional thermal energy (pressure) to support this material. In the case of a (convective) $n = 3/2$ polytrope, the star must contract (§9.3). Conversely, if $\epsilon > 1/7$, then the accretion of warm material will tend to make the star expand (e.g., Prialnik & Livio 1985). For sufficiently large accretion rates and ϵ, the accreting star may become radiative, in which case the star may expand rapidly and this treatment is invalid, as discussed below (see also §9.3).

We can relate ϵ to other physical quantities by noting that, for a perfect gas,

$$T_{acc}\,\Delta m = \frac{\Delta U}{c_V} = \frac{(\gamma - 1)\mu m_H}{k}\Delta U = \frac{2}{3}\frac{\mu m_H}{k}\Delta U, \tag{9.35}$$

where T_{acc} is the temperature of the accreting gas, k is Boltzmann's constant, m_H is the mass of the hydrogen atom, ΔU is the change in internal energy due to the accreted gas, and c_V is the ratio of specific heats at constant volume. An average internal temperature can be calculated for the $n = 3/2$ polytrope,

$$M_*\langle T \rangle \equiv \int T\,dM = \int \frac{u}{c_V}\,dM = \frac{2}{7}\frac{\mu m_H}{k}\frac{GM_*^2}{R}, \tag{9.36}$$

where u is the specific internal energy. Substituting for ΔU from (9.32) and solving for ϵ using (9.35) and (9.36),

$$\epsilon = \frac{3}{7}\frac{T_{acc}}{\langle T \rangle}.$$
(9.37)

The critical value $\epsilon_c = 1/7$ then corresponds to $T_{acc} = (1/3)\langle T \rangle$, i.e. the temperature of the accreting material is roughly comparable to the average internal temperature of the star.

As emphasized by Stahler (1988), the fusion of deuterium can play a crucial role in setting the evolutionary tracks of low-mass protostars. Deuterium energy generation is proportional to $\rho T^{14.8}$; integrating over the interior of an $n = 3/2$ polytrope,

$$L_D = 1.92 \times 10^{17} f\,[D/H] \left(\frac{M_*}{M_\odot}\right)^{13.8} \left(\frac{R_*}{R_\odot}\right)^{-14.8} L_\odot$$
(9.38)

(Stahler 1988). Here f is the fractional concentration of deuterium relative to its initial number abundance, which is taken to be $[D/H] = 2.5 \times 10^{-5}$. This equation assumes that the convection in the star instantaneously and thoroughly mixes all the deuterium, so that f is spatially constant at any time within the star. The strong dependence on stellar mass and radius arises from the steep dependence of the deuterium fusion on the central temperature, which is $\propto M_*/R_*$.

Because of the importance of L_D, evolutionary calculations must consider the variation of f,

$$\frac{df\,M_*}{dt} = \dot{M} - \frac{L_D}{\beta_D},$$
(9.39)

where β_D is the total energy available from deuterium fusion per gram of material (Stahler 1988). Rearranging this, we may write

$$\frac{df}{dt} = \frac{\dot{M}}{M_*}\left(1 - f - \frac{L_D}{L_{DSS}}\right),$$
(9.40)

where $L_{DSS} = \beta_D \dot{M}$ is the steady luminosity release that would be produced if the deuterium in the accreted material were instantaneously fused; numerically, for the assumed deuterium abundance,

$$L_{DSS} = 1.5 \left(\frac{\dot{M}}{10^{-6}\,M_\odot\,\mathrm{yr}^{-1}}\right) L_\odot.$$
(9.41)

The resulting evolution of the protostellar core can be calculated if the photospheric radiative losses are known. In this case, the surface photospheric radiative losses are the same as for an isolated pre-main-sequence star because accretion effects are limited to only a small area of the photosphere (§6.4). The detailed form of $L_*(M_*, R_*)$ can be taken in principle from the results of stellar structure calculations. For illustrative purposes one may adopt the following fit to the 'CMA' stellar evolutionary tracks of D'Antona & Mazzitelli (1994):

$$L_* = 1 \left(\frac{M_*}{0.5\,M_\odot}\right)^{0.9} \left(\frac{R_*}{2\,R_\odot}\right)^{2.34} L_\odot$$
(9.42)

(Hartmann *et al.* 1997). This fit is only approximate, and appropriate in the regime $0.3\,M_\odot \le M_* \le 1\,M_\odot$.

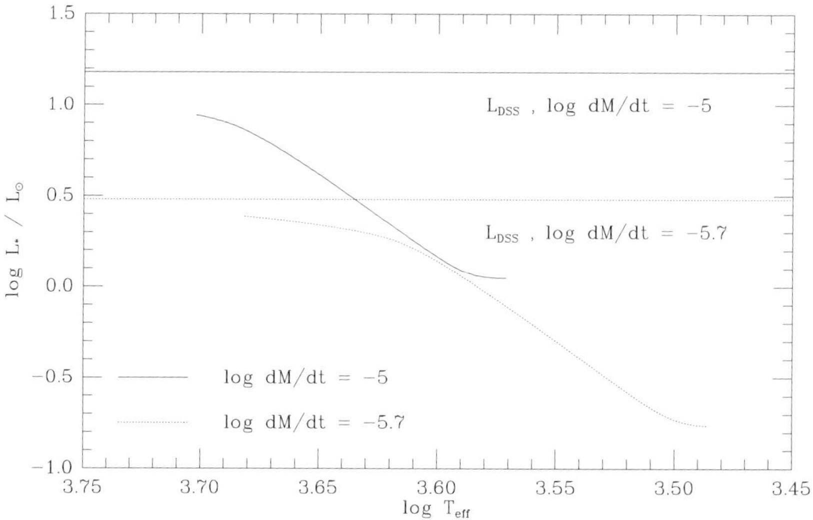

Fig. 9.4. HR diagram tracks for the 'surface cooling' calculations of Figure 9.2. The evolutionary tracks initially follow the $R_* \propto M_*$ relation when the deuterium fusion luminosity is high; however, once the (proto)star approaches the steady-state luminosity L_{DSS} the evolutionary tracks become more horizontal, because the energy represented by the fresh deuterium being accreted is now balanced by the photospheric radiative losses (see text). From Hartmann *et al.* (1997).

Figure 9.2 shows evolutionary tracks calculated for two different disk mass accretion rates. The curves have been calculated by simultaneously solving (9.34), (9.38), (9.40), and (9.42), starting with a low-mass core of small radius, and assuming that accretion is 'cold' ($\epsilon = 0$). These disk accretion curves depart only modestly from the spherical accretion results of Stahler (1988; solid dots in Figure 9.2, 9.3). In part, the differences between the two treatments derive from Stahler's use of more detailed stellar interior calculations and a careful treatment of the accreting material. However, it appears that the main reason for the differences between disk and spherical accretion is the difference in outer boundary conditions. Stahler assumed much smaller stellar photospheric radiative energy losses L_* (see discussion in Hartmann *et al.* (1997)). To illustrate this, the dashed lines in Figure 9.2 show the effect of setting $L_* = 0$ in the simple model calculations. The resulting $R_*(M_*)$ relations track the original calculations of Stahler more closely.

To clarify the effect of the photospheric radiative energy loss, which is assumed not to be blocked or trapped by the (magnetospherically) accreting material, it is useful to compare L_* to the steady-state luminosity that would occur if deuterium were fused as it is accreted into the protostar, L_{DSS}. If $L_* < L_{DSS}$, the demands on deuterium fusion are low, and the concentration f remains high. However, for $L_* > L_{DSS}$ the energy losses require fusing deuterium at a rate faster than is supplied through accretion, and f must drop. This behavior is shown explicitly in Figure 9.4. The evolutionary tracks turn over slightly before $L_* = L_{DSS}$ because some energy must be invested in heating the star to help support the weight of the cold accreting material. The corresponding deuterium concentrations are shown in Figure 9.3. As the stellar luminosity becomes

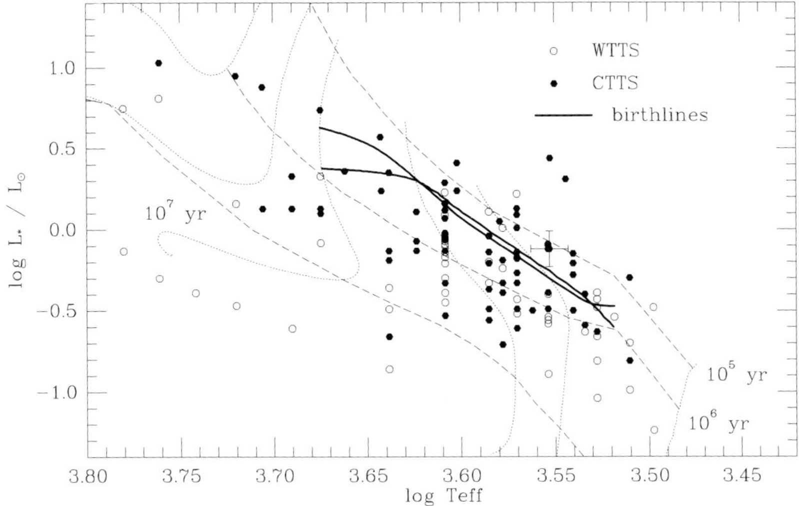

Fig. 9.5. Stellar evolutionary tracks for low-temperature ($\epsilon = 0$) disk accretion, compared with observed HR diagram positions of T Tauri stars in the Taurus–Auriga star-forming region. The evolutionary tracks are calculated as the simple evolutionary models shown in Figure 9.3, except interpolating in the D'Antona & Mazzitelli (1994) CMA tracks to provide the calibration of $L_*(M_*, R_*)$, producing small departures from the power-law calibration results shown in previous figures. The light solid lines show the CMA pre-main-sequence evolutionary tracks of D'Antona & Mazzitelli for masses of 0.1, 0.3, 0.5, 1, 1.5, and 2.5 M$_\odot$. The dashed lines show isochrones for 1×10^5 yr, 1×10^6 yr, and 1×10^7 yr. The HR diagram positions of T Tauri stars in the Taurus-Auriga molecular cloud, segregated between weak-emission (WTTS) and strong-emission (CTTS) stars, are taken from Kenyon & Hartmann (1995). The heavy solid lines correspond to birthline accretion rates of $\dot{M} = 2 \times 10^{-6}$ M$_\odot$ yr^{-1} (lower) and $\dot{M} = 10^{-5}$ M$_\odot$ yr^{-1} (upper). Modified from Hartmann *et al.* (1997).

increasingly large, the deuterium becomes depleted, allowing the star to have a higher central temperature and thus contract below the deuterium main sequence.

9.6 Comparison with observations

Figure 9.5 shows a comparison of the cold disk accretion birthlines (which are fairly similar to the spherical accretion birthlines; Figure 9.2) with the HR diagram positions of Taurus stars (e.g., Figure 1.2). In this figure the evolutionary tracks have been calculated using the detailed $L_*(M_*, R_*)$ relation from D'Antona & Mazzitelli (1994) CMA tracks. Comparison with the Taurus sample suggests that the low-temperature birthline is in rough, though not exact, agreement with the positions of the most luminous optically-visible stars. There are a few objects which appear to be above the birthline, but it is possible that these discrepancies arise from observational problems.

Note that there is nothing in the birthline calculations which forbids accreting protostars to lie above the deuterium main sequence in the HR diagram. The birthline calculations, either in the spherical or disk accretion limit, do not address conditions of core formation (§9.3). If the initial protostellar core has a radius well above the birthline, the accreting core will then descend until it reaches the deuterium main sequence; but the region above the birthline is by no means forbidden.

In principle, the best test of birthline calculations would be to compare directly with accreting protostars. At the moment, however, masses and radii for these heavily-extincted objects are not generally available. The median Class I luminosity in Taurus of $\sim 0.5\,L_\odot$ is reasonably consistent with the calculations shown in Figure 9.5, assuming a reasonable distribution of protostellar masses. However, as discussed in §4.1, there is a problem with total Class I source luminosities. With the birthline calculations, either from Stahler (1988) or the disk accretion calculations, of Figure 9.5, the protostellar radius–mass relation is constrained to be

$$R_*/\,\mathrm{R}_\odot \;\approx\; 6M_*/\,\mathrm{M}_\odot. \tag{9.43}$$

This implies an accretion luminosity of

$$L_{acc} \;\approx\; 10\,\mathrm{L}_\odot(\,\dot{M}/2 \times 10^{-6}\,\mathrm{M}_\odot\,\mathrm{yr}^{-1}), \tag{9.44}$$

which is not consistent with the observed Class I luminosity function in Taurus (Figure 4.2). If the birthline $R_*(M_*)$ is much larger than predicted, then the accretion luminosity might be lowered; but then the stellar luminosity would become unacceptably large (Figure 9.5; equation (9.42)). These considerations seem to indicate that protostellar mass accretion rates generally are much lower than assumed.

If disk accretion is highly episodic, so that much of the mass is accreted during (FU Ori) outbursts at high accretion rates, the protostar would then spend most of its time accreting relatively slowly from its disk, and so have a low accretion luminosity most of the time. However, in this case birthline calculations would need to be modified to include the effects of finite thermal energy in the accreted material. As discussed in Chapter 7, the inner disk midplane temperatures of FU Ori disks can be 10^5 K or more, which begins to approach the average internal temperature of the central star. Since the disk is extremely optically thick, one must consider the possibility that substantial amounts of thermal energy are advected into the star during FU Ori accretion (Popham *et al.* 1996; §7.6).

Evolutionary tracks calculated using the simple disk accretion model of the previous section, but assuming $\epsilon \sim 0.1$–0.2, show modest deviations from the tracks in the 'cold' accretion limit $\epsilon = 0$ (Hartmann *et al.* 1997), supporting the idea that the greatest difference between the spherical and disk accretion results lies in the photospheric boundary conditions adopted, not in the details of the treatment of the thermal energy in the accreting material. A much more significant issue is whether the accreting star remains convective. For spherically-symmetric accretion, Prialnik & Livio (1985) showed that rapid accretion of material with finite thermal energy content onto a fully-convective, main-sequence, low-mass star can cause the star to become partially radiative and expand substantially. This can occur even when $\epsilon < 1/7$, for which our assumption of a convective polytropic star would result in contraction (in the absence of deuterium fusion). If the timescale for changing the thermal energy of the star is short in comparison with the thermal equilibrium timescale, thermal equilibrium cannot be established, which is necessary for the star to remain convective. The constraint that the thermal time of the star be shorter than the thermal energy addition timescale,

$$t_{th} \;=\; E/L_* \;<\; t_U \;=\; E/\epsilon\,GM\dot{M}/R \tag{9.45}$$

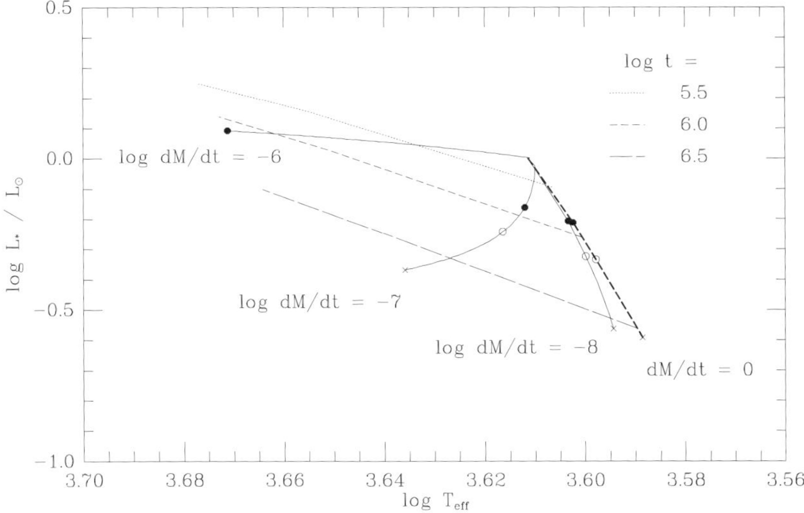

Fig. 9.6. Stellar evolutionary tracks in the low-temperature accretion limit for low accretion rates typical of T Tauri disks. In all cases the inital mass and radius are $0.5\,M_\odot$ and $2\,R_\odot$, typical of T Tauri stars. The filled circles, open circles, and crosses denote positions in the HR diagram for accretion at the indicated rates (in log $M_\odot\,yr^{-1}$) after elapsed times of 3×10^5 yr, 1×10^6 yr, and 3×10^6 yr. The bold dashed line indicates evolution for zero mass accretion. The light dashed lines show non-accreting Hayashi track isochrones for ages of 3×10^5 yr, 1×10^6 yr, and 3×10^6 yr. The effect of cold accretion is to push the convective stars lower in the HR diagram, making them appear slightly older than they really are (see text). From Hartmann *et al.* (1997).

corresponds to

$$\epsilon\, GM\dot{M}/R \;\; < \;\; L_* . \tag{9.46}$$

This condition states that convection could be suppressed if ϵ is greater than the ratio of intrinsic stellar luminosity to the accretion luminosity. The disk models of Popham *et al.* (1993) suggest that, depending upon the assumed disk viscosity, ϵ might be as large as 0.02 for $\dot{M} \sim 10^{-5}\,M_\odot\,yr^{-1}$, and $\epsilon \sim 0.1$ for $\dot{M} \sim 10^{-4}\,M_\odot\,yr^{-1}$. The relatively large values of ϵ at the highest accretion rates make it seem unlikely that the central stars in FU Ori objects, which are thought to be typical T Tauri stars with luminosities $L \sim 1\,L_\odot$ (Chapter 7), can remain convective during the outburst of rapid disk accretion. Indeed, this suggested stellar expansion, as found by Prialnik & Livio (1985) in the case of convective main sequence stars, may explain why inner disk radii for FU Ori objects seem to be a factor of two larger than the typical radii of T Tauri stars (§7.6). Under these circumstances the simple calculations of birthline evolution are not valid. It is possible that even with FU Ori accretion, protostars might spend much of their time near the cold accretion track if the time to recover to the convective state is short compared to the time between outbursts.

The cold accretion limit is much more likely to be applicable to disk accretion during the main T Tauri phase, where disk accretion rates are much lower and the accreting disk material is far cooler and less opaque than in the FU Ori outburst state. Figure 9.6 exhibits evolutionary tracks for several disk accretion rates and

$\dot{M} = 0$ in the limit that $f = 0$. The Hayashi track ages which would be inferred from the standard evolutionary tracks differ little from the elapsed times after the birthline for ages $\gtrsim 10^6$ yr. At a given age, the accreting HR diagram positions tend to fall slightly below the positions for the non-accreting stars, because of the extra contraction produced by the accretion of cold material (9.34), but the difference between the inferred ages and actual elapsed times is not larger than a factor of two.

Thus, it appears that T Tauri disk accretion does not affect age estimates very much, and that conventional Hayashi tracks are adequate to estimate ages for objects well below the birthline. Uncertainties in birthline positions remain, making it difficult to determine ages for the youngest stars. It must be emphasized again that the absolute calibration of pre-main-sequence ages is still uncertain because of uncertainties in stellar masses, which should be reduced in the near future from studies of binary T Tauri stars.

9.7   Disk evolution

We conclude with a simple model of how the circumstellar disk might evolve along with its central T Tauri star, and make some approximate comparisons with observations. To do this we make use of the similarity solution discussed in §5.2 under the assumption that the temperature structure of the disk is determined mostly by irradiation from the central star, and that the disk is nearly isothermal vertically. As discussed in Chapter 6, these approximations are probably not too bad for the outer disk, which tends to control the evolution. We further make the extremely simplifying assumption that the outer disk is reasonably 'flared' (§5.7, 6.2), so that the disk temperature varies as

$$T(R) \sim 10 \, (100 \text{AU}/R)^{1/2} \, \text{K}, \tag{9.47}$$

where the normalization is roughly what would be expected for a typical $1 \, \text{L}_\odot$ T Tauri star. Finally, we assume that the viscosity can be represented with a constant α formalism (§5.5). With these assumptions, $v_v \propto R$, and the similarity solutions of §5.2 hold. (The effect of the fading of the central star on the disk heating is neglected.) In particular, for an 'initial' disk radius R_1 (see §5.2, equations (5.47) - (5.49)), the similarity solution yields

$$\Sigma \sim 1.4 \times 10^3 \, \frac{e^{-R/(R_1 t_d)}}{(R/R_1) \, t_d^{3/2}} \left(\frac{M_d(0)}{0.1 \, \text{M}_\odot} \right) \left(\frac{R_1}{10 \, \text{AU}} \right)^{-2} \text{g cm}^{-2}, \tag{9.48}$$

and

$$\dot{M} \sim 6 \times 10^{-7} \, \frac{e^{-R/(R_1 t_d)}}{t_d^{3/2}} \left(1 - \frac{2R}{R_1 \, t_d} \right) \left(\frac{M_d(0)}{0.1 \, \text{M}_\odot} \right)$$

$$\times \left(\frac{R_1}{10 \, \text{AU}} \right)^{-1} \left(\frac{\alpha}{10^{-2}} \right) \left(\frac{M_*}{0.5 \, \text{M}_\odot} \right)^{-1/2} \left(\frac{T_{100}}{10 \, \text{K}} \right) \, \text{M}_\odot \, \text{yr}^{-1}, \tag{9.49}$$

where $M_d(0)$ is the initial disk mass, M_* is the stellar mass, and the time t_d is related to the true elapsed time by

$$t_d = 1 + t/t_s \, ; \tag{9.50}$$

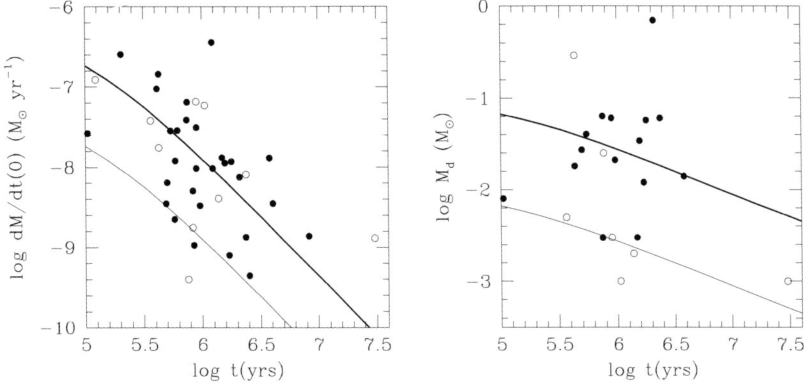

Fig. 9.7. Comparison of the central accretion rates of disk similarity solutions with observations of accreting stars in Taurus, as described in the text. The heavy solid line corresponds to an initial disk mass of $0.1\,M_\odot$ and the light solid line to an initial mass of $0.01\,M_\odot$. Modified from Gullbring *et al.* (1997) and Hartmann *et al.* (1998).

the scaling time is

$$t_s \ \sim \ 8 \times 10^4 \left(\frac{R_1}{10\,\mathrm{AU}}\right) \left(\frac{\alpha}{10^{-2}}\right)^{-1} \left(\frac{M_*}{0.5\,M_\odot}\right)^{1/2} \left(\frac{T_{100}}{10\,\mathrm{K}}\right)^{-1} \mathrm{yr}, \qquad (9.51)$$

where T_{100} is the temperature at 100 AU.

In this model, the expansion of the transition radius R_t (§5.2), which can be considered as a characteristic disk size scale, is

$$R_t \ \sim \ 5\,t_d \left(\frac{R_1}{10\,\mathrm{AU}}\right) \mathrm{AU}. \qquad (9.52)$$

Figure 9.7 compares this similarity solution using the fiducial parameters, and another solution with an initial mass of $0.01\,M_\odot$, with observations of Taurus pre-main-sequence stars (Gullbring *et al.* 1997; Hartmann *et al.* 1998). The accretion rates were estimated from measurements of hot continuum radiation as described in §6.4. The ages are taken from the Hayashi tracks of Figure 9.1, neglecting the possibility of birthline evolution. The models scaled with the above parameters suggests reasonable agreement for initial disk masses between 0.2 and $0.01\,M_\odot$. The disk masses predicted by the models at ages of ~ 1 Myr are $\sim 0.01\,M_\odot$, comparable to the disk masses estimated from mm-wave measurements (§6.3, Figure 6.5).

These solutions assume that $R_t \sim 60\,\mathrm{AU}$ at $t \sim 10^6$ yr; the 'outer' disk radius measured in various ways, such as mm emission, might be a factor of two or so larger than this. The adopted scaling is reasonably consistent with observed disk radii (§6.1). Note that the results are not very sensitive to the assumed starting radius R_1 when $t \gg t_s$. Note also that adopted fiducial viscosity parameter $\alpha = 10^{-2}$ is reasonably consistent with current estimates of the BH instability (§5.3).

These similarity solutions for viscous disk evolution are undoubtedly too simplistic in detail. For example, accreting stars show substantial variability (§6.4), and such variability is obviously beyond the power of the simple viscous evolution model to explain; possibly some of the high-accretion-rate objects are undergoing 'EXor' outbursts (§7.5).

Two qualitative aspects of the similarity solutions are worth noting. First, one expects in general that the mass accretion rates through the inner disk will generally decay with increasing age, and this might have important effects on the actual accretion of mass onto the central star. For sufficiently low mass accretion rates, the magnetospheric radius might move beyond the co-rotation radius (§8.11), especially as signatures of magnetic activity suggest that the magnetic fields of young stars do not decay very rapidly through the first 10^8 yr (Walter *et al.* 1988). This would halt accretion onto the central star, and the usual diagnostics of T Tauri accretion – hot continuum, strong emission lines – would disappear, and as the inner disk hole increases in size, perhaps even near-infrared excess emission would become difficult to detect (Clarke *et al.* 1995). All this could happen even if a substantial circumstellar disk remains around the star.

The second point to note is that simple viscous disk models take a very long time to empty out their mass; the accretion rate decreases with increasing time because the disk spreads out, and thus evolves on ever increasing viscous times. This behavior is very likely to be modified by other processes such as gravitational interactions with binary companions, disk mass loss driven by stellar heating or winds, etc. Nevertheless, it is interesting to speculate that, at least in some cases, the absence of strong infrared emission from older stars might be due to dust coagulation (or planet formation) in the disk rather than any major decrease in disk mass. This speculation takes us to the extreme frontier of research at the present time; further mid-infrared and far-infrared surveys with *ISO* and *SIRTF*, along with improved sensitivity in radio-wavelength interferometry, should go a long way to help understand the later stages of pre-main-sequence disk evolution.

Appendix I

Basic hydrodynamic and MHD equation

Here we list some basic MHD and gas-dynamic equations for reference. Derivations and further discussion may be found, for example, in Priest (1984) and Shu (1992).

We assum a non-relativistic, one-component fluid of density ρ and velocity $\mathbf{v}$. The equation of continuity or mass conservation,

$$\frac{\partial \rho}{\partial t} + \nabla \cdot (\rho \mathbf{v}) = 0, \tag{A1.1}$$

is sometimes written as

$$\frac{D\rho}{Dt} + \rho \nabla \cdot \mathbf{v} = 0, \tag{A1.2}$$

where the total time derivative

$$\frac{D}{Dt} = \frac{\partial}{\partial t} + \mathbf{v} \cdot \nabla \tag{A1.3}$$

is sometimes called the convective time derivative, or the derivative following a fluid element; $\partial/\partial t$ is the time derivative at a fixed location in space.

The momentum equation is

$$\rho \frac{D\mathbf{v}}{Dt} = -\nabla P - \rho \nabla \Phi + \mathbf{F}, \tag{A1.4}$$

where P is the gas pressure, Φ is the gravitational potential, and $\mathbf{F}$ includes other forces on the volume. The gravitational potential is determined by Poisson's equation

$$\nabla^2 \Phi = 4\pi G \rho. \tag{A1.5}$$

If magnetic fields are present then the momentum equation becomes (using Gaussian cgs units)

$$\rho \frac{D\mathbf{v}}{Dt} = -\nabla P - \rho \nabla \Phi + \frac{1}{c} \mathbf{J} \times \mathbf{B}, \tag{A1.6}$$

where $\mathbf{J}$ is the current density and $\mathbf{B}$ is the magnetic field strength, and the net charge is assumed to be zero. If the fluid velocity is much smaller than the speed of light c, then it is possible to neglect the 'displacement current' term in Maxwell's equation

$$\nabla \times \mathbf{B} - \frac{1}{c} \frac{\partial \mathbf{E}}{\partial t} = \frac{4\pi}{c} \mathbf{J}. \tag{A1.7}$$

Then

$$\frac{1}{c} \mathbf{J} \times \mathbf{B} = \frac{1}{4\pi} (\nabla \times \mathbf{B}) \times \mathbf{B}, \tag{A1.8}$$

and with a vector identity one may then write the momentum equation as

$$\rho \frac{d\mathbf{v}}{dt} = -\nabla P - \rho \nabla \phi - \frac{1}{8\pi} \nabla B^2 + \frac{1}{4\pi} (\mathbf{B} \cdot \nabla) \mathbf{B}. \qquad (A1.9)$$

Ohm's law gives the current density $\mathbf{J}$ in terms of the electric and magnetic fields,

$$\mathbf{J} = \sigma \left(\mathbf{E} + \frac{\mathbf{v}}{c} \times \mathbf{B} \right), \qquad (A1.10)$$

where σ is the conductivity of the fluid. In the simple ideal MHD limit that we consider in this book, the conductivity is assumed to be effectively infinite, so that

$$\mathbf{E} + \frac{\mathbf{v}}{c} \times \mathbf{B} = 0. \qquad (A1.11)$$

The first law of thermodynamics can be written as

$$du = -P d(1/\rho) + T \, ds, \qquad (A1.12)$$

where u is the internal energy of the fluid, T is the temperature, and s is the entropy. For an ideal polytropic gas, $u = c_V T$, where c_V is the specific heat at constant volume. The generalized energy equation is

$$\rho \frac{D}{Dt} \left(u + v^2/2 \right) = -\Lambda_r - \nabla \cdot \mathbf{q} + H_e + \mathbf{v} \cdot \left(\frac{1}{c} \mathbf{J} \times \mathbf{B} - \nabla P + \mathbf{F} \right), \quad (A1.13)$$

where Λ_r are the radiative energy losses, $\mathbf{q}$ is the conductive flux, and H_e represents the sum of all other heating sources, including viscous dissipation (see, e.g., Priest (1984)).

Appendix 2

Jeans masses and fragmentation

The Jeans mass is a concept frequently used to discuss gravitational instabilities, although the background state assumed is not usually realizable (cf. Binney & Tremaine), i.e. the assumption of constant background density does not generally represent a solution consistent with hydrostatic equilibrium.

Assume that the velocity and density can be represented by a mean state plus a small fluctuation, so that $\mathbf{v} = \mathbf{v}_\circ + \delta\mathbf{v}$, $\rho = \rho_\circ + \delta\rho$. Then linearizing the equations of mass conservation and momentum for the case of no magnetic pressure (Appendix 1),

$$\frac{\partial}{\partial t}\delta\mathbf{v} = -\frac{1}{\rho_\circ}\nabla c_s^2 \delta\rho - \nabla\Phi, \tag{A2.1}$$

$$\frac{\partial}{\partial t}\delta\rho + \delta\mathbf{v}\cdot\nabla\rho_\circ = -\rho_\circ\nabla\cdot\delta\mathbf{v}, \tag{A2.2}$$

where it is assumed that $\mathbf{v}_\circ = \mathbf{0}$, and that the equilibrium state is

$$\frac{c_s^2}{\rho_\circ}\nabla\rho_\circ = -\nabla\Phi_\circ. \tag{A2.3}$$

Using the Poisson equation relating the gravitational potential to the density,

$$\nabla^2\Phi = -4\pi G\rho. \tag{A2.4}$$

Making the further simplification (obviously inconsistent with the above) that the background density $\rho_\circ$ is constant, one can combine the equations to arrive at

$$\frac{\partial^2\delta\rho}{\partial t^2} = -\rho_\circ\left(\frac{c_s^2}{\rho_\circ}\nabla^2\delta\rho - 4\pi G\delta\rho\right). \tag{A2.5}$$

Taking a plane wave of the form

$$\delta\rho = \delta\rho_\circ e^{i(\omega t - kx)}, \tag{A2.6}$$

the resulting dispersion equation is

$$\omega^2 = c_s^2(k^2 - k_J^2), \quad k_J^2 = 4\pi G\rho_\circ/c_s^2. \tag{A2.7}$$

When $k < k_J$, ω is imaginary; thus, there are exponentially decaying and exponentially increasing solutions; the latter correspond to gravitational instability, i.e. the density grows without limit.

It is usual to define the Jeans mass by

$$M_J = \left(\frac{2\pi}{k_J}\right)^3 \rho_\circ = \left(\frac{\pi c_s^2}{G}\right)^{3/2} \rho_\circ^{-1/2}. \tag{A2.8}$$

For a given temperature and density, masses greater than the Jeans mass will be unstable to gravitational collapse, based on this analysis. However, this solution has been called the 'Jeans swindle' (e.g., Spitzer 1978; Binney & Tremaine 1987) because the background state does not correspond to any physical equilibrium. In certain cases the Jeans mass is a useful guide to the likely scales of gravitational instability (Chapter 2).

One situation for which the assumption of a constant uniform background density as an equilibrium state is correct is the special case of a uniform, infinite, and infinitely-thin sheet of matter (or a thin sheet where gradients in the perpendicular (z) direction can be neglected). Then the conservation equations can be integrated vertically (in z), and the surface density Σ takes the place of the volume density;

$$\frac{\partial}{\partial t}\Sigma + \nabla \cdot (\Sigma \delta \mathbf{v}) = 0, \tag{A2.9}$$

$$\Sigma \frac{\partial}{\partial t}\delta \mathbf{v} = -\nabla \mathbf{P} - \Sigma \nabla \Phi, \tag{A2.10}$$

where the pressure has been vertically integrated. Linearizing as before, and starting from rest,

$$\frac{\partial}{\partial t}\delta\Sigma + \Sigma_\circ \nabla \cdot \delta \mathbf{v} = \mathbf{0}, \tag{A2.11}$$

$$\Sigma_\circ \frac{\partial}{\partial t}\delta \mathbf{v} = -c_s^2 \nabla \delta\Sigma - \Sigma_\circ \nabla \Phi. \tag{A2.12}$$

We assume here implicitly that the gradient in the background state pressure balances the gradient of the background potential. These gradients must be in the z direction by symmetry. We can use this fact with Gauss' law to derive the potential, by integrating the surface field $-\nabla\Phi$ over a pillbox of area $2dA$ encompassing the sheet at $z = 0$; then

$$2(-\nabla\Phi)\,dA = -4\pi G\Sigma\,dA, \tag{A2.13}$$

so that

$$\nabla\Phi = 2\pi G\Sigma. \tag{A2.14}$$

Therefore,

$$\lim_{\epsilon\to\infty}\int_{-\epsilon}^{+\epsilon} dz\nabla^2\Phi = \lim_{\epsilon\to\infty}\left[\nabla\Phi\,|_{+\epsilon} - \nabla\Phi\,|_{-\epsilon}\right] = 4\pi G\Sigma. \tag{A2.15}$$

From basic definitions, this implies

$$\nabla^2\Phi = 4\pi G\Sigma\delta(z), \tag{A2.16}$$

where δ is the Dirac delta function.

As before, we analyze behavior for a plane wave perturbation in the x direction,

$$\delta\Phi = \delta\Phi_a\, e^{i(kx - \omega t)}, \tag{A2.17}$$

where $\delta\Phi_a$ is a (constant) amplitude. This equation can only apply in the plane $z = 0$; for $z \neq 0$,

$$\nabla^2 \delta\Phi = 0. \tag{A2.18}$$

There is only one continuous function which satisfies both these constraints, and also goes to zero at large z,

$$\delta\Phi = \delta\Phi_a\, e^{i(kx - \omega t) - |kz|}. \tag{A2.19}$$

Returning to the Gauss' law evaluation,

$$\lim_{\epsilon \to \infty} \left[\frac{\partial}{\partial z}\delta\Phi\,|_{+\epsilon} - \frac{\partial}{\partial z}\delta\Phi\,|_{-\epsilon} \right] = -2\,|\,k\,|\,\delta\Phi_a = 4\pi G\delta\Sigma, \tag{A2.20}$$

so that

$$\Phi_a = -2\pi G\delta\Sigma/\,|\,k\,|\,. \tag{A2.21}$$

Substitution of this result into the linearized equations leads to the dispersion relation

$$\omega^2 = c_s^2 k^2 - 2\pi G\Sigma_\circ\,|\,k\,|\,. \tag{A2.22}$$

Again, when $\omega^2 < 0$, exponential growth occurs, so the sheet is gravitationally unstable for

$$|\,k_J\,| < 2\pi G\Sigma_\circ/c_s^2\,. \tag{A2.23}$$

Thus, just as in the basic Jeans mass calculation, wavelengths greater than $\lambda_J > 2\pi/k_J$ are unstable. However, the sheet dispersion relation differs in an important way from the basic Jeans mass result. Defining a growth rate $\Gamma = -i\omega$, and taking only real positive k,

$$\Gamma^2 = 2\pi G\Sigma_\circ k - c_s^2 k^2, \tag{A2.24}$$

so that $\Gamma \to 0$ as $k \to 0$. The growth rate has a maximum at $k_{max} = k_J/2$, unlike the basic Jeans mass calculation, where the growth rate is largest for $k \to 0$.

Appendix 3

Basic radiative transfer

Here we briefly review some basic equations of radiative transfer. A detailed discussion of radiation transport is beyond the scope of this work; the most thorough treatment can be found in Mihalas (1978).

The specific intensity $I_\nu(\theta, \phi)$ is defined to be the amount of energy in a unit frequency interval around ν passing through a unit area into unit solid angle in the direction (θ, ϕ) per unit time. Dealing with this quantity has the advantage that it is invariant except for sources and sinks of radiation. The total flux of energy F_ν through an area dA is then

$$F_\nu = \int_0^{2\pi} d\phi \int_0^{\pi} d\theta \cos\theta \, I_\nu, \qquad (A3.1)$$

where θ is the polar angle measured from the normal to the surface and ϕ is the azimuthal angle. The total luminosity L_ν from an object is then the integral of F_ν over the entire surface of the object, and the total radiative energy loss L is $\int_0^\infty L_\nu d\nu$.

A special case of interest is that of a flat uniform surface which radiates as a blackbody. That is, the intensity emitted into space from this surface at all angles is the Planck function B_ν. Then the flux emitted by the surface, using the assumption of axisymmetry, is

$$F_\nu = \int_0^{2\pi} d\phi \int_0^{\pi/2} d\theta \cos\theta B_\nu = 2\pi \int_0^1 d\mu \mu B_\nu = \pi B_\nu, \qquad (A3.2)$$

where $\mu = \cos\theta$ and the total flux per unit area integrated over all frequencies is $\int_0^\infty d\nu \pi B_\nu = \sigma T^4$.

The time-independent radiative equation for the change of the intensity over a path s is

$$\frac{dI_\nu}{ds} = -k_\nu I_\nu + j_\nu, \qquad (A3.3)$$

where k_ν is the opacity per unit volume, and j_ν is the emissivity. The term $k_\nu I_\nu$ represents the absorption of radiation by the medium, while the emissivity represents sources of additional radiation.

In plane-parallel geometry, with vertical height z, and with θ measured from the vertical, the transfer equation can be written as

$$\mu \frac{dI_\nu}{d\tau_\nu} = I_\nu - S_\nu, \qquad (A3.4)$$

216

where

$$d\tau_v = -k_v dz, \qquad S_v = j_v/k_v. \qquad (A3.5)$$

The formal solution of the plane-parallel transfer equation for the intensity emerging from the surface of the medium (where $\tau_v = 0$) is

$$I_v(0) = \int_o^{\tau_v} S_v(t')e^{-t'/\mu}\,dt'/\mu + I_v(\tau_v)e^{-\tau_v/\mu}. \qquad (A3.6)$$

The total intensity along the ray is the sum of the underlying intensity $I_v(\tau_v)$, attenuated by the total absorption through the medium of optical depth τ_v, plus the contribution from the emission of the medium, represented by the source function S_v.

The formal solution is generally not very useful because the source function is not known *a priori*; usually S_v depends upon the optical depth and other properties of the medium, as well as the radiation fields present. In the general case, the transfer solution is obtained iteratively.

One special case of particular interest is that of extinction (by dust, for example) in an absorbing slab of material far from the source. In this case $S_v = 0$ and only one angle of inclination needs to be taken into account, so that the observed flux

$$F_v(obs) = I_v\,d\Omega\,\exp[-\tau_v(total)] = F_v(true)\,\exp[-\tau_v(total)], \qquad (A3.7)$$

where $d\Omega$ is the solid angle subtended by the source at the observer and $\tau_v(total)$ is the total optical depth along the path to the source. Usually extinction at a given wavelength or frequency is expressed in magnitudes. The magnitude scale is related to fluxes F by $m_1 - m_2 = -2.5\log(F_1/F_2)$; then the extinction A_v in magnitudes (the reduction in brightness due to the absorbing screen) is related to the extinction optical depth by $A_v = 1.086\tau_v(total)$.

For all cases considered in detail here we may take azimuthal symmetry. Thus, the integral over solid angle becomes

$$\oint \Omega/4\pi = (1/2)\int_{-1}^{1} d\mu. \qquad (A3.8)$$

We define the moments

$$J_v = (1/2)\int_{-1}^{1} d\mu\,I_v\,; \quad H_v = (1/2)\int_{-1}^{1} d\mu\,\mu\,I_v\,; \quad K_v = (1/2)\int_{-1}^{1} d\mu\,\mu^2\,I_v. \quad (A3.9)$$

The quantity J_v is the mean intensity, and $4\pi H_v$ is the total flux of energy per unit area, while $4\pi K_v/c$ is the radiation pressure (Mihalas 1978). (Note that all of these quantities are per unit interval frequency at frequency v.)

For simplicity assume that S_v is isotropic. Then the first two moments of the transfer equation yield

$$\frac{dH_v}{d\tau_v} = J_v - S_v\,; \qquad \frac{dK_v}{d\tau_v} = H_v. \qquad (A3.10)$$

Recasting the first equation as

$$\frac{dH_v}{dz} = -k_v(J_v - S_v), \qquad (A3.11)$$

and if there is no net energy production in the atmosphere, as is often the case,

$$\frac{d}{dz} \int_0^\infty H_v dv = 0, \tag{A3.12}$$

and so

$$\int_0^\infty k_v (J_v - S_v) dv = 0, \tag{A3.13}$$

which is the condition of radiative equilibrium. For example, the models of dusty infalling envelopes discussed in §4.2 generate little energy except at the center; therefore the enforcement of radiative equilibrium provides the solution for the envelope source function and therefore the emergent spectrum.

To see the general nature of the solutions, we make the approximation that the opacity $k_v = k$ is constant as a function of frequency. Then the two moments of the transfer equation can be integrated over frequency,

$$\frac{dH}{d\tau} = 0, \qquad \frac{dK}{d\tau} = H, \tag{A3.14}$$

which can be easily integrated to yield

$$K = H\tau + \text{constant}, \quad J = S. \tag{A3.15}$$

To complete the solution one requires a closure relation between J and K and a boundary condition. One simple approximation is to assume that, in this plane-parallel region, the radiation is restricted to streaming only in two opposing directions,

$$I = I_+ \delta(\mu - 1/\sqrt{3}) + I_- \delta(\mu + 1/\sqrt{3}), \tag{A3.16}$$

where δ is the Dirac delta function. With this choice of mean angle, $J = 3K$, which is also the result for the nearly isotropic radiation fields expected at large optical depth (cf. Mihalas 1978). Assuming that the incoming radiation field $I_- = 0$ at the outer boundary $\tau = 0$ of the atmosphere, it is straightforward to show that $J(\tau = 0) = \sqrt{3}H$. The solution of the transfer equation for this case becomes

$$J = S = 3H(\tau + 1/\sqrt{3}). \tag{A3.17}$$

If one further assumes that the source function is equal to the Planck function $B = \int_0^\infty B_v \, dv = \sigma T^4/\pi$, and defines the effective temperature by $4\pi H = \sigma T_{eff}^4$, the temperature structure of the atmosphere is given by the standard gray atmosphere equation

$$T^4 = (3/4) T_{eff}^4 (\tau + 1/\sqrt{3}). \tag{A3.18}$$

Thus, the atmosphere can be taken to be approximately radiating as a blackbody, with effective temperature T_{eff}, with the radiation arising from a surface where $\tau \sim 0.76$. (The equation for the gray plane-parallel atmosphere can be solved to arbitary accuracy, but the exact solution is not of great interest because the opacity is never constant with frequency.)

For dusty protostellar envelopes, one is generally interested in situations in which the atmosphere cannot be approximated as plane-parallel. We therefore also consider the transfer equation in spherical coordinates,

$$\mu \frac{\partial I_v}{\partial r} + \frac{(1 - \mu^2)}{r} \frac{\partial I_v}{\partial \mu} = -k_v (I_v - S_v). \tag{A3.19}$$

Taking moments as before,

$$\frac{\partial H_v}{\partial r} + \frac{2H_v}{r} = -k_v(J_v - S_v), \tag{A3.20}$$

$$\frac{\partial K_v}{\partial r} + \frac{(3K_v - J_v)}{r} = -k_v H_v, \tag{A3.21}$$

where the total energy emitted in dv at v is $L_v = 16\pi^2 r^2 H_v$.

We consider first the case in which the envelope is optically thin, which is relevant to dusty envelopes surrounding YSOs at low frequencies. In this case clearly $H_v \propto r^{-2}$. If we further assume that all of the radiation is provided by a central 'core' with uniform intensity I_c, then

$$J_v = \frac{1}{2} \int_{\mu_{min}}^{1} d\mu I_c = \frac{I_c}{2}\left[1 - \left(1 - \frac{R_c^2}{r^2}\right)^{1/2}\right], \tag{A3.22}$$

where μ_{min} corresponds to the tangent angle to the stellar surface and R_c is the radius of the core. It is simple to show that in the limit $r \gg R_c$,

$$J_v \to I_c R_c^2 / 4r^2 \to H_v \to K_v. \tag{A3.23}$$

At the other extreme, suppose that the optical depth is very large. In this circumstance one expects the radiation field to be nearly isotropic and to approach LTE. This implies $J_v \simeq 3K_v$ and $S_v \simeq B_v \simeq J_v$. Equation (A3.21) becomes

$$\frac{1}{k_v}\frac{dB_v}{dr} = \frac{1}{k_v}\frac{dB_v}{dT}\frac{dT}{dr} = -3H_v. \tag{A3.24}$$

Defining the Rosseland mean opacity

$$\frac{1}{k_R} = \int_0^\infty \frac{1}{k_v}\frac{dB_v}{dT}dv \Big/ \int_0^\infty \frac{dB_v}{dT}dv, \tag{A3.25}$$

and integrating over frequency, one arrives at the well-known diffusion equation from the theory of stellar structure (cf. Schwarzchild 1958)

$$L = -\frac{64\pi\sigma r^2 T^3}{3k_R}\frac{dT}{dr}. \tag{A3.26}$$

Finally, we consider an approximate solution for the gray (constant) opacity case where sphericity is important. In this case the radiation field tends to be peaked in the radial direction. For illustrative purposes we take $K = J$. Then

$$\frac{dJ}{dr} + \frac{2J}{r} = -kH = -\frac{kL}{16\pi^2 r^2}. \tag{A3.27}$$

Suppose that the opacity per unit volume has a power-law dependence, $k = k_o(r/r_o)^{-n}$ $(n > 1)$. Then

$$\frac{dJ}{dr} + \frac{2J}{r} = \frac{k_o L}{16\pi^2 r_o^2}\left(\frac{r_o}{r}\right)^{2+n}. \tag{A3.28}$$

This has the solution

$$J = J_1\left(\frac{r_o}{r}\right)^{1+n} + J_2\left(\frac{r_o}{r}\right)^2, \tag{A3.29}$$

where J_1 and J_2 are constants; the second term is the solution to the homogeneous equation. Solving in terms of the radial optical depth

$$\tau_r = \int_r^\infty k\,dr = \frac{k_\circ r_\circ}{n-1}\left(\frac{r}{r_\circ}\right)^{1-n},$$
(A3.30)

and using the boundary condition that as $\tau \to 0$, $J \to H$, the result is

$$J = \frac{L}{16\pi^2 r^2}(\tau_r + 1),$$
(A3.31)

which has obvious similarities to the plane-parallel case.

List of symbols

A	Atomic species (Chapter 8)
A_{ij}	Einstein spontaneous emission probability
B	magnetic field
B_p	poloidal magnetic field (Chapter 8)
B_ν	Planck function
B_ϕ	toroidal magnetic field (Chapter 8)
c	speed of light
c_s	isothermal sound speed
c_V	ratio of specific heats at constant volume
C_{ij}	collisional transition rate between (atomic) levels
dj/dt	angular momentum flux per unit magnetic flux tube (Chapter 8)
$\dot{e}_w$	wind energy flux per unit magnetic flux tube (Chapter 8)
E	energy
f	fractional deuterium abundance (Chapter 9)
f	oscillator strength
F	radiant energy flux per unit area
g	disk annulus torque
g_i	statistical weight of atomic level i
G	gravitational constant
h	Planck constant
h	specific angular momentum (Chapter 5)
H	disk scale height
J	angular momentum
J	current density (Chapter 8)
k	Boltzmann constant, wavenumber
l	angular momentum constant on magnetic field line (Chapter 8)
l_B	angular momentum on field line carried by magnetic field (Chapter 8)
L	luminosity
L_{acc}	accretion luminosity
L_b	bolometric luminosity
L_d	disk luminosity
L_D	deuterium fusion luminosity
L_{DSS}	steady-state deuterium fusion luminosity

L_J	stellar bolometric luminosity estimated at J (Chapter 4)
m_H	mass of hydrogen atom
M_*	stellar mass
M_d	disk mass
M_J	Jeans mass
M_r	mass interior to radius r
M_{cl}	mass of cloud
$\dot{M}$	mass infall or accretion rate
$\dot{M}_w$	wind mass loss rate
n	number density
n	power-law index of infall density distribution (Chapter 4)
n	polytropic index (Chapter 9)
N_{H_2}	number density of hydrogen molecules
p	impact parameter
p	power-law dependence of surface density on radius (Chapter 6)
P	gas pressure
Q	Toomre parameter for gravitational instability
r	radial distance
r_c	centrifugal radius (Chapter 3)
r_λ	photospheric radius at wavelength λ (Chapter 4)
R	cylindrical radius distance
R_c	protostellar core radius (Chapter 9)
R_{cl}	radius of cloud
R_{co}	disk co-rotation radius
R_m	magnetospheric (disk truncation) radius
R_s	sonic point radius (Chapter 8)
R_T	disk truncation radius
R_*	stellar radius
t	time
t_{ff}	free-fall time
t_{kh}	Kelvin–Helmholtz timescale
T	temperature (K)
$<T>$	mean internal stellar temperature
T_d	disk surface or effective temperature
T_{eff}	(stellar) effective temperature
T_g	non-dimensional time for disk similarity solution
T_{max}	disk peak temperature
u	velocity, r/r_c (Chapter 3)
U	internal thermal energy of star
$UBVRIJHKL$	Standard optical-near infrared, broad-band photometric magnitudes
v	velocity
v_p	poloidal velocity (Chapter 8)
v_{th}	thermal velocity width
v_ϕ	azimuthal velocity
v_ϕ	toroidal velocity (Chapter 8)
V	volume, velocity
w	turbulent velocity

w_D	drift velocity
W	gravitational potential energy of star
x	non-dimensional radial distance (Chapter 3)
α	orbital angle (Chapter 3), viscosity parameter
α	overdensity (Chapter 2), non-dimensional density (Chapter3)
α_m	line absorption cross-section (Chapter 4)
α_v	line absorption cross-section at frequency v
β	exponent of dust opacity vs. frequency
γ	angle between incoming ray and disk normal
$\delta(x)$	Dirac delta function
ϵ	thermal energy content of accreted material (Chapter 9)
η	parameter for flattened collapse models (Chapter 3)
η	ratio between mass and magnetic flux (Chapter 8)
θ	polar angle in cylindrical coordinates
$\theta_\circ$	streamline polar angle of rotating collapse solution at infinity (Chapter 3)
κ	epicyclic frequency (Chapter 5)
κ	scalar relation between **v** and **B** (Chapter 8)
κ_g	wavenumber for disk torque decomposition (Chapter 5)
κ_R	Rosseland mean opacity
κ_λ	opacity per unit mass at wavelength λ
κ_v	opacity per unit mass at frequency v
λ	wavelength, mean free path (Chapter 5)
λ_J	Jeans length
μ	mean molecular weight in hydrogen masses
v	frequency
v_v	viscosity
ρ	gas mass density
σ	cross-section for collisions (Chapter 2)
σ	Stefan–Boltzmann constant
Σ	surface density
τ	optical depth
ϕ	potential; gravitational, effective
Φ_B	magnetic flux
ω	angular frequency
Ω	angular velocity

Bibliography

Adams, F.C., Lada, C.J., & Shu, F.H. 1987, *Ap.J.*, **321**, 788 (ALS)
Adams, F.C., Lada, C.J., & Shu, F.H. 1988, *Ap.J.*, **326**, 865
Adams, F.C., Ruden, S.P., & Shu, F.H. 1989, *Ap.J.*, **347**, 959
Adams, F.C., & Shu, F.H. 1986, *Ap.J.*, **308**, 836
Ali, B. & Depoy, D. 1995, *A.J.*, **109**, 709
Aly, J.J., & Kuijpers, J. 1990, *A&Ap.*, **227**, 473
Ambartsumian, J.A. 1947, *Stellar Evolution and Astrophysics* (Yerevan: Acad. Sci. Armenian SSR).
André, P., Deeney, B.D., Phillips, R.B., & Lestrade, J.-F. 1992, *Ap.J.*, **401**, 667
André, P., & Montmerle, T. 1994, *Ap.J.*, **420**, 837
André, P., Montmerle, T., & Feigelson, E.D. 1987, *A.J.*, **93**, 1192
André, P., Phillips, R.B., Lestrade, J.-F., & Klein, K.-L. 1991, *Ap.J.*, **376**, 630
André, P., Ward-Thompson, D., & Barsony, M. 1993, *Ap.J.*, **406**, 122
Anglada, G., Rodriguez, L.F., Canto, J., Estalella, R., & Lopez, R. 1987, *A&Ap.*, **186**, 280
Appenzeller, I., Jankovics, I., & Östreicher, R. 1984, *A&Ap.*, **141**, 108
Appenzeller, I., & Tscharnuter, W.M. 1974, *A&Ap.*, **30**, 423
Armitage, P.J., & Clarke C.J. 1996, *M.N.R.A.S.*, **280**, 458
Artymowicz, P., & Lubow, S.H. 1996a, in *Disks and Outflows around Young Stars*, eds. S.V.W. Beckwith, J. Staude, A.M. Quetz, & A. Natta (Berlin: Springer), p. 115
Artymowicz, P., & Lubow, S.H. 1996b, *Ap.J.*, **467**, L77
Bachiller, R. 1996, *Ann. Rev. Astr. Ap.*, **34**, 111
Balbus, S.A., & Hawley, J.F. 1991, *Ap.J.*, **376**, 214
Balbus, S.A., & Hawley, J.F. 1997, *Rev. Mod. Phys.*, in press
Bally, J., & Devine, D. 1994, *Ap.J.*, **428**, L65
Bally, J., Stark, A.A., Wilson, R.W., & Langer, W.D. 1987, *Ap.J. Lett.*, **312**, L45
Basri, G., & Bertout, C. 1989, *Ap.J.*, **341**, 340
Basri, G., Marcy, G.W., & Valenti, J.A. 1992, *Ap.J.*, **390**, 622
Bastien, P., & Ménard, F. 1988, *Ap.J.*, **326**, 334
Bastien, P., & Ménard, F. 1990, *Ap.J.*, **364**, 232
Beckwith, S.V.W., & Sargent, A.I. 1991, *Ap.J.*, **381**, 250
Beckwith, S.V.W., Sargent, A.I., Chini, R.S., Güsten, R. 1990, *A.J.*, **99**, 924 (BSCG)
Beckwith, S.V.W., Sargent, A.I., Koresko, C.D., & Weintraub, D.A. 1989, *Ap.J.*, **343**, 393
Beichman, C.A., Boulanger, F., & Moshir, M. 1992, *Ap.J.*, **386**, 248
Beichman, C.A., Myers, P.C., Emerson, J.P., Harris, S., Mathieu, R., Benson, P.J., & Jennings, R.E. 1986, *Ap.J.*, **307**, 337
Bell, K.R., & Lin, D.N.C. 1994, *Ap.J.*, **427**, 987
Bell, K.R., Lin, D.N.C., Hartmann, L.W., & Kenyon S.J. 1995, *Ap.J.*, **444**, 376
Benson, P.J., & Myers, P.C. 1989, *Ap.JSupp.*, **71**, 89
Bertout, C. 1989, *Ann. Rev. Astr. Ap.*, **27**, 351
Bertout, C., Basri, G., & Bouvier, J. 1988, *Ap.J.*, **330**, 350
Binney, J., & Tremaine, S. 1987, *Galactic Dynamics* (Princeton: Princeton University Press).
Blaes, O.M., & Balbus, S.A. 1994, *Ap.J.*, **421**, 163
Blandford, R.D., & Payne, D.G. 1982, *M.N.R.A.S.*, **199**, 883

Blitz, L. 1991, in *The Physics of Star Formation and Early Stellar Evolution*, eds. C.J. Lada & N.D. Kylafis (Dordrecht:Kluwer), p. 3

Bodenheimer, P. 1978, *Ap.J.*, **224**, 488

Böhm, K.H. 1990, in *Structure and Dynamics of the Interstellar Medium*, eds. G. Tenorio-Tagle, M. Moles, & J. Melnick (Berlin: Springer), 282

Bok, B.J., & Reilly, E.F. 1947, *Ap.J.*, **105**, 255

Boland, W., & DeJong, T. 1984, *A&Ap.*, **134**, 87

Bonnell, I., & Bastien, P. 1992, *Ap.J.*, **401**, 654

Bonnell, I., & Bate, M.R. 1994, *M.N.R.A.S.*, **271**, 999

Bonnell, I.A., Bate, M.R., Clarke, C.J., & Pringle, J.E. 1996b, *M.N.R.A.S.*, **285**, 201

Bonnell, I.A., Bate, M.R., & Price, N.M. 1996a, *M.N.R.A.S.*, **279**, 121

Bonnell, I., Martel, H., Bastien, P. & Arcoragi J.-P., & Benz, W. 1991, *Ap.J.*, **377**, 553

Bonnor, W.B. 1956, *M.N.R.A.S.*, **112**, 195

Bontemps, S., André, P., Terebey, S., & Cabrit, S. 1996, *A&Ap.*, **311**, 858

Boss, A.P. 1993, *Ap.J.*, **410**, 157

Boss, A.P. 1995, in *Circumstellar Disks, Outflows, and Star Formation*, eds. S. Lizano, & J.M. Torrelles, *Rev. Mex. Astr. Astrof., Serie de Conferencias*, **1**, 165

Boss, A.P., & Black, D.C. 1982, *Ap.J.*, **258**, 270

Boss, A.P., & Yorke, H.W. 1995, *Ap.J.*, **439**, 55L

Bouvier, J., & Bertout, C. 1992, *A&Ap.*, **263**, 113

Bouvier, J., Bertout, C., Benz, W., & Mayor, M. 1986, *A&Ap.*, **165**, 110

Bouvier, J., Cabrit, S., Fernandez, M., Martin, E.L., & Matthews, J.M. 1993, *A&Ap.*, **272**, 176

Brandenburg, A., Nordlund, A., Stein, R.F., & Torkelsson, U. 1996, *Ap.J.*, **458**, L45

Briceño, C., Hartmann, L., Stauffer, J.R., Gagné, M., Stern, R.Q., & Caillault, J.-P. 1997, *A.J.*, **113**, 740

Burrows, C.J., Stapelfeldt, K.R., Watson, A.M., Krist, J.E., Ballester G.E., Clarke, J.T. Crisp, D., Gallagher III, J.S., Griffiths, R.E., Hester, J.J., Hoessel, J.G., Holtzman, J.A., Mould, J.R., Scowen, P.A., Trauger, J.T., & Westphal, J.A. 1996, *Ap.J.*, **473**, 437

Butner, H.M., Evans, N.J. II, Lester, D.F., Levreault, R.M., & Strom, S.E. 1991, *Ap.J.*, **376**, 636

Butner, H.M., Lada, E.A., & Loren, R.B. 1995, *Ap.J.*, **448**, 207

Butner, H.M., Natta, A., & Evans, N.J. II 1994, *Ap.J.*, **420**, 326

Cabot, W., & Pollack, J.B. 1992, *Geophys. Astrophys. Fluid Dyn.*, **64**, 97

Cabrit, S., & Bertout, C. 1992, *A&Ap.*, **261**, 274

Cabrit, S., Guilloteau, S., Andre, P., Bertout, C., Montmerle, T., Schuster, K. 1996, *A&Ap.*, **305**, 527

Calvet, N. 1997, in *Herbig-Haro Flows and the Birth of Low-Mass Stars*, IAU Symposium 182, eds. B. Reipurth & C. Bertout, p. 417

Calvet, N., Basri, G., & Kuhi, L.V. 1983, *Ap.J.*, **268**, 739

Calvet, N., & Hartmann, L. 1992, *Ap.J.*, **386**, 239

Calvet, N., Hartmann, L., & Kenyon, S.J. 1991, *Ap.J.*, **383**, 752

Calvet, N., Hartmann, L., & Kenyon, S.J. 1993, *Ap.J.*, **402**, 623

Calvet, N., Hartmann, L., Kenyon, S.J., & Whitney, B.A. 1994, *Ap.J.*, **434**, 330

Calvet, N., Hartmann, L., & Strom, S.E. 1997, *Ap.J.*, **481**, 912

Camenzind, M. 1990, *Reviews in Modern Astronomy* (Springer-Verlag: Berlin), **3**, p. 234

Cameron, A.C., & Campbell, C.G. 1993, *A&Ap.*, **274**, 309

Cantó, J., D'Alessio, P., & Lizano, S. 1995, in *Circumstellar Disks, Outflows, and Star Formation*, eds. S. Lizano, & J.M. Torrelles, *Rev. Mex. Astr. Astrof., Serie de Conferencias*, **1**, 217

Carr, J.S., Harvey, P.M., Lester, D.F. 1987, *Ap.J.*, **321**, L71

Carr, J.S., Tokunaga, A.T., Najita, J., Shu, F.H., & Glassgold, A.E. 1993, *Ap.J.*, **411**, L37

Caselli, P., & Myers, P.C. 1995, *Ap.J.*, **446**, 665

Cassen, P., & Moosman, A. 1981, *Icarus*, **48**, 353

Cernicharo, J. 1991, in *The Physics of Star Formation and Early Stellar Evolution*, eds. C.J. Lada & N.D. Kylafis (Dordrecht: Kluwer), p. 287

Chandler, C.J., & Sargent, A.I. 1993, *Ap.J.*, **414**. L29

Chandrasekhar, S., 1960, *Proc. Nat. Acad. Sci.*, **46**, 53

Chandrasekhar, S. 1961, *Hydrodynamic and Magnetohydrodynamic Stability* (Oxford: Oxford University Press)

Chandrasekhar, S. 1967, *An Introduction to the Study of Stellar Structure* (New York: Dover)

Chernin, L.M., & Masson, C.R. 1995, *Ap.J.*, **443**, 181

Choi, M., Evans, N.J. II., Gregerson, E.M., & Wang, Y. 1995, *Ap.J.*, **448**, 742
Churchwell, E., Wood, D.O.S., Felli, M., & Massi, M. 1987, *Ap.J.*, **321**, 516
Clarke, C.J., Armitage, P.J., Smith, K.W., & Pringle, J.E. 1995, *M.N.R.A.S.*, **273**, 639
Clarke, C.J., Lin, D.N.C., & Papaloizou, J.C.B. 1989, *M.N.R.A.S.*, **236**, 495
Clarke, C.J., Lin, D.N.C., & Pringle, J.E 1990, *M.N.R.A.S.*, **242**, 439
Clarke, C.J., & Pringle, J.E. 1991, *M.N.R.A.S.*, **249**, 588
Clarke, C.J., & Syer, D. 1996, *M.N.R.A.S.*, **278**, L23
Clemens, D.P., & Barvainis, R.E. 1988, *Ap.JSupp.*, **68**, 257
Cohen, M. 1973a,b,c, *M.N.R.A.S.*, **161**, 85-105
Cohen, M., Bieging, J., & Schwartz, R.D. 1982, *Ap.J.*, **253**, 707
Cohen, M., Emerson, J.P., & Beichman, C.A. 1989, *Ap.J.*, **339**, 455
Cohen, M., & Kuhi, L.V. 1979, *Ap.JSupp.*, **41**, 743
Cohen, M., & Schwartz, R.D. 1976, *M.N.R.A.S.*, **174**, 37
Cohen, M., & Woolf, N.J. 1971, *Ap.J.*, **169**, 543
Covino, E., Terranegra, L., Vittone, A.A., & Russo, G. 1984, *A.J.*, **89**, 1868
Cram, L.E. 1979, *Ap.J.*, **234**, 949
Cram, L.E., Giampapa, M.S., & Imhoff, C.L. 1980, *Ap.J.*, **238**, 905
Cram, L.E., & Mullan, D.J. 1985, *Ap.J.*, **294**, 626
Croswell, K., Hartmann, L., & Avrett, E.H. 1987, *Ap.J.*, **312**, 227
Crutcher, R.M., Mouschovias, T. Ch., Troland, T.H., & Ciolek, G.E. 1994, *Ap.J.*, **427**, 839
Curry, C., Pudritz, R.E., & Sutherland, P.G. 1994, *Ap.J.*, **434**, 206
D'Alessio, P. 1996, PhD Thesis, Universidad Nacional Autonoma de Mexico
D'Alessio, P., Calvet, N., & Hartmann, L. 1997, *Ap.J.*, **474**, 397
Dame, T.M., Ungerechts, H., Cohen, R.S., DeGeus, E.J., Grenier, I.A., May, J., Murphy, D.C.,
    Nyman, L.-A., & Thaddeus, P. 1987, *Ap.J.*, **322**, 706
D'Antona, F., & Mazzitelli, I. 1994, *Ap.JSupp.*, **90**, 467
DeCampli, W.M. 1981, *Ap.J.*, **244**, 124
DeGeus, E.J., Bronfman, L., & Thaddeus, P. 1990, *A&Ap.*, **231**, 137
Dopita, M.A., Schwartz, R.D., & Evans, I. 1982, *Ap.J.*, **263**, L73
Draine, B.T., & Lee, H.M. 1984, *Ap.J.*, **285**, 89
Duquennoy, A., & Mayor, M. 1991, *A&Ap.*, **248**, 485
Dutrey, A., Guilloteau, S., Duvert, G., Prato, L., Simon, M., Schuster, K., & Menard, E. 1995,
    A&Ap., **309**, 493
Dyson, J.E. 1987, in *Circumstellar Matter, IAU Symposium 122*, eds. I. Appenzeller & C. Jordan
    (Dordrecht: Reidel), 159
Ebert, R. 1955, *Zs. Ap.*, **37**, 217
Ebert, R., von Hoerner, S., & Temesvary, S. 1960, *Die Enstehung von Sternen durch Kondensation
    diffuser Materie (Berlin: Springer-Verlag)*, p. 184.
Edwards, S. 1995, in *Circumstellar Disks, Outflows, and Star Formation*, eds. S. Lizano, & J.M.
    Torrelles, *Rev. Mex. Astron. Astrofis., Serie de Conferencias*, **1**, 309
Edwards, S., Cabrit, S., Strom, S., Heyer, I., Strom, K., & Anderson, E. 1987, *Ap.J.*, **321**, 473
Edwards, S., Hartigan, P., Ghandour, L., & Andrulis, C. 1994, *A.J.*, **108**, 1056
Edwards, S., Mundt, R., & Ray, T. 1993, in *Protostars and Planets III*, eds. E.H. Levy & J.I. Lunine
    (Tucson: University of Arizona Press), p. 567
Edwards, S., Strom, S.E., Hartigan, P., Strom, K.M., Hillenbrand, L.A., Herbst, W., Attridge, J.,
    Merrill, K.M., Probst, R., & Gatley, I. 1993, *A.J.*, **106**, 372
Efstathiou, A., & Rowan-Robinson, M. 1991, *M.N.R.A.S.*, **252**, 528
Elmegreen, B.G. 1991, in *The Physics of Star Formation and Early Stellar Evolution*, eds. C.J. Lada
    & N.D. Kylafis (Dordrecht:Kluwer), p. 35
Feigelson, E.D., Casanova, S., Montmerle, T., Guibert, J. 1993, *Ap.J.*, **416**, 623
Feigelson, E.D., & DeCampli, W.M. 1981, *Ap.J.*, **243**, L89
Feigelson, E.D., Jackson, J.M., Mathieu, R.D., Myers, P.C., & Walter, F.M. 1987, *A.J.*, **94**, 1251
Feigelson, E.D., & Kriss, G.A. 1981, *Ap.J.*, **248**, L35
Feigelson, E.D., & Kriss, G.A. 1989, *Ap.J.*, **338**, 262
Fiedler, R.A., & Mouschovias, T. Ch. 1993, *Ap.J.*, **415**, 680
Folha, D., Emerson, J., & Calvet, N. 1997, Poster proceedings of IAU Symposium 182, eds. F.
    Malbet & A. Castets (Grenoble: Observatoire de Grenoble)
Foster, P.N., & Chevalier, R.A. 1993, *Ap.J.*, **416**, 303

Frank, J., King, A., & Raine, D. 1992, *Accretion Power in Astrophysics*, 2nd edition (Cambridge University Press)

Fricke K. 1969, *A&Ap.*, **1**, 388

Fukui, Y., Iwata, T., Mizuno, A., Bally, J., & Lane, A.P. 1993, in *Protostars and Planets III*, eds. E.H. Levy & J.I. Lunine (Tucson: University of Arizona Press), p. 603

Fuller, G.A. 1994, in *Clouds, Cores, and Low-Mass Stars*, eds. D.P. Clemens and R. Barvainis, Astronomical Society of the Pacific Conference Series, **65**, p. 3

Fuller, G.A., & Myers, P.C. 1992, *Ap.J.*, **384**, 523

Fuller, G.A., & Myers, P.C. 1993, *Ap.J.*, **418**, 273

Gahm, G.F. 1981, *Ap.J.*, **242**, L163

Gahm G.F., 1994, in *Flares and Flashes*, eds. J. Greiner, H.W. Duerbeck, & R.E. Gershberg, IAU Coll. No. 151 (Berlin: Springer), p. 203

Gahm, G.F., Fredga, K., Liseau, R., & Dravins, D. 1979, *A&Ap.*, **73**, L4

Gahm G.F., Lodén K., Gullbring E., & Hartstein D., 1995, *A&Ap.***301**, 89

Galli, D., & Shu, F.H. 1993a,b, *Ap.J.*, **417**, 220, 243

Gammie, C.F. 1996a, *Ap.J.*, **457**, 355

Gammie, C.F. 1996b, *Ap.J.*, **462**, 725

Gammie, C.F., & Ostriker, E.C. 1996, *Ap.J.*, **466**, 814

Genzel, R., & Stutzski, J. 1989, *Ann. Rev. Astr. Ap.*, **27**, 41

Ghez, A.M., Neugebauer, G., Gorham, P.W., Haniff, C.A., Kulkarni, S.R., Matthews, K., Koresko, C., & Beckwith, S. 1991, *A.J.*, **102**, 2066

Ghez, A.M., Neugebauer, G., & Matthews, K. 1993, *A.J.*, **106**, 2005

Ghosh, P. 1995, *M.N.R.A.S.*, **272**, 763

Ghosh, P., & Lamb, F.K. 1979, *Ap.J.*, **234**, 296

Giampapa, M.S., Basri, G.S., Johns, C.M., & Imhoff, C. 1993, *Ap.JSupp.*, **89**, 321

Giampapa, M.S., Calvet, N., Imhoff, C.L., & Kuhi, L.V. 1981, *Ap.J.*, **251**, 113

Goldreich, P., & Lynden-Bell, D. 1965, *M.N.R.A.S.*, **130**, 125

Goldsmith, P.F. 1988, in *Molecular Clouds in the Milky Way*, eds. R.L Dickman, R.L. Snell, & J.S. Young, Lecture Notes in Physics (Berlin: Springer), **315**, 1

Gomez, M., Hartmann, L., Kenyon, S.J., & Hewett, R. 1993, *A.J.*, **105**, 1927

Goodman, A.A., Benson, P.J., Fuller, G.A., & Myers, P.C. 1993, *Ap.J.*, **406**, 528

Goodman, A.A., Crutcher, R.M., Heiles, C., Myers, P.C., & Troland, T.H. 1989, *Ap.J.*, **338**, L61

Goodrich, R.W. 1987, *Pub. Astr. Soc. Pac.*, **99**, 116

Grasdalen, G.L. 1973, *Ap.J.*, **182**, 781

Grasdalen, G.L., Sloan, G., Stout, M., Strom, S.E., & Welty, A.D. 1989, *Ap.J. Lett.*, **339**, L37

Greene, T.P., & Meyer, M.R. 1995, *Ap.J.*, **450**, 233

Greene, T.P., Wilking, B.A., Andre, P., Young, E.T., & Lada, C.J. 1994, *Ap.J.*, **434**, 614

Gullbring E., 1994, *A&Ap.*, **287**, 131

Gullbring, E., Barwig, H., Chen, P.S., Gahm, G., & Bao, M.X. 1996, *A&Ap.*, **307**, 791

Gullbring, E., Hartmann, L., Briceño, C., & Calvet, N. 1997, *Ap.J.*, in press

Haas, M., Christou, J.C., Zinnecker, H., Ridgway, S.T., Leinert, Ch. 1993, *A&Ap.*, **269**, 282

Haro, G. 1952, *Ap.J.*, **115**, 572

Haro, G. 1953, *Ap.J.*, **117**, 73

Haro, G., Iriarte, B., & Chavira, E. 1953, *Bol. Obs. Tonantzintla y Tacubaya*, **8**, 3

Hartigan, P., Edwards, S., & Ghandour, L. 1995, *Ap.J.*, **452**, 736

Hartigan, P., Hartmann, L., Kenyon, S.J., Strom, S.E., & Skrutskie, M.F. 1990, *Ap.J.*, **354**, 25L

Hartigan, P., Morse, J.A., & Raymond, J. 1994, *Ap.J.*, **346**, 125

Hartigan, P., Raymond, J., & Hartmann, L. 1987, *Ap.J.*, **316**, 323

Hartigan, P., Strom, S., Edwards, S., Kenyon, S., Hartmann, L., Stauffer, J., & Welty, A. 1991, *Ap.J.*, **382**, 617

Hartmann, L., Avrett, E.H., Loeser, R., & Calvet, N. 1990, *Ap.J.*, **349**, 168

Hartmann, L., Boss, A., Calvet, N., & Whitney, B. 1994b, *Ap.J. Lett.*, **430**, L49

Hartmann, L., & Calvet, N. 1995, *A.J.*, **109**, 1846

Hartmann, L., Calvet, N., & Boss, A. 1996, *Ap.J.*, **464**, 387

Hartmann, L., Calvet, N., Gullbring, E., & D'Alessio, P. 1998, *Ap.J.*, in press

Hartmann, L., Cassen, P., & Kenyon, S.J. 1997, *Ap.J.*, **475**, 770

Hartmann, L., Hewett, R., Stahler, S., & Mathieu, R.D. 1986, *Ap.J.*, **309**, 275

Hartmann, L., Hewett, R., & Calvet, N. 1994a, *Ap.J.*, **426**, 669

Hartmann, L., Jones, B.F., Stauffer, J.R., & Kenyon, S.J., 1991, *A.J.*, **101**, 1050
Hartmann, L., & Kenyon, S.J. 1996, *Ann. Rev. Astr. Ap.*, **34**, 207
Hartmann, L., & Kenyon, S.J. 1985, *Ap.J.*, **299**, 462
Hartmann, L., & Kenyon, S.J. 1987a, *Ap.J.*, **312**, 243
Hartmann, L., & Kenyon, S.J. 1987b, *Ap.J.*, **322**, 393
Hartmann, L., & MacGregor, K.B. 1982, *Ap.J.*, **259**, 180
Hartmann, L., Soderblom, D.R., & Stauffer, J.R. 1988, *A.J.*, **93**, 907
Hawley, J.F., Gammie, C.F., & Balbus, S.A. 1995, *Ap.J.*, **440**, 742
Hawley, J.F., Gammie, C.F., & Balbus, S.A. 1996, *Ap.J.*, **464**, 690
Hayashi, C. 1966, *Ann. Rev. Astr. Ap.*, **4**, 171
Hayashi, C., Hoshi, R., & Sugimoto, D. 1962, *Prog. Theor. Phys. Suppl.*, **22**, 1
Hayashi, M., Ohashi, N., & Miyama, S. 1993, *Ap.J.*, **418**, L71
Heiles, C., Goodman, A.A., McKee, C.F., & Zweibel, E.G. 1993, in *Protostars and Planets III*, eds.
    E.H. Levy & J.I. Lunine (Tucson: University of Arizona Press), p. 279
Henning, T., & Stognienko, R. 1996, *A&Ap.*, **311**, 291
Henriksen, R., André, P., & Bontemps, S. 1997, *A&Ap.*, **323**, 549
Henyey, L.G., LeLevier, R., & Levee, R.D. 1955, *Pub. Astr. Soc. Pac.* , **67**, 154
Herbig, G.H. 1951, *Ap.J.*, **114**, 697
Herbig, G.H. 1952, *J. Roy. Astr. Soc. Canada*, **46**, 222
Herbig, G.H. 1954, *Ap.J.*, **119**, 483
Herbig, G.H. 1957, *Ap.J.*, **128**, 259
Herbig, G.H. 1960, *Ap.JSupp.*, **4**, 337
Herbig, G.H. 1962, *Adv. Astr. Ap.* , **1**, 47
Herbig, G.H. 1966, *Vistas in Astronomy*, **8**, 109
Herbig, G.H. 1970, *Mem. Roy. Soc. Scie. Liege, Ser. 5*, **9**, 13
Herbig, G.H. 1977a, *Ap.J.*, **214**, 747
Herbig, G.H. 1977b, *Ap.J.*, **217**, 693
Herbig, G.H. 1989, in *ESO Workshop on Low-Mass Star Formation and Pre-Main Sequence Objects*,
    ed. B. Reipurth (Garching: ESO), p. 233
Herbig, G.H., & Bell, K.R. 1988, *Third Catalog of Emission-Line Stars of the Orion Population*, Lick
    Observatory Bulletin No. 1111 (HBC)
Herbig, G.H., & Kuhi, L.V. 1963, *Ap.J.*, **137**, 398
Herbig, G.H., & Terndrup, D.M. 1986, *Ap.J.*, **307**, 609
Herbst, W., Herbst, D.K., Grossman, E.J., & Weinstein D., 1994, *A.J.*, **108**, 1906
Heyvaerts, J., & Norman, C.A. 1989, *Ap.J.*, **347**, 1055
Heyvaerts, J., Priest, E.R., & Bardou, A. 1996, *Ap.J.*, **473**, 403
Hildebrand, R.H. 1983, *Quart. Journ. Roy. Astr. Soc.*, **24**, 267
Hillenbrand, L. 1997, *A.J.*, **113**, 1733
Hillenbrand, L.A., & Hartmann, L.W. 1998, *Ap.J.*, in press
Hillenbrand, L.A., Strom, S.E., Vrba, F.J., & Keene, J. 1992, *Ap.J.*, **397**, 613
Hollenbach, D. 1985, *Icarus*, **61**, 36
Hollenbach, D., & Natta, A. 1995, *Ap.J.*, **455**, 133
Hoyle, F. 1953, *Ap.J.*, **118**, 513
Huenemoerder, D.P., Lawson, W.A., & Feigelson, E.D. 1994, *M.N.R.A.S.*, **271**, 967
Hunter, C. 1977, *Ap.J.*, **218**, 834
Imhoff, C.L., & Giampapa, M.S. 1980, *Ap.J.*, **239**, L115
Jackson, J.D. 1962, *Classical Electrodynamics* (New York: Wiley)
Jensen, E.L.N., Mathieu, R.D., & Fuller, G.A. 1994, *Ap.J.*, **429**, L29
Jensen, E.L.N., Mathieu, R.D., & Fuller, G.A. 1996, *Ap.J.*, **458**, 312
Johns, C.M., & Basri, G. 1995a, *A.J.*, **109**, 2800
Johns, C.M., & Basri, G. 1995b, *Ap.J.*, **449**, 341
Johns-Krull, C.M., & Basri, G. 1997, *Ap.J.*, **474**, 433
Jones, B.F., & Herbig, G.H. 1979, *A.J.*, **84**, 1872
Jones, B.F. & Walker, M.F. 1988, *A.J.*, **95**, 1755
Joy, A.H. 1945, *Ap.J.*, **102**, 168
Julian, W.H., & Toomre, A. 1966, *Ap.J.*, **146**, 810
Kawazoe, E., & Mineshige, S. 1993, *Pub. Astr. Soc. Japan*, **45**, 715
Keene, J., & Masson, C.R. 1990, *Ap.J.*, **335**, 635

Kenyon, S.J., Calvet, N., & Hartmann, L. 1993, *Ap.J.*, **414**, 676 (KCH)
Kenyon, S.J., Gomez, M., Marzke, R.O., & Hartmann, L. 1994, *A.J.*, **108**, 251
Kenyon, S.J., Hartmann, L., Strom, K.M, & Strom, S.E. 1990, *A.J.*, **99**, 869
Kenyon, S.J., & Hartmann, L. 1987, *Ap.J.*, **323**, 714
Kenyon, S.J., & Hartmann, L. 1990, *Ap.J.*, **349**, 197
Kenyon, S.J., & Hartmann, L. 1991, *Ap.J.*, **383**, 664
Kenyon, S.J., & Hartmann, L. 1995, *Ap.JSupp.*, **101**, 117
Kenyon, S.J., Hartmann, L., & Hewett, R. 1988, *Ap.J.*, **325**, 231
Kenyon, S.J., Hartmann, L., Imhoff, C.L., & Cassatella, A. 1989, *Ap.J.*, **344**, 925
Kenyon, S.J., Hartmann, L., & Kolotilov, E.A. 1991, *Pub. Astr. Soc. Pac.* , **103**, 1069
Kenyon, S.J., Yi, I., & Hartmann, L. 1996, *Ap.J.*, **462**, 439
Kepner, J., Hartigan, P., Yang, C., & Strom, S. 1993, *Ap.J.*, **415**, L119
Kolotilov, E.A., & Petrov, P.P. 1983, Pis'ma Astr. Zh., **9**, 171
Kolotilov, E.A., & Petrov, P.P. 1985, Pis'ma Astr. Zh., **11**, 846
Königl, A. 1989, *Ap.J.*, **342**, 208
Königl, A. 1991, *Ap.J. Lett.*, **370**, L39
Königl, A., & Ruden, S.P. 1993, in *Protostars and Planets III*, eds. E.H. Levy & J.I. Lunine (Tucson: University of Arizona Press), p. 641
Koresko, C.D., Beckwith, S.V.W., Ghez, A.M., Matthews, K., & Neugebauer, G. 1991, *A.J.*, **102**, 2073
Koerner, D.W., Chandler, C.J., & Sargent, A.I. 1995, *Ap.J.*, **452**, L69
Koerner, D.W., & Sargent, A.I. 1995, *A.J.*, **109**, 2138
Koerner, D.W., Sargent, A.I., & Beckwith, S.V.W. 1993, *Icarus*, **106**, 2
Kuan, P. 1975, *Ap.J.*, **202**, 425
Kuhi, L.V. 1964, *Ap.J.*, **140**, 1409
Kuiper, T.B.H., Rodriguez-Kuiper, E.N., & Zuckerman, B. 1978, *Ap.J.*, **219**, 129
Lada, C.J. 1985, *Ann. Rev. Astr. Ap.*, **23**, 267
Lada, C.J. 1987, in *Star Forming Regions*, IAU Symposium 115, eds. M. Peimbert & J. Jugaku (Dordrecht: Reidel), p. 1
Lada, C.J., & Wilking, B.A. 1984, *Ap.J.*, **287**, 610
Lada, E.A., Evans, N.J. II, Depoy, D.L., Gatley, I. 1991, *Ap.J.*, **371**, 171
Ladd, E.F., Adams, F.C., Casey, S., Davidson, J.A., Fuller, G.A., Harper, D.A., Myers, P.C., & Padman, R. 1991, *Ap.J.*, **382**, 555
Lago, M.T.V.T. 1984, *M.N.R.A.S.*, **210**, 323
Larson, R.B. 1969a,b, *M.N.R.A.S.*, **145**; 271, 297
Larson, R.B. 1972, *M.N.R.A.S.*, **157**, 121
Larson, R.B. 1980, *Rev. Mex. Astron. Astrof.*, **7**, 219
Larson, R.B. 1981, *M.N.R.A.S.*, **194**, 809
Larson, R.B. 1983, *M.N.R.A.S.*, **190**, 321
Larson, R.B. 1984, *M.N.R.A.S.*, **206**, 197
Larson, R.B. 1985, *M.N.R.A.S.*, **214**, 379
Laughlin, G., & Bodenheimer, P. 1994, *Ap.J.*, **436**, 335
Lay, O.P., Carlstrom, J.E., Hills, R.E., & Phillips, T.G. 1994, *Ap.J.*, **434**, L75
Leinert, C.H., Zinnecker, H., Weitzel, N., Christou, J., Ridgeway, S.T., Jameson, R., Haas, M., & Lenzen, R. 1993, *A&Ap.*, **278**, 129
Li, Z.Y., & McKee, C.F. 1996, *Ap.J.*, **464**, 373
Li, Z.Y., & Shu, F.H. 1996, *Ap.J.*, **468**, 261
Lin, D.N.C., & Papaloizou, J.C.B. 1980, *M.N.R.A.S.*, **191**, 37
Lin, D.N.C., & Papaloizou, J.C.B. 1985, in *Protostars and Planets II*, eds. D.C. Black and M.C. Matthews (Tucson: University of Arizona Press), p. 981
Lin D.N.C., & Pringle J.E. 1990, *Ap.J.*, **358**, 515
Livio, M. 1995, in *Cataclysmic Variables*, eds. A. Bianchini et al. (Dordrecht: Kluwer), p. 349
Lizano, S., & Shu, F.H. 1989, *Ap.J.*, **342**, 834
Lizano, S., & Torrelles, J.M. eds., 1995, *Circumstellar Disks, Outflows, and Star Formation, Rev. Mex. Astr. Astrof., Serie de Conferencias*, **1**
Loren, R.B., Wooten, A., & Wilking, B.A. 1990, *Ap.J.*, **365**, 269
Lovelace, R.V.E., Romanova, M.M., & Contoupoulos, J. 1993, *Ap.J.*, **403**, 158
Lovelace, R.V.E., Wang, J.C.L., & Sulkanen, M.E. 1987, *Ap.J.*, **315**, 504
Lubow, S., Papaloizou, J.C.B., & Pringle, J.E. 1994, *M.N.R.A.S.*, **268**, 1010

Lynden-Bell, D., & Pringle, J.E. 1974, *M.N.R.A.S.*, **168**, 603
Malbet, F., Rigaut, F., Bertout, C., & Léna, P. 1993, *A&Ap.*, **271**, L9
Mannings, V., & Emerson, J.P. 1994, *M.N.R.A.S.*, **267**, 361
Mardones, D., Myers, P.C., Tafalla, M., Wilner, D.J., Bachiller, R., & Garay, G. 1997, *Ap.J.*, in press
Masson, C.R., & Chernin, L.M. 1993, *Ap.J.*, **414**, 230
Mathieu, R.D. 1994, *Ann. Rev. Astr. Ap.*, **32**, 465
Mathis, J. 1990, *Ann. Rev. Astr. Ap.*, **28**, 37
Mazzitelli, I. 1989, in *ESO Workshop on Low-Mass Star Formation and Pre-Main Sequence Objects*, ed. B. Reipurth (ESO:Garching), 433
McCaughrean, M.J., & O'Dell, C.R. 1996, *A.J.*, **111**, 1977
McCaughrean, M.J. & Stauffer, J. 1994, *A.J.*, **108**, 1382
McCullough, P.R., Fugate, R.Q., Christou, J.C., Ellerbroek, B.L., Higgins, C.H., Spinhirne, J.M., Cleis, R.A., & Moroney, J.F. 1995, *Ap.J.*, **438**, 394
McKee, C.F. 1989, *Ap.J.*, **345**, 782
McKee, C.F., Zweibel, E.G., Goodman, A.A., & Heiles, C. 1993, in *Protostars and Planets III*, eds. E.H. Levy & J.I. Lunine (Tucson: University of Arizona Press), p. 327
Mendoza, E.E. 1966, *Ap.J.*, **143**, 1010
Mestel, L. 1961, *M.N.R.A.S.*, **122**, 473
Mestel, L. 1965, *Quart. Journ. Roy. Astr. Soc.*, **6**, 161
Mestel, L. 1968, *M.N.R.A.S.*, **138**, 359
Mestel, L. 1985, in *Protostars and Planets II*, eds. D.C. Black & M.S. Mathews (Tucson: University of Arizona Press), p. 81
Mestel, L., & Spitzer, L., Jr 1956, *M.N.R.A.S.*, **116**, 503
Mihalas, D. 1978, *Stellar Atmospheres* (San Francisco: Freeman)
Miller, G.E. & Scalo, J.M. 1979, *Ap.JSupp.*, **41**, 513
Montmerle, T., Feigelson, E.D., Bouvier, J., & André, P. 1993, in *Protostars and Planets III*, eds. E.H. Levy & J.I. Lunine (Tucson: University of Arizona Press), p. 689
Montmerle, T., Koch-Miramond, L., Falgarone, E., Grindlay, J.E. 1983, *Ap.J.*, **269**, 182
Mould, J.R., Hall, D.N.B., Ridgway, S.T., Hintzen, P., Aaronson, M. 1978, *Ap.J.*, **222**, L123
Mouschovias, T. Ch. 1976, *Ap.J.*, **206**, 753
Mouschovias, T. Ch. 1991, in *The Physics of Star Formation and Early Stellar Evolution*, eds. C.J. Lada & N.D. Kylafis (Dordrecht: Kluwer) pp 61, 449
Mouschovias, T.Ch., & Spitzer, L. 1976, *Ap.J.*, **210**, 326
Mundt, R. 1984, *Ap.J.*, **280**, 749
Mundt, R. 1985, in Protostars & Planets II, eds. D.C. Black & M.S. Mathews (Tucson: University of Arizona Press), 414
Mundt, R., Brugel, E.W., & Bührke, T. 1987, *Ap.J.*, **319,** 275
Mundt, R., & Fried, J.W. 1983, *Ap.J.*, **274**, L83
Mundt, R., Ray, T.P., Bührke, T., Raga, A.C., & Solf, J. 1990, *A&Ap.*, **232**, 37
Mundt, R., Walter, F.M., Feigelson, E.D., Finkenzeller, U., Herbig, G.H., & O'Dell, A.P. 1983, *Ap.J.*, **269**, 229
Mundy, L.G., Looney, L.W., Erikson, W., Grossman, A., Welch, W.J., Forster, J.R., Wright, M.C.H., Plambeck, R.L., Lugten, J., & Thornton, D.D. 1996, *Ap.J.*, **464,** L169
Myers, P., & Benson, P.J. 1983, *Ap.J.*, **266**, 309
Myers, P.C., Fuller, G.A., Goodman, A.A., & Benson, P.J. 1991, *Ap.J.*, **376**, 561
Myers, P.C., & Fuller, G.A. 1992, *Ap.J.*, **396**, 631
Myers, P.C., Fuller, G.A., Mathieu, R.D., Beichman, C.A., Benson, P.J., Schild, R.E., & Emerson, J.P. 1987, *Ap.J.*, **319**, 340
Myers, P.C., & Goodman, A.A. 1988a, *Ap.J. Lett.*, **326**, L27
Myers, P.C., & Goodman, A.A. 1988b, *Ap.J.*, **329**, 392
Myers, P.C., Goodman, A.A., Heiles, C., & Güsten, R. 1995, *Ap.J.*, **442**, 177
Myers, P.C., & Khersonsky, V.K. 1995, *Ap.J.*, **442**, 186
Myers, P.C., Mardones, D., Tafalla, M., Williams, J.P., & Wilner, D.J. 1995, *Ap.J.*, **449**, L65
Najita, J., Carr, J.S., Glassgold, A.E., Shu, F.H., & Tokunaga, A.T. 1996, *Ap.J.*, **462**, 919
Najita, J., Carr, J.S., & Tokunaga, A.T. 1996, *Ap.J.*, **456**, 292
Najita, J., & Shu, F.H. 1994, *Ap.J.*, **429**, 808
Nakamura, F., Hanawa, T., & Nakano, T. 1995, *Ap.J.*, **444**, 770
Nakano, T. 1984, *Fund. Cosmic Phys.*, **9**, 139

Nakano, T., & Nakamura, T. 1978, *Pub. Astr. Soc. Japan*, **30**, 671
Narayan, R., & Popham, R. 1994, in *Theory of Accretion Disks – 2*, eds. W.J. Duschl, J. Frank, Meyer, F., Meyer-Hofmeister, E., & Tschnarnuter, W.M. (Dordrecht: Kluwer), p. 293
Natta, A. 1993, *Ap.J.*, **412**, 761
Natta, A., & Giovanardi, C. 1990, *Ap.J.*, **356**, 646
Nelson, R., & Papaloizou, J.C. 1993, *M.N.R.A.S.*, **265**, 905
Neufeld, D.A., & Hollenbach, D.J. 1994, *Ap.J.*, **428**, 170
Neuhauser, R., Sterzik, M.F., Schmitt, J.H.M.M., Wichmann, R., & Krautter, J. 1995, *A&Ap.*, **297**, 391
O'Dell, C.R., & Wen, Z. 1994, *Ap.J.*, **436**, 194
O'Dell, C.R., Wen, Z., & Hu, X. 1993, *Ap.J.*, **410**, 696
Ohashi, N., Hayashi, M., Ho, P.T.P., Momose, M., & Hirano, N. 1996, *Ap.J.*, **466**, 317
Onishi, T., Mizuno, A., Kawamura, A., Ogawa, H., Fukui, Y. 1996, *Ap.J.*, **465**, 815
Ossenkopf, V., & Henning, Th. 1994, *A&Ap.*, **291**, 943
Osterloh, M., & Beckwith, S.V.W. 1995, *Ap.J.*, **439**, 288
Ostriker, E.C., & Shu, F.H. 1995, *Ap.J.*, **447**, 813
Ostriker, E.C., Shu, F.H., & Adams, F.C. 1992, *Ap.J.*, **399**, 192
Paatz, G., & Camenzind, M. 1996, *A&Ap.*, **308**, 77
Paczynski, B. 1976, cited in Trimble, V., *Quart. Journ. Roy. Astr. Soc.*, **17**, 25
Palla, F., & Stahler, S.W. 1992, *Ap.J.*, **392**, 667
Papaloizou, J.C.B., & Lin, D.N.C. 1995, *Ann. Rev. Astr. Ap.*, **33**, 505
Parker, E.N. 1963, *Interplanetary Dynamical Processes* (New York: Wiley).
Parker, E.N. 1966, *Ap.J.*, **145**, 811
Pelletier, G., & Pudritz, R.E. 1992, *Ap.J.*, **394**, 117
Penston, M.V. 1969, *M.N.R.A.S.*, **144**, 425
Petrov, P.P., Gullbring, E., Ilyin, I., Gahm, G.F., Tuominen, I., Hackman, T., & Loden, K. 1996, *A&Ap.*, **314**, 821
Petrov, P.P., & Herbig, G.H. 1992, *Ap.J.*, **392**, 209
Pollack, J.B., Hollenbach, D., Beckwith, S., Simonelli, D.P., Roush, T., & Fong, W. 1994, *Ap.J.*, **421**, 615
Popham, R. 1995, *Ap.J.*, **467**, 749
Popham, R., Kenyon, S., Hartmann, L., & Narayan, R. 1996, *Ap.J.*, **473**, 422
Popham, R., Narayan, R., Hartmann, L., & Kenyon, S. 1993, *Ap.J.*, **415**, L127
Prialnik, D., & Livio, J. 1985, *M.N.R.A.S.*, **216**, 37
Priest, E.R. 1984, *Solar Magneto-hydrodynamics* (Dordrecht: Reidel).
Pringle, J.E. 1981, *Ann. Rev. Astr. Ap.*, **19**, 137
Pringle, J.E. 1989, *M.N.R.A.S.*, **236**, 107
Pudritz, R.E., & Norman, C.A. 1983, *Ap.J.*, **274**, 677
Pudritz, R.E., & Norman, C.A. 1986, *Ap.J.*, **301**, 571
Pudritz, R.E., Pelletier, G., & Gomez de Castro, A.I. 1991, in *The Physics of Star Formation and Early Stellar Evolution*, eds. C.J. Lada & N.D. Kylafis (Dordrecht: Kluwer), p. 539
Raga, A.C. 1993, *Ap. Space Sci.*, **208**, 163
Raga, A.C. 1989, in *ESO Workshop on Low Mass Star Formation and Pre-Main Sequence Objects*, ed. B. Reipurth (Garching: ESO Conference and Workshop Proceedings No. 33), p. 281
Raga, A.C. 1995, in *Circumstellar Disks, Outflows, and Star Formation*, eds. S. Lizano, & J.M. Torrelles, *Rev. Mex. Astr. Astrof., Serie de Conferencias*, **1**, 103
Randich, S., Schmitt, J.H.M.M., Prosser, C.F., & Stauffer, J.R. 1996, *A&Ap.*, **305**, 785
Ray, T.P., Mundt, R., Dyson, J.E., Falle, S.A.G., & Raga, A.C. 1996, *Ap.J.*, **468**, L103
Reipurth, B. 1989, in *ESO Workshop on Low Mass Star Formation and Pre-Main Sequence Objects*, ed. B. Reipurth (Garching: ESO Conference and Workshop Proceedings No. 33), p. 247
Reipurth, B. 1990, in *Flare Stars in Star Clusters*, IAU Symp. 137, eds. L.V. Mirzoyan, B.R. Petterson, M.K. Tsvetkov (Dordrecht: Kluwer), p. 229
Reipurth, B. 1991 in *The Physics of Star Formation and Early Stellar Evolution*, eds. C.J. Lada & N.D. Kylafis (Dordrecht:Kluwer), p. 497
Reipurth, B., & Bertout, C. 1997, eds., *Herbig-Haro Flows and the Birth of Low-Mass Stars*, IAU Symposium 182 (Dordrecht: Kluwer)
Reipurth, B., & Heathcote, S. 1997, in *Herbig-Haro Flows and the Birth of Low-Mass Stars*, IAU Symposium 182, eds. B. Reipurth & C. Bertout (Dordrecht: Kluwer), p. 3

Reipurth, B., Pedrosa, A., & Lago, M.T.V.T. 1996, *A&AS*, **120**, 229
Reipurth, B., & Zinnecker, H. 1993, *A&Ap.*, **278**, 81
Reyes-Ruiz, M., & Stepinski, T.F. 1995, *Ap.J.*, **438**, 750
Rieke, G., Lee, T., & Coyne, G. 1972, *Pub. Astr. Soc. Pac.* , **84**, 37
Roddier, C., Roddier, F., Northcott, M.J., Graves, J.E., & Kim, J. 1996, *Ap.J.*, **463**, 326
Rodriguez, L.F., & Hartmann, L.W. 1992, *Rev. Mex. Astr. Astrof.*, **24**, 135
Rodriguez, L.F., Hartmann, L.W., & Chavira, E. 1990, *Pub. Astr. Soc. Pac.* , **102**, 1413
Rozyczka, M., & Spruit, H.C. 1993, *Ap.J.*, **417**, 677
Rucinski, S.M. 1985, *A.J.*, **90**, 2321
Ruden, S.P., & Pollack, J.B. 1991, *Ap.J.*, **375**, 740
Rybicki, G.B., & Hummer, D.G. 1978, *Ap.J.*, **219**, 654
Rydgren, A.E., Strom, S.E., & Strom, K.M. 1976, *Ap.J.Supp.*, **30**, 307
Rydgren, A.E., & Vrba, F.J. 1981, *A.J.*, **86**, 1069
Ryu, D., & Goodman, J. 1992, *Ap.J.*, **338**, 438
Safier, P. 1993, *Ap.J.*, **408**, 115
Safier P., 1995, *Ap.J.*, **444**, 818
Scalo, J.M. 1986, *Fund. Cosmic Physics*, **11**, 1
Scalo, J.M. 1990, in *Physical processes in fragmentation and star formation*, Proceedings of the Workshop, Rome, Italy (Dordrecht:Kluwer), p.151
Schwartz, R.D. 1975, *Ap.J.*, **195**, 631
Schwartz, R.D. 1983, *Ann. Rev. Astr. Ap.*, **21**, 209
Schwarzschild, M. 1958, *Structure and Evolution of the Stars* (New York: Dover)
Shakura, N.I., & Sunyaev, R.A. 1973, *A&Ap.*, **24**, 337
Shapiro, S.L., & Teukolsky, S.A. 1983, *Black Holes, White Dwarfs, and Neutron Stars*, The Physics of Compact Objects (Wiley: New York), p. 450 ff.
Shu, F.H. 1977, *Ap.J.*, **214**, 488
Shu, F.H. 1991, in *The Physics of Star Formation and Early Stellar Evolution*, eds. C.J. Lada & N.D. Kylafis (Dordrecht: Kluwer), p. 376
Shu, F.H. 1992, *The Physics of Astrophysics, vol. 2 (Gas Dynamics)* (Mill Valley: University Science Books).
Shu, F.H. 1995, in *Circumstellar Disks, Outflows, and Star Formation*, eds. S. Lizano, & J.M. Torrelles, *Rev. Mex. Astr. Astrof., Serie de Conferencias*, **1**, 375
Shu, F.H., Adams, F.C., & Lizano, S. 1987, *Ann. Rev. Astr. Ap.*, **25**, 23
Shu, F., Najita, J., Ostriker, E.C., & Shang, H. 1995, *Ap.J.*, **455**, L155
Shu, F., Najita, J., Ostriker, E., Wilkin, F., Ruden, S., & Lizano, S. 1994, *Ap.J.*, **429**, 781
Shu, F.H., Tremaine, S., Adams, F.C., & Ruden, S.P. 1990, *Ap.J.*, **358**, 495
Simon, M., Ghez, A.M., Leinert, Ch., Cassar, L., Chen, W.P., Howell, R.R., Jameson, R.F., Matthews, K., Neugebauer, G., & Richichi, A. 1995, *Ap.J.*, **443**, 625
Simon, T. 1975, *Pub. Astr. Soc. Pac.* , **87**, 317
Simon, T., Morrison, N.D., Wolff, S.C., & Morrison, D. 1972, *Pub. Astr. Soc. Pac.* , **84**, 644
Skinner, S.L. 1993, *Ap.J.*, **408**, 660
Skrutskie, M.F., Dutkevich, D., Strom, S.E., Edwards, S., Strom, K.M., & Shure, M.A. 1990, *A.J.*, **99**, 1187
Smak, J. 1984, *Pub. Astr. Soc. Pac.* , **96**, 5
Smith, B.A., & Terrile, R.J. 1984, *Science*, **226**, 1421
Snell, R.L., Loren, R.B., & Plambeck, R.L. 1980, *Ap.J. Lett.*, **239**, L17
Spitzer, L., Jr 1968, in *Nebulae and Interstellar Matter, Stars and Stellar Systems*, vol. 7, eds. B. Middlehurst and L.H. Aller (Chicago: University of Chicago Press), p. 1
Spitzer, L., Jr. 1978, *Physical Processes in the Interstellar Medium* (New York: Wiley).
Spruit, H. 1996, in *Physical Processes in Binary Stars*, eds. R.A.M.J. Wijers, M.B. Davies, & C.A. Tout (Dordrecht: Kluwer).
Stahler, S.W. 1983, *Ap.J.*, **274**, 822
Stahler, S.W. 1988, *Ap.J.*, **332**, 804
Stahler, S.W., Shu, F.H., & Taam, R.E. 1980a, *Ap.J.*, **241**, 637
Stahler, S.W., Shu, F.H., & Taam, R.E. 1980b, *Ap.J.*, **242**, 226
Stahler, S.W., & Walter, F.M. 1993, in *Protostars and Planets III*, eds. E.H. Levy & J.I. Lunine (Tucson: University of Arizona Press), p. 405

Stapelfeldt, K.R., Burrows, C.J., Krist, J.E., Trauger, J.T., Hester, J.J., Holtzman, J.A., Ballester G.E., Casertano, S., Clarke, J.T. Crisp, D., Evans, R.W., Gallagher, J.S. III, Griffiths, R.E., Hoessel, J.G., Mould, J.R., Scowen, P.A., Watson, A.M., & Westphal, J.A. 1995, *Ap.J.*, **449**, 888

Stauffer, J.R., Caillault, J.-P., Gagne, M., Prosser, C.F., & Hartmann, L.W. 1994, *Ap.JSupp.*, **91**, 625

Stepinski, T. 1995, in *Circumstellar Disks, Outflows, and Star Formation*, eds. S. Lizano, & J.M. Torrelles,*Rev. Mex. Astr. Astrof., Serie de Conferencias*, **1**, 267

Stocke, J.T., Hartigan, P.M., Strom, S.E., Strom, K.M., Anderson, E.R., Hartmann, L.W., & Kenyon, S.J. 1988, *Ap.JSupp.*, **68**, 229

Stone, J.M., & Balbus, S.A. 1996, *Ap.J.*, **464**, 364

Stone, J.M., Hawley, J.F., Gammie, C.F., & Balbus, S.A. 1996, *Ap.J.*, **463**, 656

Stone, J.M., & Norman, M.L. 1994, *Ap.J.*, **433**, 746

Strom, K.M., Strom, S.E., Edwards, S., Cabrit, S., & Skrutskie, M.F. 1989, *A.J.*, **97**, 1451

Strom, K.M., Strom, S.E., Kenyon, S.J., & Hartmann, L. 1988, *A.J.*, **95**, 534

Strom, S.E. 1972, *Pub. Astr. Soc. Pac. *, **84**, 745

Strom, S.E., Edwards, S., & Skrutski, M. 1993, in *Protostars and Planets III*, eds. E.H. Levy & J.I. Lunine (Tucson: University of Arizona Press), p. 837

Strom, S.E., Strom, K.M., Brooke, A.L., Bregman, J., & Yost, J. 1972, *Ap.J.*, **171**, 267

Suess, S.T., & Nerney, S.F. 1975, *Solar Phys.*, **40**, 487

Tamura, M., Gatley, I., Waller, W., & Werner, M.W. 1991, *Ap.J.*, **374**, L25

Terebey, S., Chandler, C.J., & Andre, P. 1993, *Ap.J.*, **414**, 759

Terebey, S., Shu, F.H., & Cassen, P. 1984, *Ap.J.*, **286**, 529 (TSC)

Troland, T.H., Crutcher, R.M., Goodman, A.A., Kazes, I., & Myers, P.C. 1996, *Ap.J.*, **471**, 302

Tomley, L., Steiman-Cameron, T.Y., & Cassen, P. 1994, *Ap.J.*, **422**, 850

Toomre, A. 1964, *Ap.J.*, **139**, 1217

Tout, C.A., & Pringle, J.E. 1992, *M.N.R.A.S.*, **259**, 604

Tscharnuter, W. 1991, in *The Physics of Star Formation and Early Stellar Evolution*, eds. C.J. Lada & N.D. Kylafis (Dordrecht:Kluwer), p. 411

Uchida, Y., & Shibata, K. 1984a, *Pub. Astr. Soc. Japan*, **36**, 105

Uchida, Y., & Shibata, K. 1984b, *Pub. Astr. Soc. Japan*, **37**, 515

Ulrich, R.K. 1976, *Ap.J.*, **210**, 377

Ulrich, R.K., & Knapp, G.R. 1979, *Ap.J.*, **230**, L99

Umebeyashi, T., & Nakano, T. 1988, *Prog. Theor. Phys. Suppl.*, **96**, 151

Umebeyashi, T., & Nakano, T. 1990, *M.N.R.A.S.*, **243**, 103

Ungerechts, H., & Thaddeus, P. 1987, *Ap.JSupp.*, **63**, 645

Valenti, J.A., Basri, G., & Johns, C.M. 1993, *A.J.*, **106**, 2024

van Ballegooijen, A.A. 1994, *Space Sci. Rev.*, **68**, 299

Velikhov, E.P. 1959, *Soviet Phys.-JETP*, **36**, 995 (*JETP Letters* **35**, 1398)

Vishniac, E.T., & Diamond, P. 1989, *Ap.J.*, **347**, 447

Vishniac, E.T., & Diamond, P. 1992, *Ap.J.*, **398**, 561

Vrba F.J., Chugainov, P.F., Weaver W.B., & Stauffer J.S., 1993, *A.J.*, **106**, 1608

Walker, C.K., Lada, C.J., Young, E.T., Maloney, P.R., & Wilking, B.A. 1986, *Ap.J. Lett.*, **309, ** L47

Walker, M.F. 1972, *Ap.J.*, **175**, 546

Walter, F.M. 1986, *Ap.J.*, **306**, 573

Walter, F.M., Brown, A., Mathieu, R.D., Myers, P.C., & Vrba, F.J. 1988, *A.J.*, **96**, 297

Walter, F.M., & Kuhi, L.V. 1981, *Ap.J.*, **250**, 254

Walter, F.M., Vrba, F.J., Mathieu, R.D., Brown, A., & Myers, P.C. 1994, *A.J.*, **107**, 692

Wang, Y., Evans, N.J. II., Zhou, S., & Clemens, D.P. 1995, *Ap.J.*, **454**, 217

Wang, Y.-M. 1995, *Ap.J.*, **449**, L153

Wang, Y.-M. 1996, *Ap.J.*, **465**, L111

Wardle, M., & Königl, A. 1993, *Ap.J.*, **410**, 218

Weber, E.J., & Davis, L., Jr. 1967, *Ap.J.*, **148**, 217

Weidenschilling, S.J., & Cuzzi, J.N. 1993, in *Protostars and Planets III*, eds. E.H. Levy & J.I. Lunine (Tucson: University of Arizona Press), p. 1031

Weintraub, D.A., Sandell, G., & Duncan, W.D. 1991, *Ap.J.*, **382**, 270

Welty, A.D., Strom, S.E., Edwards, S., Kenyon, S.J., & Hartmann, L.W. 1992, *Ap.J.*, **397**, 250

Whitney, B.A., & Hartmann, L. 1992, *Ap.J.*, **386**, 239

Whitney, B.A., & Hartmann, L. 1993, *Ap.J.*, **402**, 605

Whitworth, A.P., Bhattal, A.S., Francis, N., & Watkins, S.J. 1996, *M.N.R.A.S.*, **283**, 1061

Wilking, B.A., & Lada, C.J. 1983, *Ap.J.*, **274**, 698
Williams, J.P., DeGeus, E.J., & Blitz, L. 1994, *Ap.J.*, **428**, 693
Winkler, K.-H., & Newman, M.J. 1980a,b, *Ap.J.*, **236**, 210, & **238**, 311
Wolf, B., Appenzeller, I., & Bertout, C. 1977, *A&Ap.*, **58**, 163
Wolfire, M.G., & Cassinelli, J.P. 1987, *Ap.J.*, **315**, 315
Wolk, S.J., & Walter, F.M. 1996, *A.J.*, **111**, 2066
Wood, D.O.S., Myers, P.C., & Daugherty, D.A. 1994, *Ap.J̇Supp.*, **95**, 457
Yorke, H.W., Bodenheimer, P., & Laughlin, G. 1993, *Ap.J.*, **411**, 274
Yun, J.L., & Clemens, D.P. 1990, *Ap.J. Lett.*, **365**, L73
Zhou, S. 1994, *Ap.J.*, **442**, 685
Zhou, S., & Evans, N.J. II 1994, in *Clouds, Cores, and Low-Mass Stars*, PASP Conference Series
    Vol. 65, eds. D.P. Clemens & R. Barvainis, p. 183
Zhou, S., Evans, N.J., II, Kömpke, C., & Walmsley, C.M. 1993, *Ap.J.*, **404**, 232
Zhou, S., Wu, Y., Evans, N.J., II, Fuller, G. A., & Myers, P.C. 1989, *Ap.J.*, **346**, 168
Zinnecker, H. 1982, in *Symposium on the Orion Nebula to Honor Henry Draper*, eds. A.E. Glassgold
    et al. , (New York: New York Academy of Sciences), p. 226

Index